Gareth Guest

Electron Cyclotron Heating of Plasmas

Related Titles

d'Agostino, R., Favia, P., Kawai, Y., Ikegami, H., Sato, N., Arefi-Khonsari, F. (eds.)

Advanced Plasma Technology

479 pages with 256 figures and 13 tables
2008
Hardcover
ISBN: 978-3-527-40591-6

Smirnov, B. M.

Plasma Processes and Plasma Kinetics

580 Worked-Out Problems for Science and Technology

582 pages with 91 figures and 31 tables
2007
Softcover
ISBN: 978-3-527-40681-4

Stacey, W. M.

Fusion Plasma Physics

571 pages with 158 figures and 28 tables
2005
Softcover
ISBN: 978-3-527-40586-2

Lieberman, M. A., Lichtenberg, A. J.

Principles of Plasma Discharges and Materials Processing

approx. 800 pages
2005
Hardcover
ISBN: 978-0-471-72001-0

Woods, L. C.

Physics of Plasmas

226 pages with 69 figures
2004
Softcover
ISBN: 978-3-527-40461-2

Gareth Guest

Electron Cyclotron Heating of Plasmas

WILEY-VCH

WILEY-VCH Verlag GmbH & Co. KGaA

The Author

Dr. Gareth Guest
5433 Caminito Rosa
La Jolla, CA 92037-7234
USA

Library of Congress Card No.: applied for

British Library Cataloguing-in-Publication Data
A catalogue record for this book is available from the British Library.

Bibliographic information published by the Deutsche Nationalbibliothek
The Deutsche Nationalbibliothek lists this publication in the Deutsche Nationalbibliografie; detailed bibliographic data are available on the Internet at http://dnb.d-nb.de.

Typesetting Thomson Digital, Noida, India
Printing betz-druck GmbH, Darmstadt
Binding Litges & Dopf GmbH, Heppenheim
Cover Design

Printed in the Federal Republic of Germany
Printed on acid-free paper

ISBN: 978-3-527-40916-7

This book is dedicated with deep gratitude to Raphael A. Dandl, mentor, colleague and friend.

Contents

Electron Cyclotron Heating of Plasmas. Gareth Guest

ISBN: 978-3-527-40916-7

1
Introduction

In the late 1950s as part of the International Controlled Thermonuclear Fusion Research Program, several small independent groups started investigating the possibility of using microwave power to create magnetically confined, hot-electron plasmas. This process became known variously as *electron cyclotron heating* (ECH) or *electron cyclotron resonance heating* (ECRH) in recognition of the key role played by resonant absorption of the microwave power at the electron gyrofrequency (often called the "cyclotron frequency"). Of these, the group under R.A. Dandl at the Oak Ridge National Laboratory was unique in using continuous wave (cw) microwave power and large DC magnets to produce steady-state plasmas. Unlike pulsed discharges, this steady-state operation permitted ongoing adjustments of the gas pressure, microwave power, and magnetic field strength as well as extensive diagnostic measurements of the plasma properties. By the early 1960s, it was clear that these plasmas could be operated in regimes that exhibited some remarkable properties. Although the plasmas were confined in simple magnetic mirrors and theoretically predicted to be susceptible to large-scale plasma instabilities, it was found that if the ambient gas pressure was suitably adjusted they could be operated in completely stable, steady-state regimes. Moreover, they contained two or more distinct populations of electrons: a low-temperature group with temperatures of some 10s of electron volts together with high-temperature populations with temperatures in excess of 100 keV and kinetic pressures of at least 5% of the magnetostatic pressure of the confining magnetic field. Dandl's group devoted the next two decades to an intense study of a sequence of increasingly powerful and sophisticated embodiments of these remarkable ECH plasmas.

In the ELMO magnetic mirror device, they achieved stable, steady-state, relativistic-electron plasmas with average hot-electron temperatures in excess of 3 MeV and kinetic pressures comparable to the confining magnetostatic pressure. Thirty years after they were created, these plasmas remain unique in many respects, particularly as regards their copious emission of neutrons apparently resulting from the electron dissociation of deuterium nuclei, as well as the plasma diamagnetic modification of the confining magnetic field to yield substantial localized depressions in the magnetic intensity. The relativistic-electron shells produced in ELMO were subsequently used successfully by Dandl to stabilize toroidal plasmas confined in the ELMO Bumpy Torus.

Electron Cyclotron Heating of Plasmas. Gareth Guest

ISBN: 978-3-527-40916-7

At roughly the same time, a group under T. Consoli in Saclay, France, was investigating, among other things, the possible use of ECH plasmas to achieve the collective acceleration of plasma ions through space-charge electric fields. This group later moved to Grenoble and continued an active research program lasting over two decades. The goal of collective acceleration remained elusive, but the ECH techniques developed within this effort were to be influential in the design of sources of multiply charged ions and ultimately in sources of high-density plasma of interest for commercial applications in plasma processing. In particular, Consoli's group pioneered the use of whistler-wave heating to produce high-density albeit low-temperature plasmas. This technique, which later came to be known as high-field launch, coupled microwave power into the plasma electrons via whistler waves launched in the high-field region of the magnetic-mirror fields to propagate along the magnetic lines of force into the resonance region.

A vigorous ECH research program under H. Ikegami began in Nagoya, Japan, in the late 1960s, starting with magnetic-mirror experiments and subsequently progressing to experiments in the bumpy torus magnetic configuration. In the Soviet Union, ECH was investigated first in magnetic mirror devices and then in tokamaks, following the advent of gyrotrons, remarkable sources of high-frequency microwave power first developed in the former Soviet Union.

More recently ECH has found widespread use as a means of providing auxiliary heating in tokamaks and stellarators, as well as a means of stabilizing particular modes of instabilities and driving noninductive currents in tokamaks. In particular, early predictions that ECH could be used to stabilize neoclassical tearing modes of plasma instability in tokamaks were subsequently confirmed experimentally, and further applications became possible with the continuing development of high-power, long-pulse, and cw sources of microwave power at frequencies well above 100 GHz and, therefore, in the electron gyrofrequency range for major tokamak installations. Contemporary tokamaks routinely use several megawatts of 140 GHz microwave power to break down the gas and initiate the plasma discharge, to ameliorate the deleterious effects of plasma instabilities, and to carry out research on plasma and energy confinement. Large stellarators now use ECH to achieve current-free operation and exploit their unique advantage as a steady-state toroidal approach to fusion.

In several large tokamak installations, most notably the Joint European Tokamak (JET), deuterium–tritium plasmas have been heated to ignition temperatures and net fusion energy has been released. Encouraged by such achievements, the major fusion research programs have undertaken a broad collaboration including the United States, Japan, Russia, and the European Community to design, construct, and operate the International Thermonuclear Experimental Reactor (ITER). It appears likely that ECH will perform several important functions in ITER, including startup, auxiliary heating, and suppression of tearing modes. It is also possible that ECH could be used to drive the noninductive plasma currents required for steady-state operation, if that type of tokamak is deemed to be advantageous. Sources of microwave power and low-loss distribution systems have been under intensive development in recent years and gyrotrons in the required frequency and power range are now in operation at several tokamak and stellarator facilities.

These government-funded ECH research programs stimulated advances in various aspects of microwave technology that were essential to fusion applications, particularly the high-power cw millimeter microwave sources mentioned earlier; but the recognition of potential commercial applications of ECH plasmas led to entirely different directions for development. Rather than seeking to create plasmas with extremely high energy density, the developers of commercial ECH technologies typically sought to create large volumes of quiescent plasma with highly uniform densities and temperatures that were typically no greater than 10 eV. Increasingly, arrays of permanent magnets were employed in ECH plasma sources to replace the water-cooled or super conducting DC magnets typical of fusion experiments. Innovative coupler designs were eventually developed to facilitate the use of the ubiquitous and inexpensive 2.45 GHz microwave power sources in commercial ECH plasma devices.

In addition to these terrestrial laboratory investigations of ECH plasmas, there have been several efforts to explore possible applications of ECH to magnetospheric plasmas using ground-based antenna arrays to launch electromagnetic waves along various trajectories into the earth's magnetosphere. One goal of these active ECH experiments in space is a means of precipitating energetic electrons out of the magnetosphere to prevent damage to satellites, astronauts, and ground-based communications networks. There have also been suggestions that ECH could be used to model phenomena of astrophysical interest by employing laboratory experiments whose results can be scaled in size to provide useful insights into the behavior of the larger cosmic systems that seem to exhibit effects of nonthermal plasmas.

Thus, over the past five decades ECH has been employed in a wide range of circumstances encompassing microwave frequencies from 2 to 200 GHz and power levels ranging from less than 1 kW to 1 MW per microwave source. Magnetic configurations utilized in these applications have included simple magnetic mirrors, various types of open-ended magnetic wells, many toroidal devices, as well as magnetic geometries intended to produce unconfined plasmas for industrial processes. Much has been learned about the fundamental aspects of ECH although, regrettably, the pressure to apply ECH in large experiments has meant that some underlying phenomena still need more detailed theoretical and experimental research to resolve outstanding issues that remain. Nonetheless, much ECH physics is relatively mature – a claim that hopefully will be supported by the present work. In the future, the body of ECH science seems likely to find an increasingly wide range of goal-oriented applications. Furthermore, the remarkable achievements of ECH, particularly in regard to the generation of steady-state high energy density plasmas, are so strikingly novel and so rich in potential for further discovery that future basic research efforts are likely to be undertaken to examine phenomena that cannot readily be produced by other means and in other media.

The goal of the present work is to collect in one place most of the basic components of the science of ECH as a resource for present and future students and researchers in the physics of high energy density, relativistic-electron plasmas, as well as for scientists and engineers who are seeking to develop more utilitarian applications of ECH. In the early chapters of the book, emphasis is given to the underlying fundamental physics that governs the outcome of any particular ECH experiment. Later chapters use the published results from various experiments to examine the ways in which these underlying phenomena work in collaboration to determine the properties of ECH plasmas.

ECH of plasmas involves a number of fundamental plasma physics phenomena whose basic properties are relatively well established. These include the motions of individual electrons in various types of static magnetic fields as well as the propagation of electromagnetic waves in low-temperature magnetized plasmas. Chapter 2 deals with the analysis of illustrative types of magnetostatic fields with special emphasis on those properties that are critical to ECH. The motions of individual electrons in these magnetostatic fields are then discussed in Chapter 3. Chapter 4 addresses the coupling of microwave power into plasmas by employing highly simplified models of the plasmas and magnetic fields. Although simplified, these models are particularly applicable to the "quasioptical" plasmas in large contemporary tokamaks and stellarators as well as the ionosphere. The dynamical response of electrons to spatially localized resonant microwave electric fields, while less thoroughly documented, has been investigated by many workers with results that are presented in Chapter 5.

ECH also involves a number of plasma physics phenomena that are not as well established but are especially important, for example, in the generation of relativistic-electron plasmas with very high energy densities. Chapter 6 deals with applicable theories of plasma equilibria based, in the first instance, on simple transport models of plasma particles and heat and, in the second instance, on somewhat ad hoc microscopic models of the anisotropic equilibria confined in magnetic mirror configurations. Chapter 7 summarizes several theories of the stability of ECH plasmas in order to provide a basis for the interpretation of experiments which illustrate the dominant observable properties of specific archetypal ECH plasmas. Several such experiments in magnetic mirror devices are summarized in Chapter 8 and interpreted as fully as possible in the context of the basic ECH physics presented in the earlier chapters.

As was mentioned earlier, many of the present generation of tokamaks and stellarators use multimegawatt ECH power levels at frequencies as high as 157 GHz for several essential aspects of their functioning. Results from several of these as well as earlier tokamak experiments are interpreted in Chapter 9 using the basic physics developed in the earlier chapters.

The ELMO Bumpy Torus employed ECH in several unique roles and the key features of these are discussed in Chapter 10. Chapter 11 discusses some of the ongoing and potential future applications of ECH to space plasma phenomena, again emphasizing aspects of ECH physics that are of unique importance to these applications. Chapter 12 presents a brief overview of some of the technological aspects of the microwave sources and distribution systems that have permitted the dramatic increase in the applications of ECH to the large fusion installations.

Finally, Chapter 13 discusses the more speculative use of frequency-modulated microwave power with steady-state current drive in tokamaks as the main illustration.

As expected in a field that has been developing over five or more decades, much of the basic material in the early chapters of this book is available in many older works. Here this type of archival material is presented in as concise a form as possible and in a uniform notation and system of units (rationalized MKS) with references to much of the earlier work, particularly works that include copious references. The choice of topics covered was largely determined by the interpretative needs of the experiments to be discussed in the later chapters and readers may notice regrettable gaps. The experiments were chosen with the aim of permitting the reader to verify for himself the applicability of the basic ECH phenomenology and obviously many important experiments could not be included. In this regard, the present work differs fundamentally from a review of ECH. Fortunately, there are several excellent such reviews available to the interested reader [1, 2]. Exercises are included at the end of each chapter to encourage students to internalize and make concrete what otherwise might remain vague and intangible.

References

1 R. Prater, *Physics of Plasmas* **11**, 2349 (2004) and works cited therein.

2 V. Erckmann and U. Gasparino, *Plasma Phys. and Control. Fusion* **36**, 1869 (1994) and works cited therein.

2
Magnetic Fields

Electron cyclotron heating (ECH) depends essentially on properties of the magnetic field configuration in ways that we will consider in subsequent chapters. Presently it is perhaps self-evident that if low-energy plasma electrons are to be heated rapidly, they must be able to pass freely through a resonant interaction region where the magnetic intensity, B, is approaching the resonant value, i.e., the value at which the local cold-electron fundamental gyrofrequency, Ω_e, equals the frequency of the applied microwave power, $\omega_\mu = 2\pi f_\mu$. Since $\Omega_e = eB/m$, the resonant magnetic intensity is given by

$$B_{res} = 2\pi(m/e)\mathbf{f}_\mu, (= 1\,\mathrm{T\ at\ 28\ GHz}),$$

where $-e$ and m are the charge and mass of the electron, respectively. We will have a detailed discussion on Doppler-shifted resonance for relativistic electrons later. The location in space where $B = B_{res}$ will generally be referred to here and in what follows as the "resonance surface".

In addition to unrestricted access to the resonance surfaces, it is also important that heated electrons be prevented from striking any nearby material surfaces or escaping rapidly from the enclosing chamber. This is especially true if ECH is used to create high energy density, hot-electron plasmas; however, good confinement of heated electrons is also essential if ECH is to be used for efficient production of dense, low-temperature plasmas. Simple magnetic-mirror configurations [1] have often been employed for a wide range of ECH applications since they provide good confinement of low-energy electrons as well as energetic electrons. The low-energy electrons are electrostatically confined by the equilibrium ambipolar electric field (see Chapter 6), while the more energetic electrons are magnetically confined by the magnetic mirror effect (see Chapter 3). We will, therefore, frequently employ the simple magnetic mirror configuration as a useful paradigm for discussing the aspects of the static magnetic field that are critical to the ECH process. We will also briefly describe other magnetic field configurations in which ECH is being used, particularly the tokamak toroidal magnetic confinement configuration as well as planetary magnetospheres.

Electron Cyclotron Heating of Plasmas. Gareth Guest

ISBN: 978-3-527-40916-7

2.1 Magnetic Mirrors: Field Calculations Using the Vector Potential

An accurate mathematical description of the magnetic field of a simple magnetic mirror is given, for example, by Morozov and Solov'ev [2]. The field is generated by two or more collinear coils made up of a suitable number of circular current loops connected in series. As derived by Smythe [3], for example, the vector potential of each coil is given in cylindrical coordinates (ρ, ϕ, z) by the single component

$$A_\phi(\rho, z) = (\mu_0 I/\pi)k^{-1}(r_c/\rho)^{1/2}[(1-k^2/2)K(k^2)-E(k^2)], \quad (2.1)$$

where

$$k^2 = 4r_c\rho[(r_c+\rho)^2+z^2]^{-1}. \quad (2.2)$$

In these expressions, $K(k^2)$ and $E(k^2)$ are the complete elliptic integrals [4], I is the current in the coil, μ_0 is the permeability of free space, and r_c is the radius of the circular current loop. Note that z is the axial field position relative to the plane of the coil. The components of the magnetic field are then obtained from $\mathbf{B} = \nabla \times \mathbf{A}$. For completeness, we include here a brief recapitulation of Smythe's derivation. Our starting point is the set of three time-independent equations:

$$\nabla \times \mathbf{H} = \mathbf{j}$$
$$\nabla \cdot \mathbf{B} = 0$$
$$\mathbf{B} = \mu_0 \mathbf{H}.$$

Since $\nabla \cdot \mathbf{B} = 0$, we can set $\mathbf{B} = \nabla \times \mathbf{A}$ and choose $\nabla \cdot \mathbf{A} = 0$. The vector potential is then given by a solution of Poisson's equation, since

$$\nabla \times (\nabla \times \mathbf{A}) = \nabla(\nabla \cdot \mathbf{A}) - \nabla^2\mathbf{A} = -\nabla^2\mathbf{A} = \mu_0\mathbf{j}$$

so that

$$\mathbf{A} = (\mu_0/4\pi)\int \mathbf{j}(\mathbf{r}')[(\mathbf{r}-\mathbf{r}')^2]^{-1/2}d^3r'.$$

The current density in a circular current-carrying loop consists solely of the azimuthal component j_ϕ, which can be represented by a product of delta-functions:

$$j_\phi = (I/2\pi)\delta(r'-r_c)\delta(z'-z_c)/r_c.$$

Here the radius of the coil is r_c and its center is at $x' = y' = 0$ and $z' = z_c$. Clearly, only the azimuthal component of the vector potential is nonvanishing. The geometry of the situation is illustrated in Figure 2.1.

The volume integral, $\int \mathbf{j}(\mathbf{r}')[(\mathbf{r}-\mathbf{r}')^2]^{-1/2}d^3r'$ reduces to a line integral around the loop, since

$$j_\phi \cos\phi \, \rho d\phi \, d\rho \, dz = I \cos\phi \, \rho d\phi$$

so that

$$A_\phi = 2(\mu_0 I/4\pi)\int r_c \cos\phi \, d\phi \, [(\mathbf{r}-\mathbf{r}')^2]^{-1/2},$$

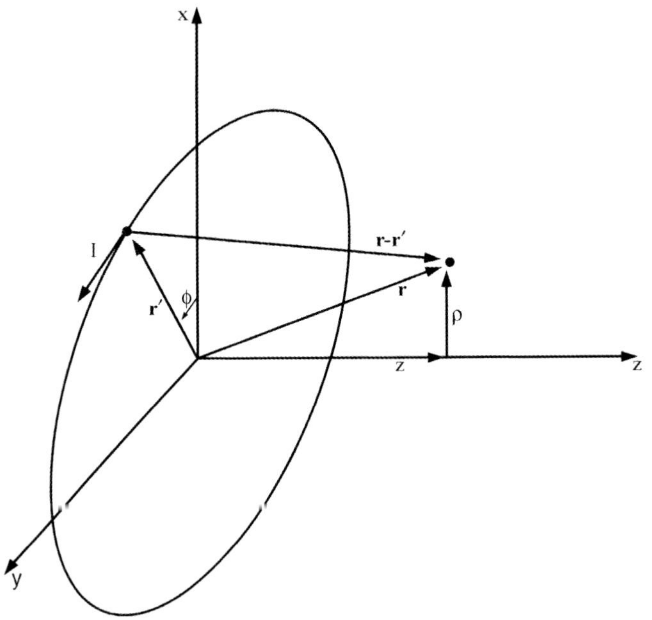

Figure 2.1 The geometry for deriving the equation for the vector potential of a single circular current loop, following Smythe [3].

where the integral runs from 0 to π. In view of the axisymmetry of the field, we may choose the point of observation, **r**, to lie in the $x-z$ plane. Then

$$\begin{aligned}(\mathbf{r}-\mathbf{r}')^2 &= (x-x')^2+(y-y')^2+(z-z')^2\\ &= (r-r_c\cos\phi)^2+(0-r_c\sin\phi)^2+(z-z_c)^2\\ &= r^2+r_c^2-2\,rr_c\cos\phi+(z-z_c)^2.\end{aligned}$$

If we now let $\phi=\pi+2\theta$, we have $d\phi=2d\theta$ and $\cos\phi=2\sin^2\theta-1$ and we then obtain the following result for A_ϕ:

$$A_\phi=(\mu_0 I r_c/\pi)\int d\theta(2\sin^2\theta-1)[r^2+r_c^2-2rr_c(2\sin^2\theta-1)+(z-z_c)^2]^{-1/2}$$

where the integral now runs from 0 to $\pi/2$. One can readily express this integral in terms of the two complete elliptic integrals [4]:

$$K(k^2)=\int(1-k^2\sin^2\theta)^{-1/2}\,d\theta$$

$$E(k^2)=\int(1-k^2\sin^2\theta)^{1/2}\,d\theta,$$

where both integrals run from 0 to $\pi/2$. In this way, we obtain Eq. (2.1) as given by Smythe [3]. One can show by tedious but straightforward differentiation that the components of the magnetic intensity are given by [3]

$$B_\rho=(\mu_0 I/2\pi)(z/\rho)[(r_c+\rho)^2+z^2]^{-1/2}\{-K+(r_c^2+\rho^2+z^2)[(r_c-\rho)^2+z^2]^{-1}E\},\tag{2.3}$$

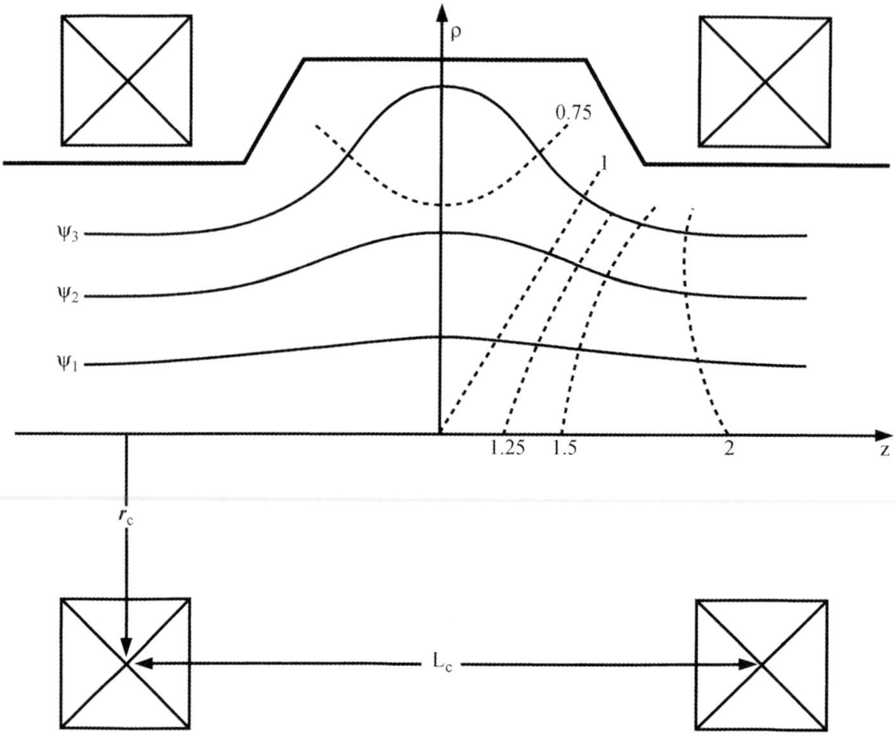

Figure 2.2 The major spatial properties of a simple magnetic mirror field with a 2 : 1 mirror ratio showing three flux surfaces and five mod-B surfaces.

and

$$B_z = (\mu_0 I/2\pi)[(r_c+\rho)^2+z^2]^{-1/2}\{K+(r_c^2-\rho^2-z^2)[(r_c-\rho)^2+z^2]^{-1}E\}. \quad (2.4)$$

We now consider a simple two-coil magnetic-mirror configuration indicated schematically in Figure 2.2.

In this rudimentary array, the distance between the coils, L_c, relative to the effective radius of the current-carrying coils, r_c, determines the magnetic mirror ratio on axis, M, defined as $M = B_{max}/B_{min}$. In the example of Figure 2.2, we have chosen $L_c = 2.4r_c$ to yield a mirror ratio on axis approximately equal to 2 : 1.

The formulas for A_ϕ, B_ρ, and B_z can be evaluated numerically using the polynomial approximations given, for example, by Abramowitz and Stegun [4]. For the axisymmetric simple magnetic mirror, the vector potential, A_ϕ, can be used to determine the locations of the surfaces on which the magnetic flux, $\Psi = \int \mathbf{B}\cdot d\mathbf{S}$ is constant everywhere. From Stokes theorem we have

$$\Psi(\rho, z) = \int \mathbf{B}\cdot d\mathbf{S} = \int (\nabla \times \mathbf{A})\cdot d\mathbf{S} = \int \mathbf{A}\cdot d\mathbf{l} = 2\pi\rho A_\phi(\rho, z). \quad (2.5)$$

The direction of the magnetic field at any point in the $\phi = \text{constant}$ plane is then given by the tangent to the corresponding flux surface through that point. We can use rudimentary numerical techniques to evaluate the flux $\Psi(\rho, z)$ and thereby map out the intersections of the (iso)flux surfaces with the (ρ, z) plane. Three such flux surfaces are shown in Figure 2.2.

With the same numerical techniques used earlier, we can evaluate the two components of the magnetic-mirror field at any point in the field and map out the surfaces on which the magnitude of the magnetic intensity is constant. These surfaces, or rather their intersections with the (ρ, z) plane, are usually referred to as "Mod-B" surfaces. The resonance surface is a particularly significant Mod-B surface for ECH, as we shall see. In Figure 2.2, we have displayed the contours on which the magnetic intensity takes on the values of 0.75, 1, 1.25, 1.5, and 2, relative to the value at the origin, $\rho = z = 0$.

2.2 Orthogonal Curvilinear Coordinates and Clebsch Representations

It will sometimes be helpful to use an orthogonal curvilinear coordinate system [5], one of whose basis vectors is parallel to the magnetic field [6]. Such a coordinate system can be constructed for our axisymmetric magnetic field by exploiting the fact that in this case the gradient of the magnetic flux, $\nabla\Psi(\rho, z)$, the ϕ-direction, $\nabla\phi$ and the magnetic field, $\mathbf{B}$, are all mutually orthogonal; since

$$\mathbf{B} = \nabla \times \mathbf{A} = -\partial A_\phi/\partial z\, \mathbf{u}_\rho + \rho^{-1}\partial(\rho A_\phi)/\partial\rho\, \mathbf{u}_z \quad \text{and}$$

$$\nabla\phi \times \mathbf{B} = \rho^{-1}[\partial(\rho A_\phi)/\partial\rho\, \mathbf{u}_\rho + \partial(\rho A_\phi)/\partial z\, \mathbf{u}_z] = \nabla\Psi(\rho, z)/(2\pi\rho).$$

In the notation of Ref. [5], we designate the three curvilinear coordinates as

$$\xi_1 = \Psi \quad \xi_2 = \phi \quad \xi_3 = s, \tag{2.6}$$

where s is the distance measured along the magnetic line of force. The three mutually orthogonal unit vectors are

$$\mathbf{u}_1 = \nabla\Psi/|\nabla\Psi|\, \mathbf{u}_2 = \nabla\phi \quad \mathbf{u}_3 = \mathbf{B}/B \equiv \mathbf{b}, \tag{2.7}$$

and the corresponding scale factors are

$$h_1 = (\rho B)^{-1} \quad h_2 = \rho \quad h_3 = HB, \tag{2.8}$$

where the Jacobian, H, is related to the curvature of the magnetic lines of force, κ, by

$$HB = \exp\left[-\int d\Psi\kappa/(\rho B)\right], \tag{2.9}$$

with the curvature given by

$$\boldsymbol{\kappa} = (\mathbf{b}\cdot\nabla)\mathbf{b} = \kappa\mathbf{u}_1 \tag{2.10}$$

This coordinate system is one example of a Clebsch representation [7] of the magnetic field:

$$\mathbf{B} = \nabla\alpha \times \nabla\beta \tag{2.11}$$

where α and β are constant on a field line. For the present case

$$\alpha = \Psi/2\pi \text{ and } \beta = \phi. \tag{2.12}$$

The curvature of the magnetic lines of force, $\kappa = (\mathbf{b} \cdot \nabla)\mathbf{b}$, has particular significance for the electron drift motions as well as for plasma stability. A useful alternative expression for this property of the magnetic lines of force can be obtained from the standard vector identity:

$$\nabla(\mathbf{A} \cdot \mathbf{B}) = \mathbf{A} \times (\nabla \times \mathbf{B}) + \mathbf{B} \times (\nabla \times \mathbf{A}) + (\mathbf{B} \cdot \nabla)\mathbf{A} + (\mathbf{A} \cdot \nabla)\mathbf{B}$$

We apply this identity to

$$\begin{aligned}\nabla(\mathbf{B} \cdot \mathbf{B}) &= 2\mathbf{B} \times (\nabla \times \mathbf{B}) + 2(\mathbf{B} \cdot \nabla)\mathbf{B} \\ &= 2\mathbf{B} \times (\nabla \times \mathbf{B}) + 2(B\mathbf{b} \cdot \nabla)B\mathbf{b} \\ &= 2\mathbf{B} \times (\nabla \times \mathbf{B}) + 2\mathbf{B}(\mathbf{b} \cdot \nabla)B + 2B^2(\mathbf{b} \cdot \nabla)\mathbf{b}\end{aligned}$$

so that

$$B^2(\mathbf{b} \cdot \nabla)\mathbf{b} = B[\nabla B - \mathbf{b}(\mathbf{b} \cdot \nabla)B] - \mathbf{B} \times (\nabla \times \mathbf{B}) = B\nabla_\perp B - \mathbf{B} \times (\nabla \times \mathbf{B}) \tag{2.13}$$

2.3 Magnetic Mirrors: Field Calculations Using the Scalar Potential

An alternative approach to describe the magnetic fields of magnetic-mirror configurations using the magnetic scalar potential, χ, has been discussed by Post [1]. This description provides a very useful approximate expression for some key properties of the magnetic field. In this approach, the magnetic intensity in current-free regions is given by $\mathbf{B} = -\nabla\chi$, where $\chi(\rho, z)$ is a solution of Laplace's equation, $\nabla^2\chi = 0$. If we require χ to be periodic in z, then the general solution is

$$\chi(\rho, z) = \sum C_n \sin(nk_0 z) I_0(nk_0\rho) \tag{2.14}$$

where $I_0(nk_0\rho)$ is the modified Bessel function, $k_0 = 2\pi/L$, L is the periodicity length in z, and the summation runs from zero to infinity. The two components of $\mathbf{B}$ are then given by

$$B_z = \sum C_n \cos(nk_0 z)\, I_0(nk_0\rho),$$

and

$$B_\rho = \sum C_n \sin(nk_0 z)\, I_1(nk_0\rho) \tag{2.15}$$

The lowest order terms in these infinite series can be written in the following form:

$$B_z = B_0\{[(M+1)/2] - [(M-1)/2]\cos(k_0 z)\, I_0(k_0\rho)\}, \tag{2.16}$$

and

$$B_\rho = -B_0[(M-1)/2]\sin(k_0 z)\, I_1(k_0\rho). \tag{2.17}$$

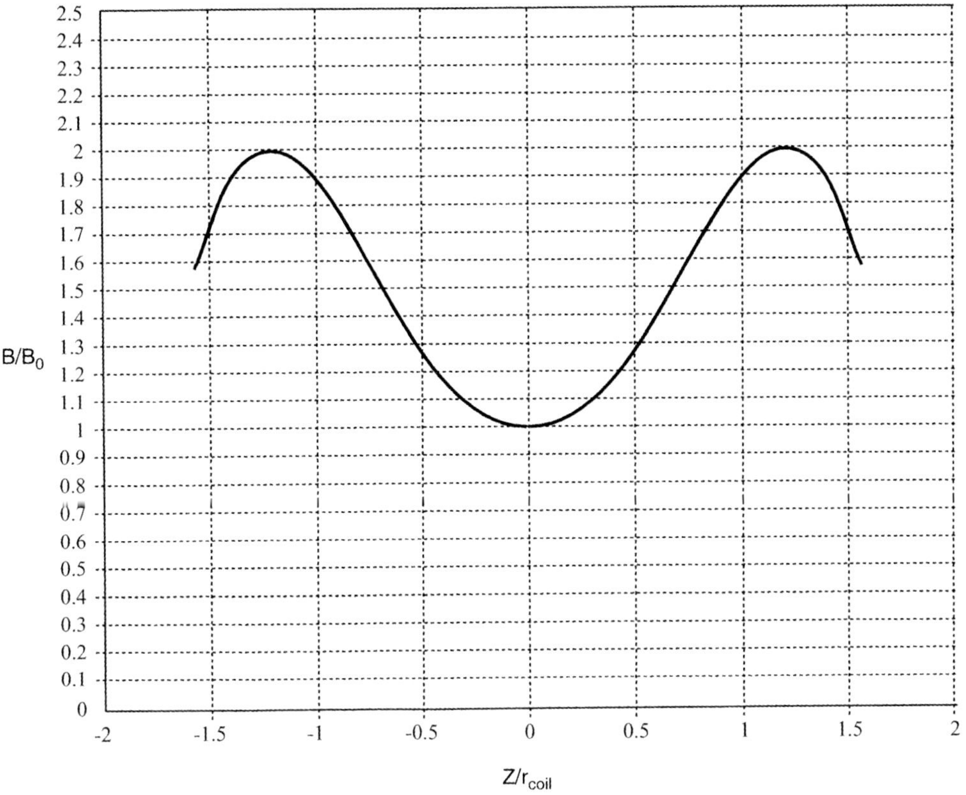

Figure 2.3 The variation of the magnetic intensity along the axis of the simple magnetic mirror field shown in Figure 2.2.

We will frequently use the approximate model for the axial variation of the magnetic intensity on the axis, $\rho = 0$, given by either of the following two equivalent versions of Eq. (2.16):

$$B(z) = B_0[1 + (M-1)\sin^2(k_0z/2)] = B_0\{[(M+1)/2]-[(M-1)/2]\cos k_0z\}, \tag{2.18}$$

where M is the mirror ratio and $k_0 = 2\pi/L_c$. For the case of the 2 : 1 mirror ratio, the simple model varies by less than 6% from the accurate mathematical model for all positions along the axis and between the coils. Figure 2.3 displays a plot of relative magnetic intensity on the axis of the simple magnetic mirror shown in Figure 2.2 as a function of the distance along the axis of the pair of coils.

The radial component of the magnetic intensity vanishes on the axis as well as on the symmetry plane midway between the coils, the so-called midplane of the configuration. In the midplane, the magnetic intensity falls with radial distance from the axis as shown in Figure 2.4. Here the accurate numerical results are shown as a solid line, while the approximate results from Eq. (2.16) are shown as a dashed line.

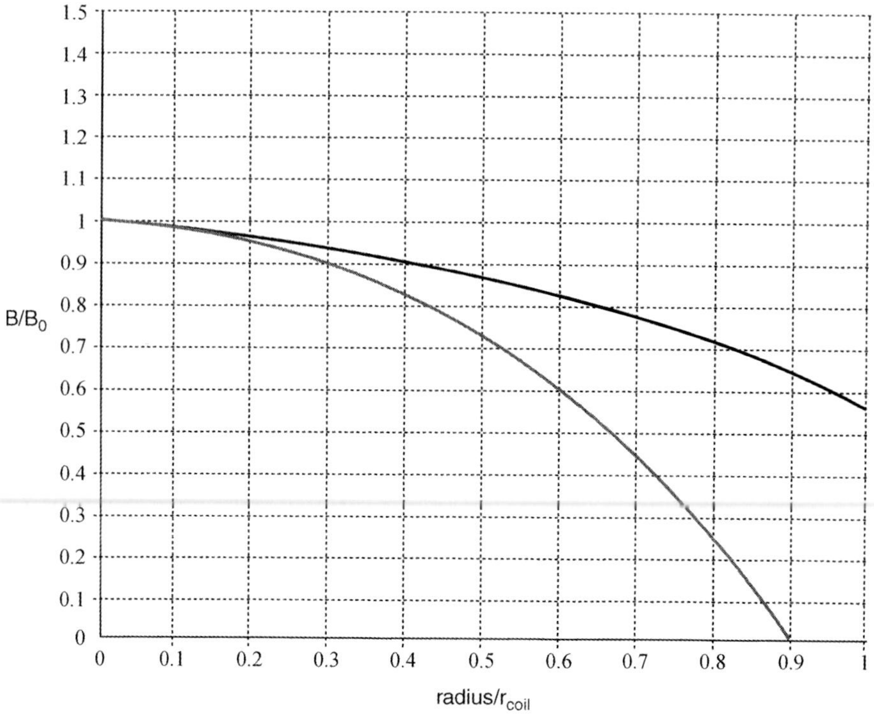

Figure 2.4 The radial dependence of the magnetic intensity on the midplane of the simple magnetic mirror field shown in Figure 2.2. The solid line is the result of numerical calculation, while the dashed line is from the approximate expression (2.17).

According to Eq. (2.16), the radial gradient of the magnetic intensity in the midplane is given to the lowest order by

$$\partial B_z/\partial \rho = -k_0 B_0[(M-1)/2] I_1(k_0\rho). \tag{2.19}$$

A convenient measure of this gradient is the characteristic length defined by $R_b = |(\partial \ln B_z/\partial \rho)^{-1}|$. Using Eq. (2.17) we find

$$R_b/L = \{[(M+1)/(M-1)] - I_0(k_0\rho)\}/[2\pi I_1(k_0\rho)], \tag{2.20}$$

where R_b/L is infinite on the axis but falls off rapidly with increasing radius. For small values of the argument $k_0\rho$, $I_0(k_0\rho) \approx 1$, and $I_1(k_0\rho) \approx k_0\rho/2$. Thus,

$$\rho R_b \approx L^2[\pi^2(M-1)]^{-1}, \tag{2.21}$$

and is roughly constant for a given simple magnetic mirror configuration. In Chapter 3, we will see that this feature of simple magnetic mirror configurations plays a major role in limiting the maximum energy of the confined electrons.

2.4
The Dipole Limit: Planetary Magnetic Fields

Planetary magnetospheres are similar in some respects to simple magnetic mirror configurations. In discussing the earth's magnetic field, for example, it is conventional to employ the dipole approximaton of Eq. (2.1) to describe the vector potential and magnetic field at distances larger than the dimensions of the currents [8]. In spherical polar coordinates, $\rho = r \sin\theta$ and $z = r\cos\theta$, the argument of the elliptic integrals is given by

$$k^2 = 4(r_c/r)\sin\theta\,[1 + 2(r_c/r)\sin\theta + (r_c/r)^2]^{-1}. \tag{2.22}$$

Clearly, if $r^2 \gg r_c^2$ then $k^2 \ll 1$ and $K(k^2)$ and $E(k^2)$ can be approximated by the following power series [9]:

$$\begin{aligned}(2/\pi)\,K(k^2) &= 1 + 2(k^2/8) + 9(k^2/8)^2 + \cdots \\ (2/\pi)E(k^2) &= 1 - 2(k^2/8) - 3(k^2/8)^2 - \cdots\end{aligned} \tag{2.23}$$

Thus, to lowest order in $(k^2/8)$

$$(1 - k^2/2)K(k^2) - E(k^2) = (2\pi)(k^2/8)^2$$

and

$$A_\phi(r,\theta) = (\mu_0 M/4\pi)\,r^{-2}\sin\theta, \tag{2.24}$$

where M is the dipole moment of the current loop, $M = \pi r_c^2 I$. The corresponding components of the magnetic intensity are given by

$$B_r = (\mu_0 M/2\pi) r^{-3}\cos\theta,$$

and

$$B_\theta = (\mu_0 M/4\pi)\,r^{-3}\sin\theta \tag{2.25}$$

2.5
Tokamaks: Rotational Transform and the "Safety Factor"

As a final example of relatively simple magnetic field configurations for which there are important applications of ECH, we consider the magnetic field in the tokamak [10]. It consists of a superposition of two separate fields, a toroidal field generated by currents carried in external windings distributed uniformly around the torus, and a poloidal magnetic field generated by toroidal currents induced in the confined plasma itself. The toroidal field can be conveniently evaluated using Ampere's law in its integral form:

$$\int \mathbf{B}_t \cdot \mathbf{dl} = \mu_0 \int \mathbf{j} \cdot \mathbf{dS} = \mu_0 I_t. \tag{2.26}$$

Here I_t is the total current carried by the external windings and threading the torus. Since the toroidal magnetic field is independent of the azimuthal angle,

$$\int \mathbf{B}_t \cdot \mathbf{dl} = 2\pi\rho B_t = \mu_0 I_t, \tag{2.27}$$

where ρ lies inside the toroidal vacuum chamber: $R_t - a \leq \rho \leq R_t + a$.

Here R_t and a are the major and minor radii, respectively, of the toroidal chamber. Clearly, the magnitude of the toroidal field varies as $1/\rho$:

$$B_t = \mu_0 I_t / 2\pi\rho \tag{2.28}$$

so that its maximum and minimum values inside the chamber are given, respectively, by

$$B_{t,max} = \mu_0 I_t / 2\pi(R_t - a) \text{ and } B_{t,min} = \mu_0 I_t / 2\pi(R_t + a). \tag{2.29}$$

The poloidal magnetic field strength, B_p, can also be obtained using Ampere's Law, provided the radial dependence of the plasma current density, j_p, is known:

$$\int \mathbf{B}_p \cdot \mathbf{dl}_p = \mu_0 \int j_p(r')2\pi\, r'\, dr'.$$

In the limit of large aspect ratios, $R_t/a \gg 1$, this reduces approximately to

$$rB_p = \mu_0 \int j_p(r') r'\, dr'$$

so that

$$aB_p(a) = \mu_0 I_p / (2\pi), \tag{2.30}$$

where I_p is the total plasma current in the discharge. Since the radial dependence of the plasma current was formerly not generally known, it became customary to describe the poloidal magnetic intensity in terms of the so-called safety factor q, where

$$q(r) = rB_t / (R_t B_p). \tag{2.31}$$

For stable tokamak operation, it is usually necessary that $q(a) \geq 1$ so that

$$B_p / B_t \leq r/R_t \ll 1.$$

The surfaces of constant B are thus approximately cylinders on which ρ = constant.

Magnetic lines of force spiral around the magnetic axis and close on themselves after q transits around the major axis of the torus if it is an integer. If q is not an integer, the spiraling lines of force trace out a magnetic flux surface. Electrons moving along the magnetic lines of force can be trapped, under suitable circumstances, between two magnetic mirrors for which the mirror ratio along a line of force with a distance r from the magnetic axis is

$$M = B_{max}/B_{min} = (R_t + r)/(R_t - r). \tag{2.32}$$

In many conventional tokamaks, the radial profile of the electron temperature, $T_e(r)$, is observed to be approximately Gaussian with $T_e(r) = T_e(0)\exp(-\alpha r^2/a^2)$. The constant $\alpha > 1$ is a "peaking factor". If the plasma parallel conductivity is proportional to $T_e^{3/2}$, the current density will also have a Gaussian profile: $j(r) = j(0)\exp(-3\alpha r^2/2a^2)$. The total plasma current is then

$$I_p = j(0)(2\pi a^2/3\alpha)[1 - \exp(-3\alpha/2)],$$

and the radial profile of the safety factor is then given by

$$q(r)/q(a) = (r^2/a^2)\{[1 - \exp(-3\alpha/2)]/[1 - \exp(-3\alpha r^2/2a^2)]\}$$

so that $q(a) = 1.5\alpha q(0)$.

References

1 R.F. Post, The magnetic mirror approach to fusion, *Nucl. Fusion*, **27**, 1579–1743 (1987).

2 A.I. Morozov and L.S. Solov'ev, The structure of magnetic fields, in *Reviews of Plasma Physics* (Acad. M.A. Leontovich, ed.), Vol. 2, p. 32, Consultants Bureau, New York (1966).

3 William R. Smythe, *Static and Dynamic Electricity*, third edition, pp. 290–291, McGraw-Hill, New York (1968).

4 *NBS Handbook of Mathematical Functions*, M. Abramowitz and I.A. Stegun, eds., U.S. Department of Commerce, National Bureau of Standards, Applied Mathematics Series 55, Washington, D.C. (1964) reprinted by Dover Publications, New York (1970), see Section 17.3, p. 590.

5 For a concise summary of curvilinear coordinate systems see David L. Book, *Revised and Enlarged Collection of Plasma Physics Formulas and Data*, pp. 16–20, NRL Memorandum Report 3332, Naval Research Laboratory, Washington, D.C., May 1977.

6 I.B. Bernstein, E.A. Frieman, M.D. Kruskal, and R.M. Kulsrud, *Proc. Roy. Soc.*, **A 224**, 17 (1058).

7 H. Grad and H. Rubin, *Proceedings of the Second United Nations Conference on Peaceful Uses of Atomic Energy*, United Nations, New York (1958) Vol. 31, p. 190.

8 For a brief description of the geomagnetic field with references to detailed studies, see Hannes Alfvén and Carl-Gunne Fälthammar, *Cosmical Electrodynamics*, Second edition, pp. 3–6, Oxford University Press, London (1963); for additional mathematical details, see George Arfken, *Mathematical Methods for Physicists*, third edition, pp. 672–676, Academic Press, New York (1985).

9 E. Jahnke, F. Emde and Friedrich Lösch, *Tables of Higher Functions*, sixth edition, p. 62, McGraw-Hill, New York (1960).

10 For an encyclopedic compendium on the tokamak, see John Wesson, *Tokamaks*, third edition, Oxford University Press, Oxford (2004).

Exercises

The following exercises require numerical evaluation of the formulas for the vector potential and the two components of the magnetic intensity in a simple, two-coil magnetic mirror configuration. A satisfactory program for carrying out the required numerical computations can readily be devised using common spreadsheet applications by employing the polynomial approximations for the elliptic integrals as cited, for example, in Abramowitz and Stegun [4].

2.1 *For the simple magnetic mirror configuration shown in Figure 2.2, determine the dependence of mirror ratio on the coil separation relative to the coil radius, L_c/r_c for $1 < L_c/r_c < 3$.*

2.2 *For the same simple magnetic mirror configuration, determine the radial location of the separatrix in the magnetic flux surfaces as a function of the separation of the two coils, L_c/r_c.*

2.3 *For the same simple magnetic mirror configuration shown in Figure 2.2, determine the magnetic intensity at the origin, $\rho = z = 0$ versus the ratio of the current in each coil to the effective radius of each coil, I/r_c.*

2.4 *Derive a formula for the curvature of the field lines near the axis of a simple magnetic mirror (a) at the midplane, and (b) in the mirror throat.*

2.5 *Verify the expression for the Jacobian for the (Ψ, ϕ, s) coordinate system, $HB = \exp[-\int d\Psi k/(B)]$, by evaluating both sides of the following vector identity: $(\nabla \times \mathbf{B}) \times \mathbf{B} = (\mathbf{B} \cdot \nabla)\mathbf{B} - \nabla B^2/2$. Use formulas from Ref. [5] or the following:*

$$\nabla f = \varrho B(\partial f/\partial \Psi)\boldsymbol{\Psi} + \varrho^{-1}(\partial f/\partial \zeta)\boldsymbol{\zeta} + (HB)^{-1}(\partial f/\partial s)\mathbf{b}$$

$$\begin{aligned}(\nabla \times \mathbf{A}) = {} & \boldsymbol{\Psi}(\varrho BH)^{-1}[(\partial(BHA_b)/\partial\zeta] - \partial(\varrho A_\zeta)/\partial\sigma] \\ & + \boldsymbol{\zeta}(\varrho/H)[\partial(A_\Psi/\varrho B)/\partial s - \partial(BHA_b)/\partial\Psi] \\ & + \mathbf{b}B[\partial(\varrho A_\zeta)/\partial\Psi - \partial(A_\Psi/\varrho B)/\partial\zeta]\end{aligned}$$

3
Electron Orbits

We now consider the unperturbed motions of individual electrons in magnetic fields such as the simple magnetic-mirror configuration discussed in Chapter 2. In most situations where electrons are confined for long times by a static magnetic field there are three well-separated time scales that can provide a convenient albeit approximate description of these motions; namely, gyration about magnetic lines of force, bounce along field lines and between magnetic mirrors, and drift across lines of force. We begin with a nonrelativistic description which will be generalized later to include relativistic effects.

3.1
Electron Gyromotion

The most rapid of the three motions is the (right-handed) electron gyration about magnetic lines of force. This gyration is caused by the Lorentz force

$$\mathbf{F} = \mathrm{m}\,\mathrm{d}\mathbf{v}/\mathrm{dt} = -\mathrm{e}(\mathbf{E} + \mathbf{v} \times \mathbf{B}). \tag{3.1}$$

In general, since the magnetic intensity varies along the trajectory of the electron, this equation of motion, Eq. (3.1), is intrinsically nonlinear. Rigorous treatments of this problem using expansions in the small parameter (m/e) have been published by several authors [1]. To obtain a useful approximate integration of this equation of motion we simply assume that within a region large enough to contain the electron for many periods of its gyromotion we may treat the local magnetic intensity, **B**, as spatially uniform and in the z-direction to write this equation as

$$\mathrm{d}\mathbf{v}/\mathrm{dt} + \Omega_e \mathbf{v} \times \mathbf{u}_z = -\mathrm{e}\mathbf{E}/\mathrm{m}. \tag{3.2}$$

Here $\mathbf{u}_z$ is a unit vector in the z-direction. This approach relies for its validity on the typical magnetic confinement situation in which the magnetic intensity varies negligibly over the extent of an electron's orbit, as discussed, for example, by Alfvén and Fälthammar as well as many others [1]. The two components of the equation of

Electron Cyclotron Heating of Plasmas. Gareth Guest

ISBN: 978-3-527-40916-7

motion perpendicular to the magnetic field are decoupled in the so-called rotating coordinates [2]:

$$v_{\pm} = (v_x \pm i v_y)/\sqrt{2}, \tag{3.3}$$

giving separate equations of motion for each component of the velocity:

$$\begin{aligned} dv_+/dt - i\Omega_e v_+ &= -eE_+/m \\ dv_-/dt + i\Omega_e v_- &= -eE_-/m \\ dv_z/dt &= -eE_{\parallel}/m. \end{aligned} \tag{3.4}$$

In the absence of electric fields, the time-varying perpendicular components of the unperturbed velocity are

$$v_{\pm}(t) = v_{\pm}(0)\exp\left(\pm i \int \Omega_e dt\right). \tag{3.5}$$

The velocity parallel to the (uniform) magnetic field is constant in time. Here we write the phase integral as $\int \Omega_e dt$ to emphasize the point that until the electron trajectory is known it is not possible to give precise values to $v_{\pm}(t)$ except in the local region where **B** is approximately constant. If we introduce a gyrophase angle, ϕ, such that at some (arbitrary) initial time

$$\begin{aligned} v_x(0) &= v_{\perp}(0)\cos\phi_0 \\ v_y(0) &= v_{\perp}(0)\sin\phi_0, \end{aligned}$$

then in the rotating coordinates the initial constants, $v_{\pm}(0)$, are given by

$$v_{\pm}(0) = v_{\perp}(0)\exp(\pm i\phi_0)/\sqrt{2},$$

and the time-dependent circular components of the velocity are simply

$$v_{\pm}(t) = v_{\perp}(0)\exp[\pm i\phi(t)]/\sqrt{2}, \tag{3.6}$$

where the time-dependent gyrophase angle is $\phi(t) = \phi_0 + \int \Omega_e dt$. In Cartesian coordinates the time-dependent perpendicular components of velocity are simply

$$v_x(t) = (v_+ + v_-)/\sqrt{2} = v_{\perp}(0)\cos\left(\phi_0 + \int \Omega_e dt\right),$$

and

$$v_y(t) = (v_+ - v_-)/i\sqrt{2} = v_{\perp}(0)\sin\left(\phi_0 + \int \Omega_e dt\right). \tag{3.7}$$

The perpendicular speed is constant in time and the electron gyration in the plane perpendicular to the magnetic field is a uniform circular motion. Since the Lorentz force provides the centripetal force that maintains the electron in uniform circular motion with gyroradius ρ, the radius of curvature of the orbit must satisfy

$$m\Omega_e^2\rho = e\,\mathbf{v} \times \mathbf{B}, \tag{3.8}$$

or $mv_{\perp}^2/\rho = ev_{\perp}B$, from which it follows that $\rho = v_{\perp}/\Omega_e$. The geometry of the orbit is schematically shown in Figure 3.1.

Provided the magnetic field is essentially constant over the orbit of the electron for times long compared to Ω_e^{-1}, Eqs. (3.7) can be integrated once more in time to obtain

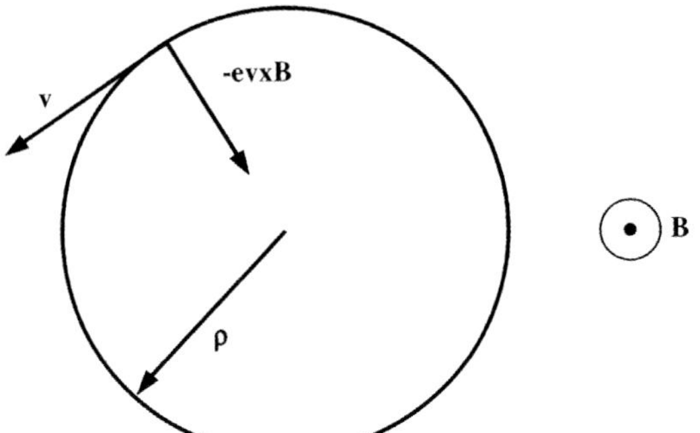

Figure 3.1 Projection of the electron gyro orbit onto the plane perpendicular to the magnetic intensity, **B**.

the following description of the time-dependent position (i.e., the "unperturbed orbit") of an electron:

$$\begin{aligned} x(t) &= x_0 + [v_\perp(0)/\Omega_e]\sin\left(\phi_0 + \int \Omega_e dt\right) - [v_\perp(0)/\Omega_e]\sin\phi_0 \\ y(t) &= y_0 - [v_\perp(0)/\Omega_e]\cos\left(\phi_0 + \int \Omega_e dt\right) + [v_\perp(0)/\Omega_e]\cos\phi_0 \\ z(t) &= z_0 + \int v_z dt, \end{aligned} \tag{3.9}$$

where Ω_e is a suitable average value of the local gyrofrequency. If we consider the particular case in which $x_0 = 0$, $y_0 = -[v_\perp(0)/\Omega_e]$, and $\phi_0 = 0$, then we have

$$\begin{aligned} x(t) &= [v_\perp(0)/\Omega_e]\sin\left(\int \Omega_e dt\right) \\ y(t) &= -[v_\perp(0)/\Omega_e]\cos\left(\int \Omega_e dt\right), \end{aligned}$$

and clearly the projection of the electron orbit onto the plane perpendicular to the magnetic field is a circle of radius $\rho = v_\perp(0)/\Omega_e$ as sketched in Figure 3.1. Thus, the unperturbed orbit of the electron is given under these assumptions by

$$\begin{aligned} x &= x_0 + \rho\sin\left(\phi_0 + \int \Omega_e dt\right) - \rho\sin\phi_0 \\ y &= y_0 - \rho\cos\left(\phi_0 + \int \Omega_e dt\right) + \rho\cos\phi_0 \\ z &= z_0 + \int v_z dt. \end{aligned} \tag{3.10}$$

The gyrating electron constitutes a current loop with a magnetic moment whose magnitude is given by

$$\mu = I\pi\rho^2 = (e\Omega_e/2\pi)\pi[v_\perp(0)/\Omega_e]^2 = mv_\perp(0)^2/2B = W_\perp(0)/B, \tag{3.11}$$

and whose direction is opposite to that of the magnetic field. Here $W_\perp = mv_\perp^2/2$ will frequently be referred to as the perpendicular kinetic energy, i.e., the kinetic energy in motion perpendicular to the magnetic field. Note that the magnetic flux through the gyrating electron's orbit is $\Psi_e = \pi\rho^2 B = 2\pi(m/e^2)\mu$. Small gradual variations in the magnetic intensity along the electron trajectory lead to additional motions of the instantaneous center of gyration, the so-called guiding center of the electron orbit. We return to this issue in subsequent sections, where we will take advantage of the adiabatic invariance of μ and Ψ_e to obtain useful descriptions of these slower motions of the electron guiding center. Since the gyrating electron can be considered as a quasisuperconducting current loop it is not surprising that the flux linking its gyro orbit should be invariant (provided that the changes in the magnetic intensity are sufficiently gradual).

3.2 Electron Bounce Motion

If the static magnetic field varies sufficiently slowly in space the magnetic moment, $\mu = W_\perp/B$, of the gyrating electron is an adiabatic invariant, an approximate constant of the motion. An empirical criterion that must be satisfied for this invariance to obtain in magnetic mirrors is [3]

$$\rho d\ln B/dr < 0.05. \tag{3.12}$$

Here again $W_\perp = mv_\perp^2/2$ is the kinetic energy in motion perpendicular to the magnetic field; i.e., the "perpendicular kinetic energy." Along the electron trajectory μ and Ψ_e are constant. Note that as an electron moves from regions of weaker magnetic intensity into regions of stronger magnetic intensity $W_\perp$ must increase proportionately to maintain μ constant. The Lorentz force cannot change the electron's total energy, $\varepsilon = W_\perp + W_\|$, since the force is perpendicular to the velocity, and therefore the kinetic energy in motion parallel to the magnetic field must decrease proportionately:

$$W_\| = \varepsilon - W_\perp = \varepsilon - \mu B. \tag{3.13}$$

Clearly $W_\|$ and thus $v_\|$ will vanish if B reaches the value ε/μ and at that point the electron's motion along the magnetic line of force will reverse; i.e., the electron will be reflected by the increasing magnetic field. In the absence of other influences electrons will bounce back and forth between the turning points where $B = B_t = \varepsilon/\mu$, provided only that $B_t = \varepsilon/\mu < B_{max}$, the maximum value of magnetic intensity in the confined region. It is customary to express this condition for magnetic-mirror confinement in terms of the electron orientation in velocity space, often called the "pitch angle," at the midplane of the magnetic mirror, $z = 0$:

$$[\varepsilon/\mu]_{z=0} = \left[\left(1 + v_\|^2/v_\perp^2\right)B\right]_{z=0} < B_{max},$$

or

$$\left[\left(v_\|^2/v_\perp^2\right)\right]_{z=0} < B_{max}/B_o - 1 = M - 1. \tag{3.14}$$

Mirror-confined electrons will bounce back and forth between the magnetic mirrors until they are scattered into the "loss cone" where $[(v_{||}/v_{\perp})]_{z=0} > (M-1)^{1/2}$ and they can promptly escape from the confined region. We can use the approximate simple magnetic mirror field from Chapter 2, Eq. (2.18), to obtain a useful expression for the period, τ_b, of the bounce motion of adiabatic electrons:

$$\tau_b = \int dz\, v_z^{-1} = \int dz[2(\varepsilon-\mu B)/m]^{-1/2}, \tag{3.15}$$

where the integral is over a complete bounce cycle. If we define M_t, the mirror ratio at the electron turning point, $M_t = \varepsilon/\mu B_o$, and let $B(z) = B_o[\,1 + (M-1)\sin^2(k_o z/2)]$, as in Eq. (2.18), we find for the bounce period,

$$\begin{aligned}\tau_b &= (8/k_o)\{2(\varepsilon/m)[(M-1)/M_t]\}^{-1/2}K(k^2)\\ &= [8/k_o v_\perp(0)](M-1)^{-1/2}K(k^2),\end{aligned} \tag{3.16}$$

where $K(k^2)$ is the complete elliptic integral whose argument is $k^2 = (M_t - 1)/(M - 1)$. A less accurate but more intuitive approach utilizes the fact that an object with magnetic moment μ, when placed in an inhomogeneous magnetic field is acted on by a force $\mathbf{F} = \mu \bullet \nabla B$. Applied to an electron moving along the axis of a simple magnetic mirror this gives

$$F_z = md^2z/dt^2 = -\mu\partial B/\partial z = -\mu B_o k_o^2[(M-1)/2]z,$$

where we have again used Eq. (2.18) to model the simple magnetic mirror field and retained only the lowest order term in the Taylor series expansion of $\sin(k_o z)$. The electron will oscillate about $z = 0$ with a frequency given by $\omega_b = k_o v_\perp(0)(M-1)^{1/2}/2$.

To compare with the more accurate result, Eq. (3.16), we rewrite Eq. (3.16) in the form

$$\omega_b = 2\pi/\tau_b = [k_o v_\perp(0)(M-1)^{1/2}/2] \times [(\pi/2)/K(k^2)]. \tag{3.17}$$

For typical conditions the bounce frequency, $\omega_b = 2\pi/\tau_b$, may be several orders of magnitude smaller than the gyrofrequency; supporting our use of the local approximation for B to describe the gyromotion.

3.3
Electron Drift Motions

In addition to the rapid gyration about the magnetic line of force and the much slower bounce along the line of force, there is a still slower motion of the electron across the lines of force arising from static electric fields perpendicular to **B** as well as any spatial gradients in the magnetic intensity. These slow drift motions are conveniently described in terms of the position of the "guiding center," $\mathbf{r}_{gc}$, indicated schematically in Figure 3.2.

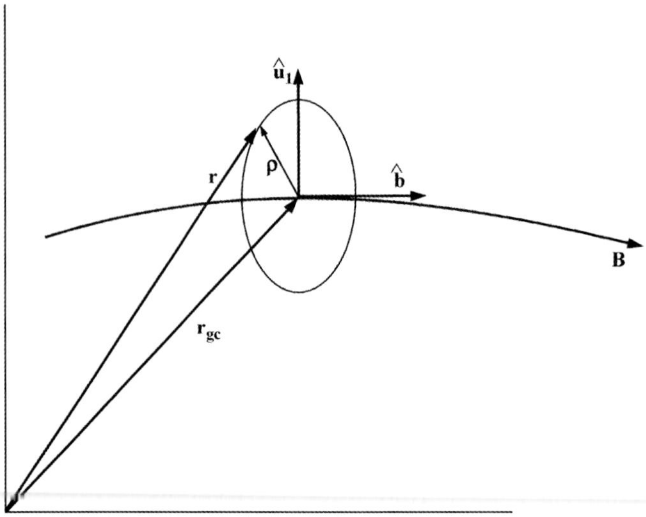

Figure 3.2 The guiding center geometry.

Here the position of the electron is given by **r** and the (vector) gyroradius by ρ. The location of the guiding center is then given by

$$\mathbf{r} = \mathbf{r}_{gc} + \rho \tag{3.18}$$

with $\rho = (m/e)\mathbf{v} \times \mathbf{B}/B^2$. In a uniform magnetic field the velocity of the guiding center is

$$\mathbf{v}_{gc} = d\mathbf{r}_{gc}/dt = d\mathbf{r}/dt - d\rho/dt = \mathbf{v} - (m/e)d\mathbf{v}/dt \times \mathbf{B}/B^2. \tag{3.19}$$

But the Lorentz force requires that $-(m/e)d\mathbf{v}/dt = \mathbf{E} + \mathbf{v} \times \mathbf{B}$ so that

$$\mathbf{v}_{gc} = \mathbf{v} + (\mathbf{E} + \mathbf{v} \times \mathbf{B}) \times \mathbf{B}/B^2 = \mathbf{E} \times \mathbf{B}/B^2 + v_{\parallel}\mathbf{b}. \tag{3.20}$$

Thus, in the presence of a transverse electrostatic field, **E**, the guiding center will drift perpendicular to both **E** and **B** with a velocity given by $\mathbf{E} \times \mathbf{B}/B^2$ while moving along the magnetic line of force with the parallel speed of the electron. Note that the $\mathbf{E} \times \mathbf{B}$ drift is independent of the sign of the charge on the particle.

In order to describe the drift motion arising from the inhomogeneities in the magnetostatic field we approximate the magnetic intensity at the position of the electron by the first two terms of a Taylor's series expansion about the position of the guiding center and, eventually, average the equations over the rapid gyromotion. We write

$$\mathbf{B}(\mathbf{r}_{gc} + \rho) = \mathbf{B}(\mathbf{r}_{gc}) + (\rho \bullet \nabla)\mathbf{B}(\mathbf{r}) \equiv \mathbf{B}(\mathbf{r}_{gc}) + \Delta\mathbf{B}(\mathbf{r}). \tag{3.21}$$

We can obtain an equation of motion for the guiding center by differentiating $\mathbf{r}_{gc}$ twice with respect to time and substituting from the Lorentz force law:

$$\mathbf{a}_{gc} = d^2\mathbf{r}_{gc}/dt^2 = \mathbf{a} - d^2\rho/dt^2. \tag{3.22}$$

In the absence of the electrostatic fields that were treated earlier,

$$\mathbf{a} = -(e/m)(\mathbf{v} \times \mathbf{B}) = -(e/m)(\mathbf{v}_{gc} + d\rho/dt) \times \mathbf{B}. \tag{3.23}$$

Using the results of Section 3.1 we can express the time-dependent vector gyroradius in terms of Cartesian components as

$$\rho/\rho = \mathbf{u}_1 \cos\phi + \mathbf{u}_2 \sin\phi,$$

where

$$\phi = \phi_o + \int \Omega dt. \tag{3.24}$$

The unit vectors $\mathbf{u}_1$ and $\mathbf{u}_2$, together with the unit vector $\mathbf{b} = \mathbf{B}/B$, comprise a local right-hand orthogonal coordinate system whose origin is at the position of the guiding center. We may choose $\mathbf{u}_1$ to be normal to the flux surface containing the magnetic line of force under examination and set $\mathbf{u}_2 = \mathbf{b} \times \mathbf{u}_1$. Then

$$d\rho/dt = \Omega\rho(-\mathbf{u}_1 \sin\phi + \mathbf{u}_2 \cos\phi) = -\Omega\rho \times \mathbf{b},$$

and

$$d^2\rho/dt^2 = -\Omega^2\rho(\mathbf{u}_1 \cos\phi + \mathbf{u}_2 \sin\phi) = -\Omega^2\rho. \tag{3.25}$$

With these substitutions and some modest amount of vector manipulation the guiding center equation of motion can be expressed as follows:

$$\begin{aligned}
m\mathbf{a}_{gc} &= md^2\mathbf{r}_{gc}/dt^2 = m\mathbf{a} - md^2\boldsymbol{\rho}/dt^2 \\
&= -e(\mathbf{v}_{gc} + d\boldsymbol{\rho}/dt) \times [\mathbf{B}(\mathbf{r}_{gc}) + \Delta\mathbf{B}(\mathbf{r})] - md^2\boldsymbol{\rho}/dt^2 \\
&= -e(\mathbf{v}_{gc} - \Omega\boldsymbol{\rho} \times \mathbf{b}) \times [\mathbf{B}(\mathbf{r}_{gc}) + \Delta\mathbf{B}(\mathbf{r})] + m\Omega^2\boldsymbol{\rho} \\
&= -e\mathbf{v}_{gc} \times \mathbf{B}(\mathbf{r}_{gc}) - e\mathbf{v}_{gc} \times \Delta\mathbf{B}(\mathbf{r}) - e\Omega\mathbf{B}(\mathbf{r}_{gc}) \times (\boldsymbol{\rho} \times \mathbf{b}) \\
&\quad + e\Omega(\boldsymbol{\rho} \times \mathbf{b}) \times \Delta\mathbf{B}(\mathbf{r}) + m\Omega^2\boldsymbol{\rho} \\
&= -e\mathbf{v}_{gc} \times \mathbf{B}(\mathbf{r}_{gc}) - e\mathbf{v}_{gc} \times \Delta\mathbf{B}(\mathbf{r}) - e\Omega\{B(\mathbf{r}_{gc})\boldsymbol{\rho} - [\mathbf{B}(\mathbf{r}_{gc}) \bullet \boldsymbol{\rho}]\mathbf{b}\} \\
&\quad + e\Omega(\boldsymbol{\rho} \times \mathbf{b}) \times \Delta\mathbf{B}(\mathbf{r}) + m\Omega^2\boldsymbol{\rho} \\
&= -e\mathbf{v}_{gc} \times \mathbf{B}(\mathbf{r}_{gc}) - e\mathbf{v}_{gc} \times \Delta\mathbf{B}(\mathbf{r}) + e\Omega(\boldsymbol{\rho} \times \mathbf{b})\Delta\mathbf{B}(\mathbf{r}),
\end{aligned}$$

giving

$$m\mathbf{a}_{gc} = -e\mathbf{v}_{gc} \times \mathbf{B}(\mathbf{r}_{gc}) - e\mathbf{v}_{gc} \times \Delta\mathbf{B}(\mathbf{r}) - e\Omega\{[\mathbf{b}\bullet(\boldsymbol{\rho}\bullet\nabla)\mathbf{B}(\mathbf{r})]\boldsymbol{\rho} - [\boldsymbol{\rho}\bullet(\boldsymbol{\rho}\bullet\nabla)\mathbf{B}(\mathbf{r})]\mathbf{b}\} \tag{3.26}$$

In "local" coordinates the operator $\boldsymbol{\rho}\bullet\nabla$ is given by $\rho(\cos\phi\partial/\partial\xi_1 + \sin\phi\partial/\partial\xi_2)$ and $\boldsymbol{\rho}\bullet\mathbf{B}(\mathbf{r}) = \rho(\cos\phi B_1 + \sin\phi B_2)$. Thus, the two entities in the braces in Eq. (3.26) are

$$\begin{aligned}
[\mathbf{b}\bullet(\boldsymbol{\rho}\bullet\nabla)\mathbf{B}(\mathbf{r})]\boldsymbol{\rho} &= \rho^2[\mathbf{u}_1(\cos^2\phi\partial B_3/\partial\xi_1 + \cos\phi\sin\phi\partial B_3/\partial\xi_2) \\
&\quad + \mathbf{u}_2(\sin\phi\cos\phi\partial B_3/\partial\xi_1 + \sin^2\phi\partial B_3/\partial\xi_2)] \quad \text{and} \\
[\boldsymbol{\rho}\bullet(\boldsymbol{\rho}\bullet\nabla)\mathbf{B}(\mathbf{r})]\mathbf{b} &= \rho^2\mathbf{b}[(\cos^2\phi\partial B_1/\partial\xi_1 + \cos\phi\sin\phi\partial B_2/\partial\xi_1) \\
&\quad + (\sin\phi\cos\phi\partial B_1/\partial\xi_2 + \sin^2\phi\partial B_2/\partial\xi_2)].
\end{aligned} \tag{3.27}$$

We can now average the guiding-center equation of motion over one gyroperiod by integrating over ϕ and dividing by 2π. The average values of the terms in Eq. (3.26) are

$\langle \Delta\,\mathbf{B}(\mathbf{r})\,\rangle = 0$ and

$$\begin{aligned}
&\langle\{\mathbf{b}\bullet(\rho\bullet\nabla)\mathbf{B}(\mathbf{r})]\rho-[\rho\bullet(\rho\bullet\nabla)\mathbf{B}(\mathbf{r})]\mathbf{b}\}\rangle \\
&\quad= (\rho^2/2)[(\mathbf{u}_1\partial B_3/\partial\xi_1+\mathbf{u}_2\partial B_3/\partial\xi_2)-\mathbf{b}(\partial B_1/\partial\xi_1+\partial B_2/\partial\xi_2)] \\
&\quad= (\rho^2/2)(\mathbf{u}_1\partial B_3/\partial\xi_1+\mathbf{u}_2\partial B_3/\partial\xi_2+\mathbf{b}\partial B_3/\partial\xi_3) \\
&\quad= (\rho^2/2)\nabla B_3(\mathbf{r}) \approx (\rho^2/2)\nabla B(\mathbf{r}).
\end{aligned} \tag{3.28}$$

We have made use of the fact that $\nabla\bullet\,\mathbf{B}=0$. Also, at the position of the electron the perpendicular components of $\mathbf{B}$ are small and we can approximate B_3 by B

$$B_3^2 = B^2-B_1^2-B_2^2 = B^2\left[1-\left(B_1^2+B_2^2\right)/B^2\right] \approx B^2$$

Our gyro-averaged guiding-center equation of motion for the electron is therefore

$$\begin{aligned}
m\mathbf{a}_{gc} &= -e\mathbf{v}_{gc}\times\mathbf{B}(\mathbf{r}_{gc})-e\Omega(\rho^2/2)\nabla B(\mathbf{r}) \\
&= -e\mathbf{v}_{gc}\times\mathbf{B}(\mathbf{r}_{gc})-\mu\nabla B(\mathbf{r})
\end{aligned} \tag{3.29}$$

The parallel component, $\mathbf{b}\bullet\, m\mathbf{a}_{gc}=\mu\bullet\,\nabla B(\mathbf{r})$, as discussed previously. The transverse component of the guiding-center velocity is obtained from the guiding-center equation of motion as follows. Since from Eq. (3.29) $\mathbf{v}_{gc}\times\mathbf{B}(\mathbf{r}_{gc})=-(m/e)\mathbf{a}_{gc}-(\mu/e)\nabla B(\mathbf{r})$ we have

$$\mathbf{B}\times(\mathbf{v}_{gc}\times\mathbf{B}) = B^2\mathbf{v}_{gc}-(\mathbf{B}\bullet\mathbf{v}_{gc})\mathbf{B} = -\mathbf{B}\times[(m/e)\mathbf{a}_{gc}+(\mu/e)\nabla B(\mathbf{r})]. \tag{3.30}$$

Note that we have dropped the explicit guiding center argument of $\mathbf{B}$, which is understood in what follows. We require the lowest order terms in the acceleration of the guiding center:

$$d\mathbf{v}_{gc}/dt = d\mathbf{v}_{gc\perp}/dt+d(\mathbf{b}v_{gc\|})/dt = d\mathbf{v}_{gc\perp}/dt+\mathbf{b}d(v_{gc\|})/dt+v_{gc\|}d\mathbf{b}/dt.$$

The final term above is the dominant contribution and is given by

$$v_{gc\|}d\mathbf{b}/dt = (v_{gc\|})^2(\mathbf{b}\bullet\nabla)\mathbf{b} = (v_{gc\|})^2\nabla_\perp\mathbf{B}(\mathbf{r})/B. \tag{3.31}$$

using the results from Chapter 2 for the curvature. With this replacement we have

$$B^2\mathbf{v}_{gc}-(\mathbf{B}\bullet\mathbf{v}_{gc})\mathbf{B} = -\mathbf{B}\times[(m/e)(v_{gc\|})^2\nabla_\perp\mathbf{B}(\mathbf{r})/B+(\mu/e)\nabla B(\mathbf{r})],$$

so that

$$\mathbf{v}_{gc\perp} = (2W_\|+W_\perp)\nabla B(\mathbf{r})\times\mathbf{B}/(eB^3). \tag{3.32}$$

We have written this in the conventional form (see, for example, Krall and Trivelpiece [4]), but it must be kept in mind that Eq. (3.32) is a local value of the drift velocity. Generally speaking it is necessary to average the expression in Eq. (3.32) along the magnetic line of force followed by the guiding center. For electrons trapped in a magnetic mirror, this average is conveniently carried out using the action invariant [5], $J = \int p_{||} ds$, where the integral is taken around a complete bounce cycle. Under typical conditions, the time required for an electron to drift once around the circumference of the plasma will be several orders of magnitude longer than the bounce time and the action invariant provides a more general and comprehensive mathematical description of this drift motion that properly averages the curvature and gradients along the magnetic line of force. We will not derive the relevant formulas here, since Lehnert [1] and Northrop [5] both provide detailed derivations and copious references to earlier work. The results are usually expressed in terms of a Clebsch representation of the magnetic field as described in Chapter 2, $\mathbf{B} = \nabla\alpha \times \nabla\beta$. Note that in this representation the magnetic gradient factor from Eq. (3.32) is

$$\nabla B(\mathbf{r}) \times \mathbf{B} = \nabla B \times (\nabla\alpha \times \nabla\beta) = (\nabla B \bullet \nabla\beta)\nabla\alpha - (\nabla B \bullet \nabla\alpha)\nabla\beta. \tag{3.33}$$

Thus, the guiding-center drift velocity in the $\nabla\alpha$ -direction is due to the component of ∇B in the $\nabla\beta$ -direction, whereas the guiding-center drift in the $-\nabla\beta$ -direction is due to the component of ∇B in the $\nabla\alpha$ - direction. The main results of the analyses are [1]

$$\begin{aligned} \langle d\alpha/dt \rangle &= (e\tau_b)^{-1}\partial J/\partial\beta \\ \langle d\beta/dt \rangle &= -(e\tau_b)^{-1}\partial J/\partial\alpha \\ \tau_b &= \partial J/\partial\varepsilon. \end{aligned} \tag{3.34}$$

Here τ_b is the bounce time for which suitable expressions were derived earlier without explicit recourse to the action invariant. If we apply these results to the simple magnetic mirror field, we find (see Exercise 3.3)

$$r\Omega_\theta = [W_\perp(0)/(eBR_{co})]\{[2E(k^2) - K(k^2)]/K(k^2)\}, \tag{3.35}$$

where k^2 is once again $(M_t - 1)/(M - 1)$ and R_{co} is the radius of curvature at the midplane of the mirror. We can use the results presented earlier in this chapter to express the azimuthal drift speed in the following form:

$$r\Omega_\theta = [2(\varepsilon - q\phi) - \mu B]/qB\langle R_c \rangle + \mathbf{E} \times \mathbf{B}/B^2. \tag{3.36}$$

Here ϕ (and $\mathbf{E} = -\nabla\phi$) is any ambipolar potential that may be present to provide for equal loss rates of ions and electrons and $\langle R_c \rangle$ is the average radius of curvature of the magnetic lines of force experienced by the charged particles as they bounce back and forth along lines of force. A convenient expression for this average radius of curvature can be derived for the simple magnetic mirror by using Eq. (3.35), which was obtained using a paraxial approximation to the magnetic fields; i.e., one that is valid near the magnetic axis. One finds

$$\langle R_c \rangle = R_{co}(r)\{[2(\varepsilon - q\phi) - \mu B]/\mu B\}K(k^2)/[2E(k^2) - K(k^2)]. \tag{3.37}$$

Again, K and E are the complete elliptic integrals whose argument k^2 is given by

$$k^2 = (M-1)^{-1}[(\varepsilon - q\phi - \mu B)/\mu B] = (M_t - 1)/(M-1),$$

and we have used the model magnetic field given in Chapter 2, Eq. (2.16):

$$\begin{aligned} B_z(r,z) &= B_0\{[(M+1)/2] - [(M-1)/2]\cos(k_0 z) I_0(k_0 r)\} \\ B_r(r,z) &= -B_0[(M-1)/2]\sin(k_0 z) I_1(k_0 r)\} \end{aligned},$$

where $R_{co}(r)$ is the radius of curvature of the line of force at its intersection with the midplane a distance r from the magnetic axis. For $k^2 = 0.826$, $2E(k^2) = K(k^2)$ and the average radius of curvature becomes infinite. For particles with this value of k^2 the negative radius of curvature near the midplane and the positive radius of curvature in the mirror throat average to zero and the azimuthal speed due to the inhomogeneity in the magnetic intensity vanishes.

3.4 Relativistic Electron Kinematics for ECH

In the absence of external electric fields so that the electron is acted on solely by a magnetostatic field, the total energy, ε, the speed, v, and the relativistic factor, $\gamma = (1 - v^2/c^2)^{-1/2}$ are all constants of the motion. Under these conditions the only difference between the relativistic and nonrelativistic equations of motion is that the electron rest mass, m, is replaced by the relativistic mass, γm. When this replacement is made all results derived earlier regarding motion in a magnetostatic field are equally valid in the relativistic case. In particular, for relativistic electrons the gyrofrequency is given by $\Omega_e = eB/(\gamma m)$ and the gyroradius by $\rho = \gamma m v_\perp/(eB) = p_\perp/(eB)$.

It will often be convenient to express the velocity in terms of the Lorentz factor using $v^2/c^2 = (\gamma^2 - 1)/\gamma^2$. The momentum per unit rest mass, $\mathbf{u} = \gamma \mathbf{v}$, thus satisfies $u^2/c^2 = (\gamma^2 - 1)$. Moreover, since the total energy, including the rest energy is γmc^2, the kinetic energy is just $\varepsilon = (\gamma - 1)mc^2$. The relativistic magnetic moment is given by [5]

$$\mu = p_\perp^2/(2mB) = (\gamma m v_\perp)^2/(2mB) = m u_\perp^2/(2B), \tag{3.38}$$

whence

$$u_\perp^2 = 2\mu B/m. \tag{3.39}$$

Since

$$u_\parallel^2/c^2 + u_\perp^2/c^2 = u_\parallel^2/c^2 + 2\mu B/mc^2 = (\gamma^2 - 1), \tag{3.40}$$

the magnetic intensity at the electron turning point is given in the relativistic case by

$$B_t = (\varepsilon/\mu)(\gamma + 1)/2. \tag{3.41}$$

3.5 The Hamiltonian Approach

In later chapters we will find it advantageous to use Hamilton's equations [6] to describe the electron dynamics. Here we simply illustrate the Hamiltonian approach by applying it to the unperturbed motion of nonrelativistic electrons in axisymmetric magnetic-mirror fields. The Hamiltonian, $H(q_i,p_i)$, a function of the generalized coordinates, q_i, and their conjugate momenta, p_i, is obtained from the Lagrangian, $\mathcal{L}(q_i,dq_i/dt)$, which for electrons in magnetic and electric fields is given by [6]

$$\mathcal{L}(q_i, dq_i/dt) = mv^2/2 - e\mathbf{A}\bullet\mathbf{v} + e\Phi. \tag{3.42}$$

Here $\mathbf{A}$ is the vector potential discussed at length in Chapter 2 and Φ is any electrostatic potential that may be present. For the moment we will neglect all electrostatic potentials. The generalized momenta are defined by

$$p_i = \partial\mathcal{L}(q_i, dq_i/dt)/\partial(dq_i/dt), \tag{3.43}$$

and the Hamiltonian is then given by

$$\mathfrak{H}(q_i, p_i) = \sum p_i dq_i/dt - \mathcal{L}(q_i, dq_i/dt). \tag{3.44}$$

As we have seen in Chapter 2, the vector potential of an axisymmetric magnetic-mirror field has only the single component A_ϕ and for this magnetic field the Lagrangian is given in cylindrical coordinates by

$$\mathcal{L} = m[(d\rho/dt)^2 + (\rho d\phi/dt)^2 + (dz/dt)^2]/2 - eA_\phi \rho d\phi/dt + e\Phi. \tag{3.45}$$

From this Lagrangian, assuming $\Phi = 0$, we obtain the following three generalized momenta:

$$\begin{aligned} p_\rho &= m(d\rho/dt) \\ p_\phi &= m\rho^2(d\phi/dt) - e\rho A_\phi \,. \\ p_z &= m(dz/dt) \end{aligned} \tag{3.46}$$

To express the Hamiltonian as a function of the generalized coordinates and their conjugate momenta we make use of the following replacements:

$$\begin{aligned} d\rho/dt &= p_\rho/m \\ d\phi/dt &= (p_\phi + e\rho A_\phi)/(m\rho^2)\,. \\ dz/dt &= p_z/m \end{aligned} \tag{3.47}$$

The resulting Hamiltonian is

$$\mathfrak{H} = (p_\rho^2 + p_z^2)/2m + (p_\phi + e\rho A_\phi)^2/(2m\rho^2). \tag{3.48}$$

The Hamiltonian ("canonical") equations of motion are

$$\begin{aligned} dq_i/dt &= \partial\mathfrak{H}/\partial p_i \\ dp_i/dt &= -\partial\mathfrak{H}/\partial q_i \\ \partial\mathfrak{H}/\partial t &= -\partial\mathcal{L}/\partial t. \end{aligned} \tag{3.49}$$

For electrons moving in an axisymmetric magnetostatic field,

$$\partial\mathfrak{H}/\partial\phi = \partial\mathfrak{H}/\partial t = 0,$$

and consequently there are two constants of the motion:

$$\begin{aligned} p_\phi &= p_o = \text{constant} \\ \mathfrak{H} &= W_o = \text{constant} \end{aligned} . \tag{3.50}$$

By employing these two constants of the motion we can rewrite the Hamiltonian as

$$\mathfrak{H} = (p_\rho^2 + p_z^2)/2m + (p_o + e\rho A_\phi)^2/(2m\rho^2) = W_o. \tag{3.51}$$

Recall from Chapter 2 that the magnetic flux, $\Psi(\rho,z) = 2\pi\rho A_\phi$ and we can therefore express the Hamiltonian as

$$\mathfrak{H} = (p_\rho^2 + p_z^2)/2m + [p_o + e\Psi(\rho,z)/2\pi]^2/(2m\rho^2) = W_o. \tag{3.52}$$

If we define $U(\rho,z) = [p_o + e\Psi(\rho,z)/2\pi]^2/(2m\rho^2)$, then we obtain

$$(p_\rho^2 + p_z^2)/2m = W_o - U(\rho,z). \tag{3.53}$$

Thus, the electrons move as though they were confined in a two-dimensional potential well, $U(\rho,z)$. Note that if we choose the constant $p_o = -e\Psi(\rho,z)/2\pi$ at some reference point (ρ_o,z_o), U will vanish at every point on the flux surface that passes through the chosen point (ρ_o,z_o).

We illustrate some of the features of this Hamiltonian description of the unperturbed electron motion using a typical magnetic mirror with a mirror ratio of 2:1. We have arbitrarily chosen $\rho_o = 5$ cm on the midplane, where $z_o = 0$. For our illustrative magnetic field, the flux surface passing through this point (and thus the magnetic line of force) then passes through the mirror throat at a radius of 3.49 cm. The magnetic intensity at the position of the flux surface in the midplane is 0.2900 T, while at the mirror throat the intensity on this field line is 0.6074 T. The two figures, Figures 3.3(a) and (b), display U (in keV) versus the radial coordinate, ρ (in cm) in the midplane and in the mirror throat, respectively.

The equations of motion have real solutions only if $|\rho - \rho_o|$ is less than or equal to the value for which $W_o - U \geq 0$. One can easily verify that at the radial positions for which $\Omega_o = U$, namely, ρ_{min} and ρ_{max}, the motion is exactly perpendicular to the magnetic field and

$$v_\perp^2 = 2W_o/m \approx (\rho_{max} - \rho_o)^2\Omega^2 \approx (\rho_o - \rho_{min})^2\Omega^2$$

Since $d\phi/dt = (p_o + e\rho A_\phi)/(m\rho^2)$, $d\phi/dt$ changes sign as ρ passes through the value ρ_o and the electrons gyrate about the field line selected by our choice of p_o. One can demonstrate analytically (see Exercise 3.4) that U is approximately equal to the perpendicular kinetic energy of electrons whose gyro orbits are centered on the chosen flux surface. The plots shown in Figures 3.3(a) and (b) provide a convenient summary in that $W_\perp$ vanishes at $(\rho,z) = (\rho_o,z_o)$, where $W_{||} = W_o$, while $W_{||}$ vanishes at the radii, ρ_{max} and ρ_{min}, where $U = W_o$. At intermediate values of ρ we have

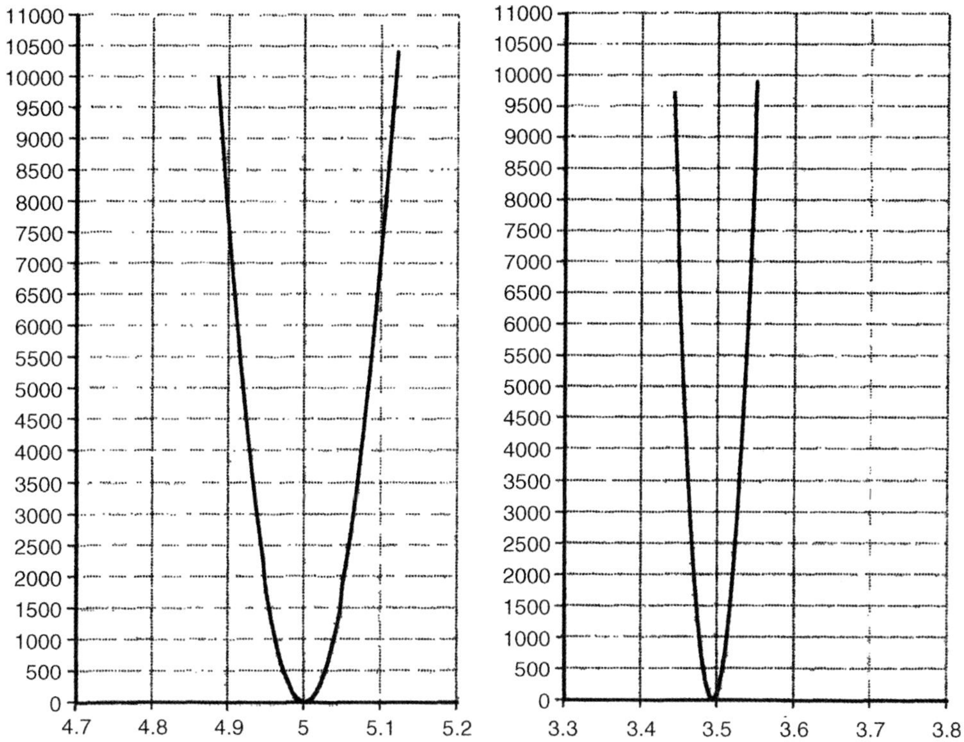

(a) (b)

Figure 3.3 (a) The function U(ρ,z_o), in keV, on the midplane, $z_o = 0$, for the magnetic line of force that intersects the midplane at ρ = 5 cm. The magnetic intensity at this point is 0.29 T. (b) The function *U*(ρ,z_o), in keV, at the mirror throat where the reference field line is at a radius of ρ = 3.49 cm and the magnetic intensity is 0.6074 T.

a simple result for the electron turning point:

$$B_t = \varepsilon/\mu = W_o/(W_\perp/B) \approx W_o B(\rho, z_o)/U(\rho, z_o) \\ \approx W_o B(\rho_o, z_o)/U(\rho, z_o) \quad (3.54)$$

The fact that U is not exactly symmetrical about ρ_o indicates that the electron gyro orbits are cycloidal, a feature which gives rise to the azimuthal drift discussed earlier. In fact, since the motion in the ρ-direction is periodic, the average over one gyroperiod of the force in this direction, F_ρ, must vanish. We can exploit this observation to exhibit a limiting case of the formula for the azimuthal drift speed cited earlier in Eq. (3.32):

$$\int F_\rho dt = -e \int \rho (d\phi/dt) B_z dt = -e \int \rho B_z d\phi = 0, \quad (3.55)$$

where the time integral is over one gyroperiod and the ϕ integral is over the corresponding nearly closed path in ϕ. Expanding B_z about the point $\rho = \rho_o$ we have

$$0 = \int \rho[B_z(\rho_o) + (\partial B_z/\partial\rho)(\rho - \rho_o) + \text{h.o.t.}]d\phi$$
$$\approx B_z(\rho_o)\Delta s_\phi + \pi(\partial B_z/\partial\rho)(\rho_{max} - \rho_o)^2.$$

Here $\Delta s_\phi = \rho_o(\phi_f - \phi_i)$ and $\pi(\rho_{max} - \rho_o)^2$ is the area of the (nearly) circular gyro orbit swept out as ϕ goes from ϕ_i to ϕ_f and back almost but not quite to ϕ_i. Since $v_\perp^2 = (\rho_{max} - \rho_o)^2\Omega^2$ and since the time interval is $\Delta t = 2\pi/\Omega$ we recover the $v_\parallel = 0$ limit of our earlier result; namely,

$$\Delta s_\phi/\Delta t = -[1/B_z(\rho_o)](\partial B_z/\partial\rho)v_\perp^2/2\Omega. \tag{3.56}$$

3.6
Drift Orbits in Toroidal Magnetic Configurations

In a simple toroidal magnetic field, formed by a toroidal array of closely spaced magnetic coils, the magnetic intensity decreases with distance from the axis as described in Chapter 2: $\mathbf{B} = B(R)\mathbf{b} = B_o(R_o/R)\mathbf{b}$. The gradient of this field is therefore $\nabla B = -\mathbf{u}_R(B_oR_o)/R^2$ and electron guiding centers will have a velocity given by Eq. (3.32):

$$\mathbf{v}_{tor} = (2W_\parallel + W_\perp)\nabla B \times \mathbf{B}/(eB^3) = (2W_\parallel + W_\perp)(-\mathbf{u}_R \times \mathbf{b})/(eB_oR_o).$$

Ions will drift in the opposite direction. This drift velocity will carry the electrons and ions toward the opposite walls of the vacuum chamber. In the early phases of the controlled fusion research program it was realized that this toroidal drift could be effectively cancelled if the magnetic field lines could be made to spiral around a magnetic axis in the center of the confinement region. In the stellarator concept this is accomplished by adding helical magnetic windings onto the exterior of the torus. In tokamaks the spiraling magnetic lines of force are produced by the plasma current flowing along the toroidal magnetic field. The velocity of charged particles flowing along the spiraling lines of force will then have a poloidal component:

$$v_{pol} = r\Omega_{pol} = rv_\parallel d\theta_{pol}/ds = v_\parallel(rd\theta_{pol}/Rd\phi_{tor}) = v_\parallel(r/Rq), \quad \text{or}$$
$$\Omega_{pol} = v_\parallel/(Rq),$$

where q is the usual safety factor. To describe the resulting drift orbits we locate a Cartesian reference frame in a cross section of the torus with its origin at the magnetic axis: $x = R - R_o = 0$ and $y = 0$ (the equatorial plane of the torus). We follow a test electron as it travels around the torus and note the point (x, y) where it passes through the reference plane after each successive transit of the torus. It is customary to visualize the electron as "puncturing" the reference plane, and after many transits of the torus these "puncture points" will trace out the drift surface we wish to describe. Note that if q is a ratio of integers the points where our test electron intersects the reference plane will not trace out a surface but will repeatedly intersect at points reflecting the fact that the field line is closing on itself after q transits of the torus. For other values of q we can describe the resulting drift surface by superposing two separate motions; namely, the poloidal motion characterized by $x = a\cos\theta(t)$ and $y = a\sin\theta(t)$ with $d\theta/dt = \Omega_{pol}$; and secondly, $dy/dt = \Omega_{pol}\, a\cos\theta(t)$ $v_{tor} = \Omega_{pol}x + v_{tor}$. Since $dx/dt = -\Omega_{pol}y$ we can readily combine the two equations to obtain $(x + v_{tor}/\Omega_{pol})dx/dt = -y dy/dt$, whence

$$(x + v_{tor}/\Omega_{pol})^2 + y^2 = \text{constant}.$$

The drift surfaces are thus circles centered at $x = -v_{tor}/\Omega_{pol}$ and $y = 0$. For $\Omega_{pol} > 0$ the center is shifted inward whereas for $\Omega_{pol} < 0$ the center is shifted outward from the magnetic axis. We have assumed to this point that the electrons are circulating freely around the torus with approximately constant $v_{\|}$, but if they are trapped in the mirror-like regions of the torus their orbits will form two arcs with centers at $x = \pm v_{tor}/\Omega_{pol}$ and with the tips of the arcs at the poloidal angle corresponding to the mirror turning point. These orbits are traditionally referred to as "banana orbits" in the tokamak literature.

It was also recognized early in the controlled fusion program that charged particles could be confined in a simple toroidal magnetic trap if the individual coils used to generate the toroidal field were spaced sufficiently far apart, relative to the minor radius of the torus, to make the magnetic field vary significantly along each magnetic line of force. In such a "bumpy" toroidal magnetic field, local gradients in magnetic field strength give energetic charged particles a poloidal drift motion which can balance the drift inherent in an ideal toroidal magnetic field. Kadomtsev [7] analyzed the single particle orbits in such a bumpy torus, and subsequently Gibson *et al.* [8], Morozov and Solov'ev [9], and others developed a rigorous picture of single-particle confinement using adiabatic constants of the motion as well as numerical studies of particle orbits.

For a particle of charge q, the drift due to the R^{-1} dependence of the magnetic intensity associated with the toroidal effect is given by

$$v_{tor} = 2(\varepsilon - q\phi - \mu B)/qBR_{tor},$$

where R_{tor} is the major radius of the torus. The poloidal drift speed is given by Eq. (3.35). Recall that the average radius of curvature and thus the poloidal speed vanishes for pitch angles such that $k^2 = 0.826$. Thus, except for those particles with pitch angles lying in narrow range around the value given by $k^2 = 0.826$, charged particles will drift on nearly circular orbits whose centers are shifted off the ring axis by

$$\delta x = v_{tor}/\Omega_{pol} = [2(\varepsilon - q\phi - \mu B)/qBR_{tor}]/\Omega_{pol}$$
$$= r[2(\varepsilon - q\phi - \mu B)/qBR_{tor}][2(\varepsilon - q\phi) - \mu B]/qB\langle R_c\rangle + \mathbf{E} \times \mathbf{B}/B^2]^{-1}$$
$$= (r\langle R_c\rangle/R_{tor})\{1 + qE_{\perp}\langle R_c\rangle/[2(\varepsilon - q\phi) - \mu B]\}^{-1}.$$

From Chapter 2 the midplane curvature of the magnetic lines of force in the simple magnetic mirror is approximately given by

$$rR_{co} \cong -4(M-1)^{-1}(L_{coil}/2\pi)^2,$$

where L_{coil} is the separation of the coils. In a bumpy torus the separation of the coils and the radius of the torus are related to the number of sectors, $N_{sectors}$ making up the complete torus: $2\pi R_{tor} = N_{sectors} L_{coil}$ Thus, in the absence of radial electrostatic fields, the centers of the drift orbits are shifted off the ring axis by an amount given in order of magnitude by $\delta x \approx [-4/(M-1)](L_{coil}/2\pi N)g(\varepsilon, \mu, \phi)$, where the pitch-angle dependence, here designated by $g(\varepsilon, \mu, \phi)$, is of order unity. Since, as we saw in Chapter 2, the coil separation $L_{coil} \approx 2.4 r_{coil}$ for a mirror ratio $M = 2$, we find that the shift of the drift surfaces is $\delta x/r_{coil} \sim O(1/N)$ and the orbits are well confined for $N \gg 1$.

References

1 Hannes Alfvén and Carl-Gunne Fälthammar, *Cosmical Electrodynamics*, second edition, Oxford University Press, London (1963), Chapter 2; Bo Lehnert, *Dynamics of Charged Particles*, North-Holland, Amsterdam (1964), Chapter 3; David J. Rose and Melville Clark, Jr., *Plasmas and Controlled Fusion*, MIT Press and John Wiley & Sons, New York (1961), Chapter 10; Donald A. Gurnett and Amitava Bhattacharjee, *Introduction to Plasma Physics*, Cambridge University Press, New York (2005) Chapter 3; A.I. Morozov and L.S. Solov'ev, in *Reviews of Plasma Physics*, Vol. 2, (M.A. Leontovich, editor) Consultants Bureau, New York (1966), pp. 201–296.

2 William P. Allis, Solomon J. Buchsbaum, and Abraham Bers, *Waves in Anisotropic Plasmas*, MIT Press, Cambridge (1963).

3 R. Kulsrud, *Phys. Rev.* **106**, 205 (1957) as well as R.H. Cohen, G. Rowlands, and J.H. Foote, *Phys. Fluids* **21**, 627 (1978) and references cited therein.

4 Nicholas A. Krall and Alvin W. Trivelpiece, *Principles of Plasma Physics*, McGraw-Hill, New York (1973).

5 T.G. Northrup and E. Teller, *Phys. Rev.* **117**, 215 (1960) andT.G. Northrup, *The Adiabatic Motion of Charged Particles*, Wiley-Interscience, New York (1963).

6 H. Goldstein, *Classical Mechanics*, pp. 19–21 and 217–225, Addison-Wesley, Reading, MA (1959); H. Goldstein, *Classical Mechanics*, second edition, pp. 21–23 and Chapter 8, Addison-Wesley: Reading, MA (1980); see also Bo Lehnert, *Dynamics of Charged Particles*, pp. 18–23 and 152–156, North-Holland, Amsterdam (1964).

7 B.B. Kadomtsev, in *Plasma Physics and the Problem of Controlled Thermonuclear Reactions* (M.A. Leontovich and J. Turkevich,eds.) Vol. 3, Pergamon Press, Oxford. (1959) p. 340 ff.

8 G. Gibson, W.C. Jordan, and E.J. Lauer, *Phys. Rev. Lett.* **4**, 217 (1960).

9 A.I. Morozov and L.S. Solov'ev, in *Review of Plasma Physics* (M.A. Leontovich, Ed.) Vol. III, Consultants Bureau, New York (1966) pp. 267–272.

Exercises

3.1 *Verify Eq. (3.16) for the period of the bounce motion on the axis of a simple magnetic mirror.*

Solution: $\tau_b = 4\int dz/v_{||}(z)$ *where the integral ranges from* $z = 0$ *to* $z = z_t$, *and* $mv_{||}(z)^2/2 = \varepsilon - B(z)$. *Use the approximate expression for the field near the axis,* $B(z) = B_o[1 + (M - 1)I_0(k_o r)\sin^2 k_o z/2]$. *Note that at the turning point* $\sin^2 k_o z_t/2 = (M_t - 1)/(M - 1)$ *and therefore* $mv_{||}(z)^2/2 = \varepsilon - B(z)$ *can be written as* $mv_{||}(z)^2/2 = (\varepsilon - B_o)[1 - (\sin^2 k_o z/2)/(\sin^2 k_o z_t/2)]$. *Let* $\sin\phi = (\sin k_o z/2)/(\sin k_o z_t/2)$. *Then* $dz = (2/k_o)(\sin k_o z_t/2)\cos\phi d\phi\{1 - [(M_t - 1)/(M - 1)]\sin^2\phi\}^{-1/2}$ *where the integral over* ϕ *ranges from* $\phi = 0$ *to* $\phi = \pi/2$. *The result follows.*

3.2 *Derive a formula for the action integral for electrons moving near the axis of a simple magnetic mirror, where the action is defined as* $J \equiv \int p_{||}ds$ *and the integral is over one complete bounce period. Show that the expression for the bounce time obtained by evaluating* $\partial J/\partial\varepsilon$ *agrees with Eq. (3.16).*

3.3 *Use the formula for J obtained in Exercise 3.2 to obtain an expression for the poloidal speed of an electron moving along field lines near the axis of a simple magnetic mirror.*

3.4 *Demonstrate analytically that the potential function, U, is approximately equal to the perpendicular kinetic energy of electrons whose gyro orbits are centered on a given flux surface.*

3.5 *In deriving the formula for the bounce frequency we found that, for electrons turning near the midplane, it was reasonably accurate to represent the electron bounce as simple harmonic motion. For simple harmonic motion the maximum (parallel) kinetic energy is given by* $m(\omega_B z_t)^2/2$. *Assume* $W_o = 2$ *keV and evaluate this estimate of* $W_{||}(0)$ *for the case shown in Figure 3.2(a) at the point* $\rho = 4.95$ *cm, where* $U = 1.87$ *keV, and again for* $\rho = 4.96$ *cm, where* $U = 1.19$ *keV. Compare these two estimates with the corresponding values obtained from* $W_{||}(0) = W_o - U$.

4
Wave Propagation and Cyclotron Damping in Magnetized Plasmas

In order for microwave power to provide direct illumination of the resonance surfaces, it is necessary to couple power from an external source into electromagnetic waves that can propagate through the intervening plasma to the resonance surfaces. The effectiveness of this coupling between the external source and the RF electric fields at the resonance surfaces depends sensitively on a number of factors, including the type of coupler, its orientation relative to the magnetic field, the shape and location of the conducting walls of the vacuum chamber, and the density and temperature of the plasma. To date it has been a common practice to employ conventional or dielectrically loaded waveguides to launch microwave power into plasma chambers, but other coupling approaches have also been used. These have ranged in complexity from simple bare conductors fed by coaxial cables to phased arrays of antennas radiating specific polarizations within particular directional patterns. Open resonators have also been employed to build up the amplitude of the RF fields in specific locations [1].

It has been clearly established [2] that highly overdense, low-temperature plasmas could be created in magnetic mirrors by launching whistler waves that propagate parallel to the magnetic field from regions where B exceeded B_{res} into the resonance surface. This approach is aptly named "high-field launch." By contrast, the production of relativistic-electron plasmas apparently requires a very different approach in which weakly damped microwave power is introduced in such a way that the resulting electromagnetic waves would seemingly have to propagate through the plasma and undergo multiple polarization-changing reflections from the chamber walls to reach resonance surfaces with a suitable polarization [3] to be absorbed. The success of various ECH applications in tokamaks depends sensitively on the propagation and absorption properties of the waves in ways that we will explore in Chapter 9. In this chapter, we summarize the basic optical properties of magnetized plasmas for waves with frequencies around the electron gyrofrequency and the ways in which these optical properties affect the illumination of resonance surfaces by microwave power. In tokamak experiments where plasma dimensions are much greater than the ECH wavelengths, the predicted optical properties have generally been well confirmed. However, in the smaller magnetic-mirror experiments, the situation is often less clear. The general subject of plasma waves has been treated extensively in the literature [4] and the present summary will, therefore, be as concise as possible.

Electron Cyclotron Heating of Plasmas. Gareth Guest

ISBN: 978-3-527-40916-7

4.1 The Cold-Plasma Dispersion Relation

In order to identify the features of plasma wave propagation that are particularly relevant to ECH, we adopt a highly idealized model of the plasma in which a spatially homogeneous plasma is immersed in a static, uniform magnetic field. The propagation of electromagnetic waves in this model plasma can be examined by employing Maxwell's equations and assuming that the electric charge and current densities entering them are solely those that arise in response to the electromagnetic fields of the wave [4]. The ion response to these fields is usually negligible at frequencies in the electron gyrofrequency range and will generally not be included here except in unusual circumstances that will be clearly identified. The electron response to the fields of the wave may be obtained from a suitable dynamical equation, such as the following Langevin equation:

$$d\mathbf{v}/dt = -(e/m)[\mathbf{E} + \mathbf{v} \times (\mathbf{B}_0 + \mathbf{B})] - \nu\mathbf{v}. \tag{4.1}$$

Here ν is the frequency with which electrons undergo momentum changing collisions, $\mathbf{E}$ is the electric field of the wave, $\mathbf{B_0}$ is the uniform static magnetic field, and $\mathbf{B}$ is the fluctuating magnetic field due to the wave itself. We consider the propagation of waves whose frequency and propagation vector are ω and $\mathbf{k}$, respectively, so that all wave properties vary in space and time as $\exp(i\mathbf{k}\bullet\mathbf{r} - i\omega t)$. In reality, we are using Fourier transforms in space and Laplace transforms in time. The full power of these transform techniques will become important later in discussions of unstable waves.

Without loss of generality, we can choose coordinates so that the static magnetic field is in the z-direction and the propagation vector lies in the x–z plane and makes an angle θ with the z-axis: $\mathbf{k} = k\sin\theta\mathbf{u}_x + k\cos\theta\mathbf{u}_z$. Convenient forms of Maxwell's equations are

$$\begin{aligned}
&\nabla \times \mathbf{E} = -\partial\mathbf{B}/\partial t && \text{(Faraday's Law)}\\
&\nabla \times \mathbf{H} = \mathbf{j} + \varepsilon_0\partial\mathbf{E}/\partial t && \text{(Maxwell-Ampere Equation)}\\
&\varepsilon_0\nabla \cdot \mathbf{E} = \rho && \text{(Poisson's Equation)}\\
&\nabla \cdot \mathbf{B} = 0 &&
\end{aligned}$$

We set $\mathbf{B} = \mu_0\mathbf{H}$ and assume that the current density is a linear function of the wave electric field through a suitable conductivity tensor, σ, so that $\mathbf{j} = \sigma\bullet\mathbf{E}$. We eliminate $\mathbf{B}$ from Faraday's Law by taking the curl of both sides and then use the Maxwell-Ampere Equation to replace $\nabla \times \mathbf{B}$: $\nabla \times (\nabla \times \mathbf{E}) = -\mu_0\partial(\mathbf{j} + \varepsilon_0\partial\mathbf{E}/\partial t)/\partial t$. Since the fields of the wave are assumed to vary in space and time as $\exp(i\mathbf{k}\bullet\mathbf{r} - i\omega t)$, our partial differential wave equation becomes the algebraic vector equation

$$\begin{aligned}
&-\mathbf{k} \times (\mathbf{k} \times \mathbf{E}) = -\mu_0(-i\omega)[\mathbf{j} + \varepsilon_0(-i\omega)\mathbf{E}]\\
\text{so that}\quad &k^2\mathbf{E} - (\mathbf{k} \cdot \mathbf{E})\mathbf{k} = \omega^2\mu_0\varepsilon_0(i\sigma \cdot \mathbf{E}/\varepsilon_0\omega + \mathbf{E}) = \omega^2/c^2\mathbf{K} \cdot \mathbf{E}, \qquad (4.2)\\
\text{and finally}\quad &(n^2\mathbf{1} - \mathbf{nn} - \mathbf{K}) \cdot \mathbf{E} = 0
\end{aligned}$$

Here $\mathbf{K}$ is the dielectric tensor whose elements will be calculated further for the model plasma using the Langevin equation, Eq. (4.1), to describe the dynamical response of electrons to the electromagnetic fields of the wave, and $n^2 = k^2c^2/\omega^2$ is the square of the (vector) index of refraction, $\mathbf{n} = \mathbf{k}c/\omega$. We note in passing that if ionization and recombination are negligible on the time scales associated with wave propagation, the charge and current densities must satisfy a continuity condition:

$$\partial\rho/\partial t + \nabla\cdot\mathbf{j} = -i\omega\rho + i\mathbf{k}\cdot\mathbf{j} = 0$$

so that $\rho = \mathbf{k}\bullet\sigma\bullet\mathbf{E}/\omega$. Because of the form of the Langevin equation we are using here to describe electron dynamics, it is convenient to employ the rotating coordinates that were used earlier in describing electron gyration in the static magnetic field. These may be obtained from a Cartesian representation through the unitary transformation, **U**:

$$\sqrt{2}U = \begin{pmatrix} 1 & i & 0 \\ 1 & -i & 0 \\ 0 & 0 & \sqrt{2} \end{pmatrix}. \tag{4.3}$$

Recall that the inverse of **U** is its Hermitian adjoint, i.e., the complex conjugate of the transposed matrix. If $v \ll v_{th}$ the Langevin equation, Eq. (4.1) can be simplified for "temperate" electrons with temperatures $T_e = mv_{th}^2/2$ by neglecting the second-order term, $\mathbf{v}\bullet\nabla\mathbf{v}$ in $md\mathbf{v}/dt = m(\partial\mathbf{v}/\partial t + \mathbf{v}\bullet\nabla\mathbf{v})$. And if the electron velocities are much less than the phase velocities of the waves of interest, we may neglect $\mathbf{v}\times\mathbf{B}$. We will address this term later. Setting $eB_o/m = \Omega_o$, we then obtain for our dynamical equation $-i\omega\mathbf{v} = -e\mathbf{E}/m - \Omega_o\mathbf{v}\times\mathbf{u}_z - \nu\mathbf{v}$. This vector equation yields the following three scalar equations in Cartesian components:

$$\begin{aligned} -i\omega v_x + \Omega_o v_y + \nu v_x &= -(e/m)E_x \\ -i\omega v_y - \Omega_o v_x + \nu v_y &= -(e/m)E_y \\ -i\omega v_z + \nu v_z &= -(e/m)Ez. \end{aligned}$$

We write this in matrix form as $\mathbf{M}\bullet\mathbf{v} = -(e/m)\mathbf{E}$ where the matrix **M** is given by

$$\mathbf{M} = \begin{pmatrix} -i\omega + \nu & \Omega_o & 0 \\ -\Omega_o & -i\omega + \nu & 0 \\ 0 & 0 & -i\omega + \nu \end{pmatrix}. \tag{4.4}$$

Transforming **M** into circular coordinates, we diagonalize **M** and obtain

$$\mathbf{UMU}^{-1} = \begin{pmatrix} -i\omega + \nu - i\Omega_o & 0 & 0 \\ 0 & -i\omega + \nu + i\Omega_o & 0 \\ 0 & 0 & -i\omega + \nu \end{pmatrix}, \tag{4.5}$$

resulting in the following three decoupled scalar dynamical equations:

$$\begin{aligned} (-i\omega + \nu - i\Omega_o)v_+ &= -(e/m)E_+ \\ (-i\omega + \nu + i\Omega_o)v_- &= -(e/m)E_- \\ (-i\omega + \nu)v_z &= -(e/m)E_z. \end{aligned} \tag{4.6}$$

For the present cold-plasma model, the perturbed current density, $\mathbf{j} = -en_o\mathbf{v}$, and we obtain for the circular components of the current density

$$\begin{aligned} j_+ &= i\varepsilon_0\omega_{pe}{}^2(\omega + \Omega_o + i\nu)^{-1}E_+ = \sigma_{11}E_+ \\ j_- &= i\varepsilon_0\omega_{pe}{}^2(\omega - \Omega_o + i\nu)^{-1}E_- = \sigma_{22}E_- \\ j_z &= i\varepsilon_0\omega_{pe}{}^2(\omega + i\nu)^{-1}E_z = \sigma_{33}E_z. \end{aligned} \tag{4.7}$$

Since from Eq. (4.2), $\mathbf{K} = 1 + i\sigma/\varepsilon_0\omega$, the dielectric tensor for a cold plasma is diagonalized in the circular coordinates with diagonal elements given by [4]

$$\begin{aligned}
\mathbf{K}_{11} &= 1-({\omega_{pe}}^2/\omega^2)/(\Omega/\omega+1+i\nu/\omega)\\
\mathbf{K}_{22} &= 1+({\omega_{pe}}^2/\omega^2)/(\Omega/\omega-1+i\nu/\omega)\\
\mathbf{K}_{33} &= 1-({\omega_{pe}}^2/\omega^2)/(1+i\nu/\omega).
\end{aligned} \tag{4.8}$$

In these expressions, ω_{pe} is the electron plasma frequency, $\omega_{pe} = (e^2 n_e/m\varepsilon_0)^{1/2}$ where n_e is the electron density and ε_0 is the permittivity of free space. As mentioned earlier, at the frequencies of interest to ECH the ions can usually be treated as an immobile neutralizing background, although an important exception will arise later.

We next transform the remainder of our wave equation into circular coordinates (the dielectric tenor, **K**, is already expressed in circular coordinates). The quantity to be transformed is $\mathbf{E} - (\mathbf{k}\bullet \mathbf{E})\mathbf{k}/k^2$. Since we have chosen **k** to lie in the x–z plane, we have for the Cartesian components

$$\begin{aligned}
\mathbf{E}-(\mathbf{k}\cdot\mathbf{E})\mathbf{k}/k^2 &= \mathbf{u_x}[E_x-\sin\theta(\sin\theta\, E_x+\cos\theta\, E_z)]\\
&\quad+\mathbf{u_y}E_y+\mathbf{u_z}[E_z-\cos\theta(\sin\theta\, E_x+\cos\theta\, E_z)],
\end{aligned}$$

or, in matrix form,

$$\mathbf{E}-(\mathbf{k}\cdot\mathbf{E})\mathbf{k}/k^2 = \begin{pmatrix} \cos^2\theta & 0 & -\sin\theta\cos\theta \\ 0 & 1 & 0 \\ -\sin\theta\cos^{\theta} & 0 & \sin^2\theta \end{pmatrix}\begin{pmatrix} E_x \\ E_y \\ E_z \end{pmatrix}$$

We again employ the unitary transformation matrix **U** and find the complete wave equation transformed into circular components to be

$$\mathbf{K}\cdot\mathbf{E}/n^2 = \begin{pmatrix} \cos^2\theta+(\sin^2\theta)/2 & -(\sin^2\theta)/2 & -\sin\theta\cos\theta/\sqrt{2} \\ -(\sin^2\theta)/2 & \cos^2\theta+(\sin^2\theta)/2 & -\sin\theta\cos\theta/\sqrt{2} \\ -\sin\theta\cos\theta/\sqrt{2} & -\sin\theta\cos\theta/\sqrt{2} & \sin^2\theta \end{pmatrix}\begin{pmatrix} E_+ \\ E_- \\ E_z \end{pmatrix} \tag{4.9}$$

Combining the two matrices gives the following wave equation:

$$\mathbf{D}\cdot\mathbf{E} = \begin{pmatrix} n^2\mathbf{K}_{11}-\cos^2\theta-(\sin^2\theta)/2 & (\sin^2\theta)/2 & \sin\theta\cos\theta/\sqrt{2} \\ (\sin^2\theta)/2 & n^{-2}\mathbf{K}_{22}-\cos^2\theta-(\sin^2\theta)/2 & \sin\theta\cos\theta/\sqrt{2} \\ \sin\theta\cos\theta/\sqrt{2} & \sin\theta\cos\theta/\sqrt{2} & n^{-2}\mathbf{K}_{33}-\sin^2\theta \end{pmatrix}\begin{pmatrix} E_+ \\ E_- \\ E_{z=0} \end{pmatrix} \tag{4.10}$$

The condition for the existence of nontrivial solutions to this Fourier–Laplace transformed wave equation is that the determinant of the matrix of the coefficients of the electric field, $|\mathbf{D}|$, should vanish. This condition reduces to a special case of the Booker quartic equation for the square of the index of refraction [5], $n^2 = (kc/\omega)^2$:

$$\begin{aligned}
&n^4[\mathbf{K}_{33}\cos^2\theta+(\mathbf{K}_{11}+\mathbf{K}_{22})(\sin^2\theta)/2]\\
&\quad-n^2\{\mathbf{K}_{33}(\mathbf{K}_{11}+\mathbf{K}_{22})[1-(\sin^2\theta)/2]+\mathbf{K}_{11}\mathbf{K}_{22}\sin^2\theta\}\\
&\quad+\mathbf{K}_{11}\mathbf{K}_{22}\mathbf{K}_{33}=0.
\end{aligned} \tag{4.11}$$

This dispersion relation describes the properties of waves propagating through an infinite, homogeneous cold plasma immersed in a uniform magnetic field with the three elements of the dielectric tensor given by Eq. (4.8). In general, it will have two solutions for n^2, giving the indices of refraction of the two waves as functions of the angle of propagation, θ, and the three frequencies, ω, ω_{pe}, and Ω. The polarization of waves corresponding to a given solution of Eq. (4.11) is obtained directly from Eq. (4.10) in terms of the relative amplitudes of the electric field components. One of the roots of the dispersion relation will vanish if the coefficient of n^4 vanishes. This occurs when the angle of propagation equals θ_{crit} given by

$$\theta_{crit} = \arctan[-2\mathbf{K}_{33}/(\mathbf{K}_{11}+\mathbf{K}_{22})]^{1/2.}$$

We can identify other major critical conditions for these solutions by considering the limiting cases of propagation exactly parallel and exactly perpendicular to the magnetic field.

4.2 Critical Conditions for Parallel Propagation

For waves propagating parallel to the magnetic field, $\theta = 0$, Eq. (4.11) becomes

$$n^4\mathbf{K}_{33} - n^2\mathbf{K}_{33}(\mathbf{K}_{11}+\mathbf{K}_{22}) + \mathbf{K}_{11}\mathbf{K}_{22}\mathbf{K}_{33} = 0. \tag{4.12}$$

Provided $\mathbf{K}_{33} \neq 0$, this equation describes two waves with indices of refraction given by $n^2 = \mathbf{K}_{11}$ and $n^2 = \mathbf{K}_{22}$. The first of these two solutions, the left-hand circularly polarized "fast" wave corresponding to $n^2 = \mathbf{K}_{11}$, satisfies the following dispersion relation:

$$(kc/\omega)^2 = 1 - \left(\omega_{pe}^2/\omega^2\right)/(\Omega/\omega+1). \tag{4.13}$$

The phase velocity of this wave exceeds the speed of light in vacuum, $(kc/\omega)^2 < 1$; and the propagation of this wave is cut off ($k = 0$) if $\omega_{pe}^2/\omega^2 = 1 + \Omega/\omega$. A conventional linear turning point analysis [4] shows that this left-hand circularly polarized wave is completely reflected at cutoff with no change of polarization. The left-hand polarization of this wave can be verified by substituting $n^2 = \mathbf{K}_{11}$ into Eq. (4.10). The second of these solutions, corresponding to $n^2 = \mathbf{K}_{22}$, satisfies the dispersion relation for right-hand circularly polarized waves:

$$(kc/\omega)^2 = 1 + \left(\omega_{pe}^2/\omega^2\right)/(\Omega/\omega-1). \tag{4.14}$$

Evidently these waves have phase velocities that are less than the speed of light in vacuum, since $(kc/\omega)^2 > 1$, and hence are sometimes called the "slow waves" or, more frequently, "whistlers." They propagate in plasmas of arbitrarily high density provided the wave frequency remains below the electron gyrofrequency, $\omega < \Omega$. As the wave frequency approaches the gyrofrequency from below, the phase velocity approaches zero and, in the presence of a dissipative mechanism, the waves are strongly damped at resonance, as we will see later in this chapter. These waves have a cutoff when $\omega_{pe}^2 = \omega^2(1-\Omega/\omega)$ and $\Omega < \omega$.

4.3
Critical Conditions for Perpendicular Propagation

For waves propagating perpendicular to the magnetic field, $\theta = \pi/2$, Eq. (4.11) becomes $n^4(K_{11} + K_{22})/2 - n^2\,[K_{33}(K_{11} + K_{22})/2 + K_{11}K_{22}] + K_{11}K_{22}K_{33} = 0$. Provided $(K_{11} + K_{22}) \neq 0$, the two solutions are

$$n^2 = K_{33}, \quad \text{and} \quad n^2 = 2K_{11}K_{22}/(K_{11} + K_{22}). \tag{4.15}$$

Waves corresponding to the first solution, $n^2 = K_{33}$, are usually called "ordinary" modes or O-modes; their dispersion relation is independent of the magnetic intensity. They have a cutoff when $\omega_{pe} = \omega$. Waves corresponding to the second solution are called "extra-ordinary" modes, or X-modes. The X-modes have two cutoffs given by $\omega_{pe}^2 = \omega^2 \pm \Omega\omega$, and the wave has a resonance, $(n \to \infty)$, the "upper hybrid resonance," at $\omega_{pe}^2 = \omega^2 - \Omega^2$. The upper hybrid resonance can play an important role in ECH and we will consider it in greater detail later.

4.4
Clemmow–Mullaly–Allis Diagrams

Organizing the implications of Eq. (4.11) is a daunting task, undertaken with considerable success by P.C. Clemmow and R.F. Mullaly in 1955 with later refinements added by Allis [6]. Their graphical display of the wave properties in particular regions of parameter space, the Clemmow–Mullaly–Allis or CMA diagram, is in widespread use and provides an economical way of describing the wave propagation issues affecting the coupling of microwave power from external sources to the resonance surfaces. A variant of the CMA diagram that meets our needs has ω_{pe}^2/ω^2, proportional to electron density, as its x-axis; and Ω/ω, proportional to the magnetic intensity, as its y-axis. In this two-dimensional space, we plot the zeros of the three diagonal elements of the dielectric tensor, K_{11}, K_{22}, and K_{33}, corresponding, respectively, to the cutoff conditions for the left-hand and right-hand circularly polarized modes ($\theta = 0$ and $\pi/2$) and the O-mode ($\theta = \pi/2$):

$$K_{11} = 0 \quad \text{if} \quad \omega_{pe}^2/\omega^2 = 1 + \Omega/\omega$$
$$K_{22} = 0 \quad \text{if} \quad \omega_{pe}^2/\omega^2 = 1 - \Omega/\omega$$
$$K_{33} = 0 \quad \text{if} \quad \omega_{pe}^2/\omega^2 = 1.$$

In addition to these three cutoff conditions, we plot the two resonances:

$\Omega/\omega = 1$, the gyroresonance, and

$\omega_{pe}^2/\omega^2 = 1 - (\Omega/\omega)^2$, the upper hybrid resonance.

The resulting modified CMA diagram is shown in Figure 4.1.

Using this diagram we can distinguish between the two plasma density regimes, "underdense" and "overdense," depending on whether $\omega_{pe} < \omega$ or $\omega_{pe} > \omega$, respectively, and the two magnetic field regimes, "low field" and "high field," depending on

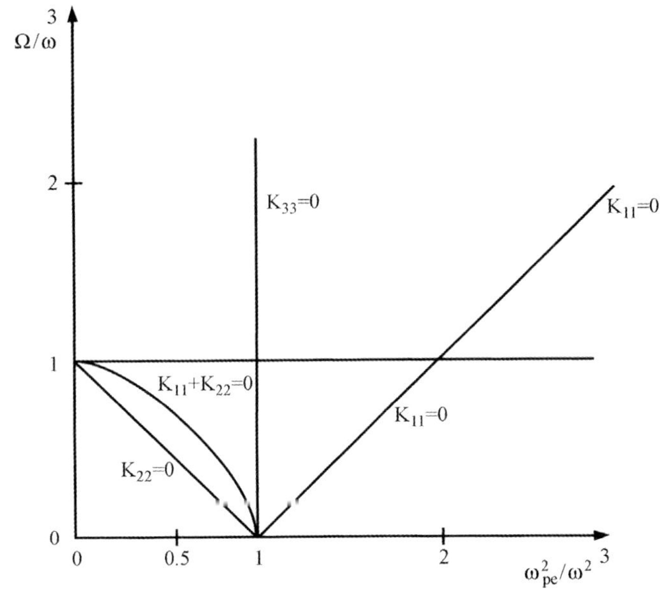

Figure 4.1 A "reduced" CMA diagram suitable for ECH; all ion dynamics have been neglected.

whether $\Omega < \omega$ or $\Omega > \omega$, respectively. In the underdense regime, $\mathbf{K}_{33}$ is positive and vanishes at a critical density given by the O-mode cutoff condition, $\omega_{pe} = \omega$, corresponding to a critical density, $n_{ec} = \omega^2 m\varepsilon_0/e^2$. This critical density increases with the square of the frequency from $1.24 \times 10^{10}\,\text{cm}^{-3}$ at a frequency of 1 GHz to $1.24 \times 10^{14}\,\text{cm}^{-3}$ at 100 GHz.

4.5
The High-Field Regime

We first consider the "high-field" regime, which has two cutoffs and one resonance:

$\omega_{pe} = \omega$ the O-mode cutoff

$\omega = \Omega$ electron gryoresonance

$\omega_{pe} = \omega(1+\Omega/\omega)^{1/2}$ the left-hand cutoff.

If the plasma is underdense (i.e., for densities below the O-mode cutoff), the waves represented by the two solutions of Eq. (4.11) propagate at all angles with respect to the magnetic field, as indicated in Figure 4.2.

Here we display schematically the indices of refraction of both waves as functions of the angle of propagation for a case in which the wave frequency is less than the electron gyrofrequency ("high field") but greater than the plasma frequency ("underdense"). As the density approaches the critical value for O-mode cutoff $\mathbf{K}_{33}$ approaches zero and $2\mathbf{K}_{11}\mathbf{K}_{22}/(\mathbf{K}_{11} + \mathbf{K}_{22})$ approaches unity. The indices

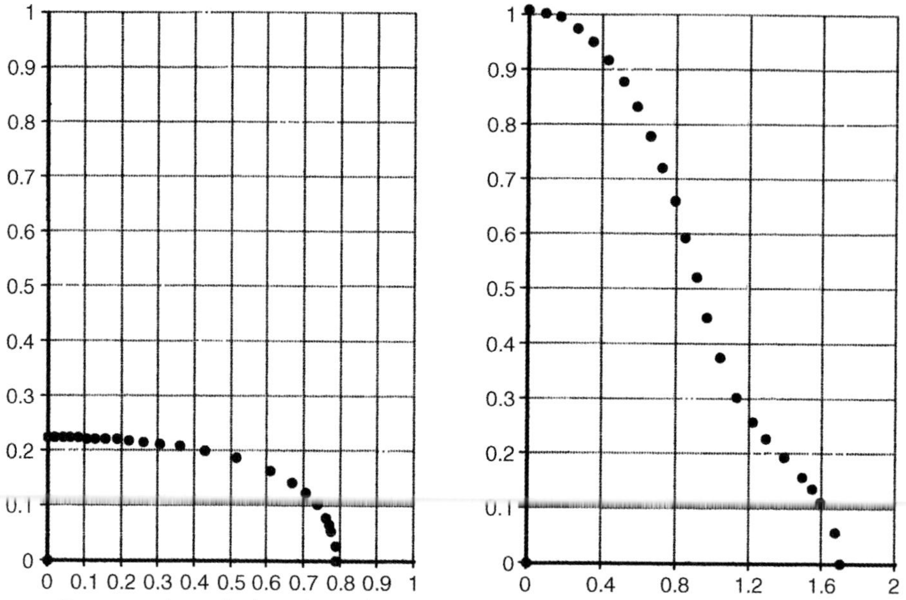

Figure 4.2 The angle-dependent indices of refraction for propagation in the high-field regime with density below the O-mode cutoff: $\omega = 1$, $\Omega = 1.5$, and $\omega_{pe} = \sqrt{0.95}$.

of refraction of the two waves in this transitional case are indicated schematically in Figure 4.3.

If the plasma is overdense, the slow wave propagates only within a cone centered on the direction of the magnetic field with its half angle given by θ_{crit}, the angle for which the coefficient of n^4 in Eq. (4.11) vanishes: $\theta_{crit} = \arctan[-2\mathbf{K}_{33}/(\mathbf{K}_{11} + \mathbf{K}_{22})]^{1/2}$.

The index of refraction of the fast wave varies continuously with angle in this overdense high-field case, as indicated in Figure 4.4.

Finally, in Figure 4.5 the indices are shown when the density is just below the value for the left-hand cutoff. For densities above this value, the left-hand circularly polarized wave is evanescent while the right-hand circularly polarized whistler continues to propagate until the wave reaches gyroresonance.

4.6
The Low-Field Regime

We next consider the low-field regime ($\Omega < \omega$) in which the plasma dispersion exhibits one additional cutoff:

$$\omega_{pe} = \omega(1 - \Omega/\omega)^{1/2} \quad \text{the right-hand cutoff,}$$

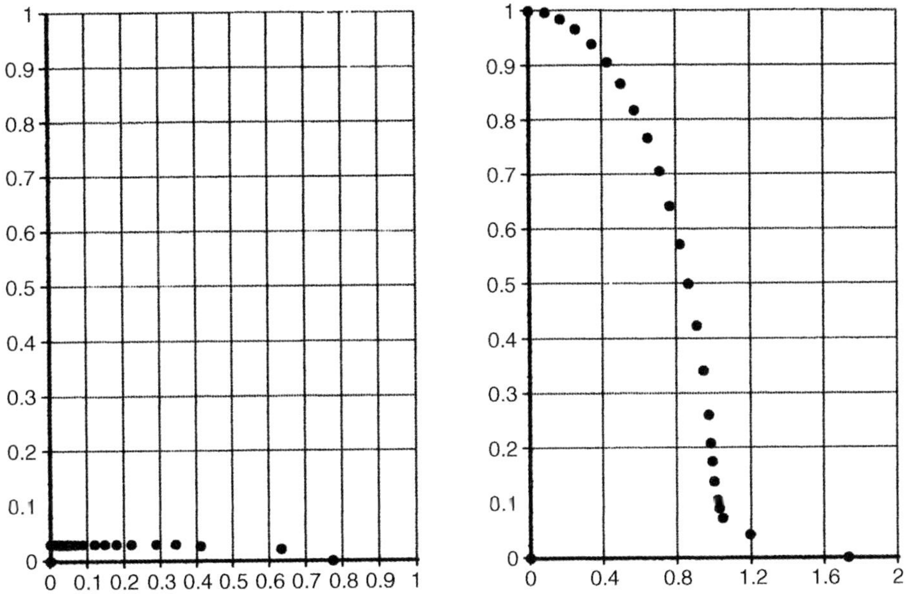

Figure 4.3 The angle-dependent indices of refraction for propagation in the high-field regime with density very slightly below the O-mode cutoff: $\omega = 1$, $\Omega = 1.5$, and $\omega_{pe} = \sqrt{0.999}$.

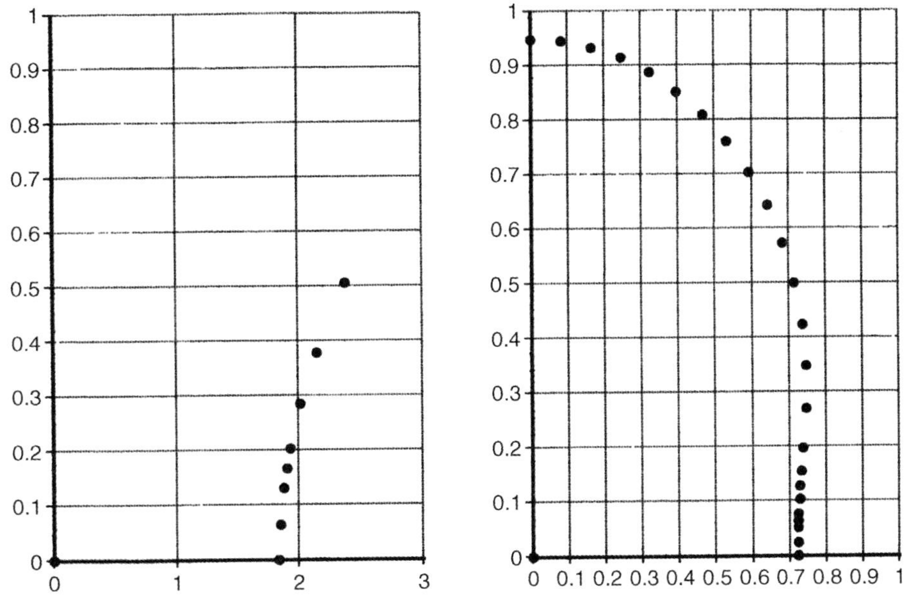

Figure 4.4 The angle-dependent indices of refraction for propagation in the high-field regime with density above the O-mode cutoff: $\omega = 1$, $\Omega = 1.5$, and $\omega_{pe} = \sqrt{1.2}$.

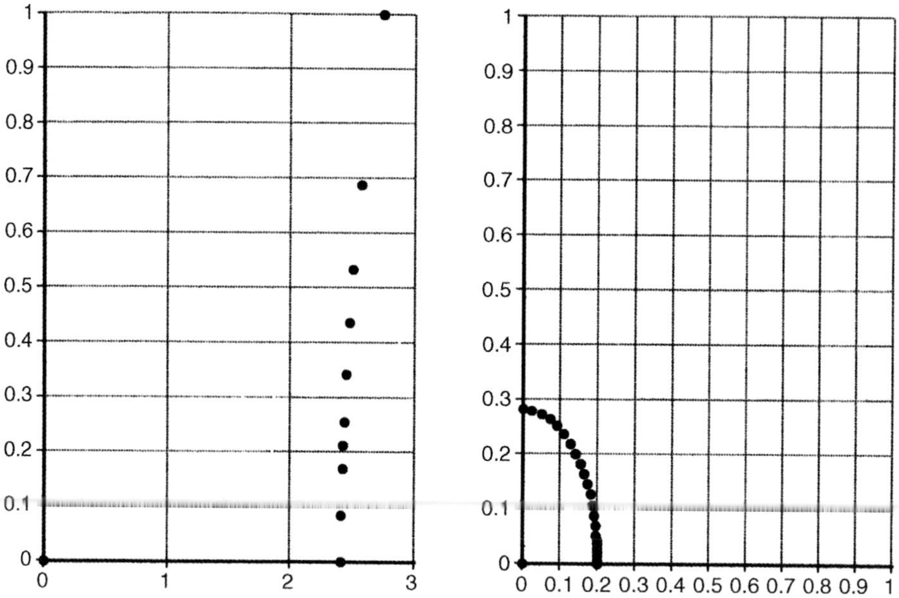

Figure 4.5 The angle-dependent indices of refraction for propagation in the high-field regime with density approaching the LH cutoff: $\omega = 1$, $\Omega = 1.5$, and $\omega_{pe} = \sqrt{2.4}$.

as well as the previous two cutoffs:

$$\omega_{pe} = \omega \qquad \text{the O-mode cutoff}$$
$$\omega_{pe} = \omega(1+\Omega/\omega)^{1/2} \qquad \text{the left-hand cutoff.}$$

In addition to these cutoffs, there is the upper hybrid resonance, given by

$$\omega_{pe} = [(\omega+\Omega)(\omega-\Omega)]^{1/2}.$$

The angle-dependent indices of refraction for plasma densities just below the right-hand cut-off are shown in Figure 4.6.

Both waves have phase velocities greater than the speed of light, and the propagation of the right-hand circularly polarized wave is clearly nearing cutoff. For densities above the right-hand cutoff but below the upper hybrid resonance only the left-hand circularly polarized wave propagates, but above the density for upper hybrid resonance the right-hand circularly polarized wave propagates perpendicular to the magnetic field and for all angles greater than the critical value $\theta_{crit} = \arctan[-2K_{33}/(K_{11} + K_{22})]^{1/2}$. This situation is illustrated in Figure 4.7.

As the density approaches the O-mode cutoff, the index of refraction of the right-hand circularly polarized wave gradually changes shape somewhat, as shown in Figure 4.8, so that just below the O-mode cutoff it propagates at almost all angles with respect to the magnetic field. Then for densities above the O-mode cutoff, the right-hand wave ceases to propagate, and for still higher densities above the left-hand cutoff all propagation ceases in this low-field regime and both waves are evanescent.

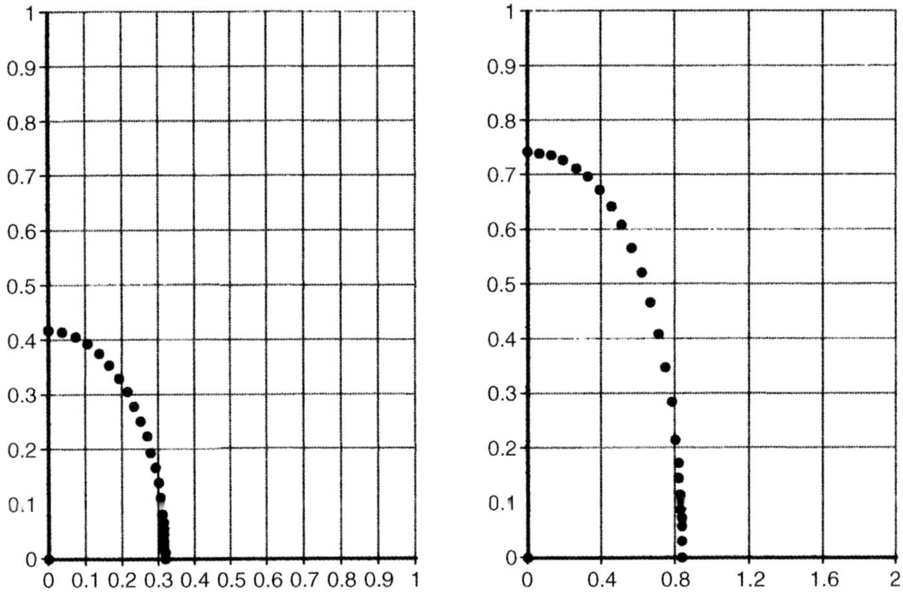

Figure 4.6 The angle-dependent indices of refraction for propagation in the low-field regime with density slightly below the RH cutoff: $\omega = 1$, $\Omega = 0.5$, and $\omega_{pe} = \sqrt{0.45}$.

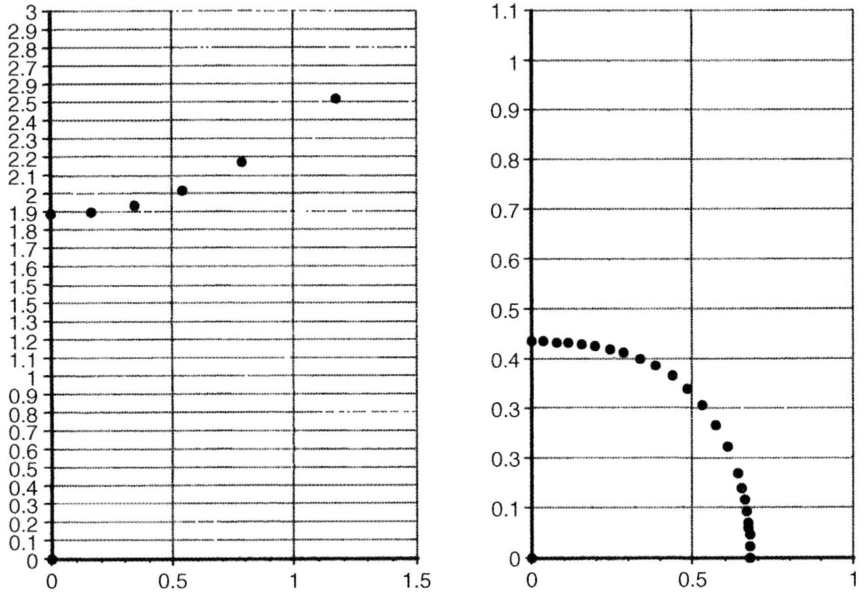

Figure 4.7 The angle-dependent indices of refraction for propagation in the low-field regime with density above the upper hybrid resonance but below the O-mode cutoff: $\omega = 1$, $\Omega = 0.5$, and $\omega_{pe} = \sqrt{0.81}$.

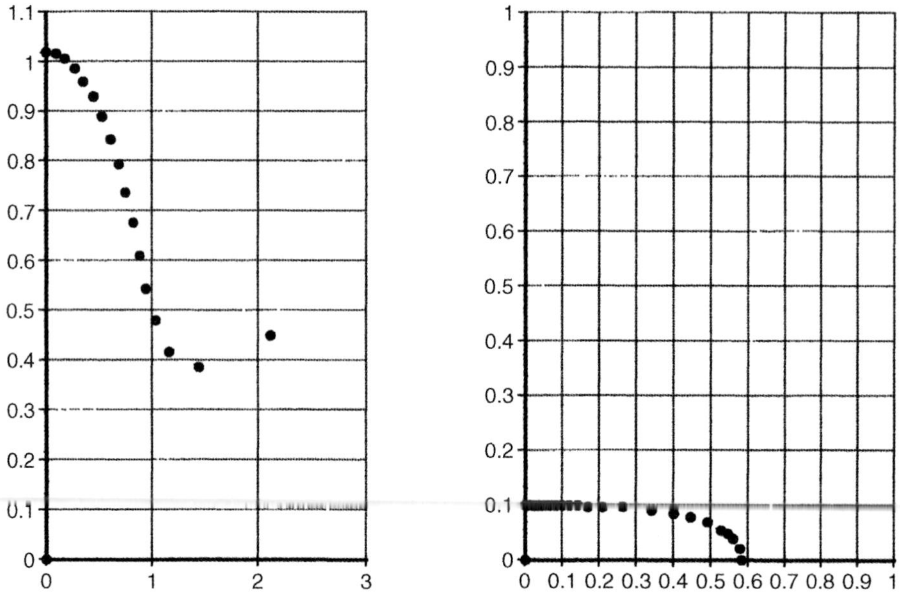

Figure 4.8 The angle-dependent indices of refraction for propagation in the low-field regime with density just below the O-mode cutoff: $\omega = 1$, $\Omega = 0.5$, and $\omega_{pe} = \sqrt{0.99}$.

4.7
A Few Preliminary Implications for ECH Experiments

Some practical consequences for ECH of these optical properties of magnetized cold plasmas are immediately apparent; others will be discussed at length in later chapters. Note especially that the optical properties of high-field launch and low-field launch are fundamentally different. First consider the high-field regime.

Right-hand circularly polarized whistler waves launched in the high-field region of a magnetic mirror field, for example, can propagate along magnetic lines of force to the resonance surface in plasmas of arbitrarily high density. Since the propagation is within a cone whose half-angle diminishes as the wave approaches resonance, the microwave power is concentrated along the magnetic field lines. Any left-hand circularly polarized component of the microwave power coupled via high-field launch will be internally reflected at the left-hand cutoff. If this wave is subsequently reflected from metallic surfaces (for example, the coupler itself or the vacuum chamber walls or suitable mirrors), it can change polarization and subsequently be absorbed at resonance.

With regard to the low-field regime, O-mode waves launched in the midplane of a magnetic mirror field, "low-field launch," can propagate directly to the resonance surfaces only in underdense plasmas. X-mode waves will encounter the right-hand cutoff followed by the upper hybrid resonance before being able to reach the resonance surface. Even so, as has been discussed by Weitzner and Batchelor [7],

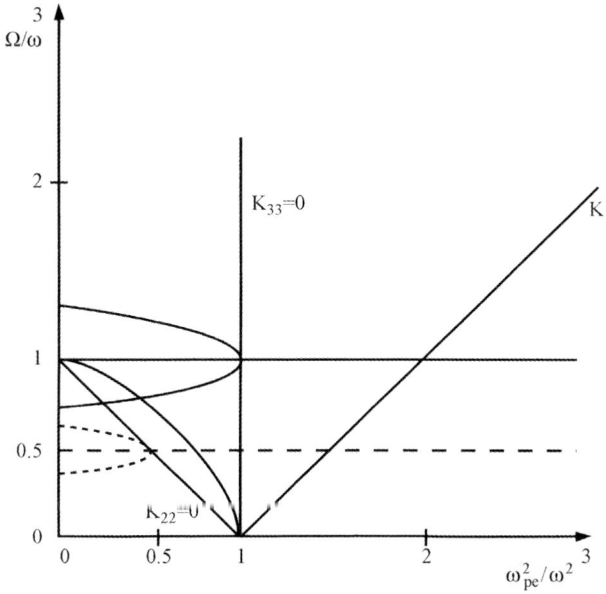

Figure 4.9 The modified CMA diagram of Figure 4.1 with typical tokamak values of relative density and magnetic intensity $(\omega_{pe}^2/\omega^2, \Omega/\omega)$, assuming fundamental (solid line) and second-harmonic (dotted line) resonance at the magnetic axis where the peak density is at the O-mode cutoff.

the microwave power may reach the resonance surfaces by indirect routes that involve Budden tunneling and mode conversion and depend sensitively on the detailed shape of the magnetic field, the plasma density profile and the metallic walls of the plasma chamber. We will return to this issue later in discussing the production of relativistic-electron plasmas which has employed "O-mode heating" or "low-field launch" with striking success.

Concerning ECH in tokamaks, it is clear from a suitably modified CMA diagram, such as the one shown in Figure 4.9 that illumination of the fundamental resonance surface in the high-field regime is limited solely by the O-mode cutoff.

Ray paths in the space spanned by our CMA diagram have been sketched for a typical tokamak with $R/a = 4$ assuming that the resonance surfaces are on the magnetic axis and the density profile is parabolic with a peak value equal to the critical density. In the low-field regime, the right-hand cutoff and the upper hybrid resonance prevent direct X-mode illumination of the fundamental resonance but do not affect the propagation of O-modes as long as the plasma is underdense. For heating at the second harmonic of the gyrofrequency, $\Omega/\omega = 0.5$, the accessibility for O-modes is again limited only by the density constraint, $\omega_{pe}^2 < \omega^2$. Direct accessibility to the resonance surface for X-modes at the second harmonic is possible in the low-field regime if the electron density is below the RH cutoff, $\omega_{pe}^2/\omega^2 < 0.5$ in Figure 4.9, but in tokamaks as in magnetic mirrors, it is possible for X-modes to tunnel through the evanescent region between the RH cutoff and the upper hybrid resonance. Here the microwave power can be converted into electrostatic plasma waves that can then propagate to the resonance surface. We shall return to this topic in considerable detail later.

4.8
Wave Damping

We next consider ways in which microwave power at the electron gyrofequency, launched into a chamber partially filled with plasma, is absorbed by the collective response of the plasma electrons. Two very different regimes are of experimental interest. The first regime applies to ECH configurations in which the plasma chamber is small enough relative to the wavelength of the incident power to function as a multimode cavity. Some have called this the "microwave oven regime." The second regime applies to plasma configurations that are sufficiently large relative to the wavelength of the incident microwave power for the propagating microwave fields to behave quasioptically and thus be described in terms of the plasma waves discussed earlier in this chapter. The small configurations are exemplified by many older magnetic-mirror devices and some ECH plasma sources, while the larger configurations are exemplified by many present-day and especially the next-generation large tokamaks and stellarators. In all cases, we will see that damping results from kinetic, microscopic properties of the plasma that are absent from the collisionless cold-fluid model of the plasma that we have considered to this point.

4.8.1
A Collisional Model of Damping

In the first regime, it is not generally possible to trace the rays corresponding to individual plasma waves excited by the incident microwave power. Instead, it may be more realistic to assume that the plasma chamber, viewed as a microwave cavity of irregular shape but high microwave integrity, is filled with electric fields having all three spatial components [8]. For low-temperature plasmas, the currents that flow in response to these fields are given as before by Eqs. (4.7), where collisions of frequency, ν, provide a formal basis for evaluating the damping. In the circular coordinates, we have used earlier the three components of the currents driven by the microwave power given by the following three equations:

$$
\begin{aligned}
j_+ &= i\varepsilon_0\omega_{pe}^2(\omega+\Omega+i\nu)^{-1}E_+ = \sigma_{11}E_+ \\
j_- &= i\varepsilon_0\omega_{pe}^2(\omega-\Omega+i\nu)^{-1}E_- = \sigma_{22}E_- \\
j_z &= i\varepsilon_0\omega_{pe}^2(\omega+i\nu)^{-1}E_z = \sigma_{33}E_z.
\end{aligned}
\tag{4.7}
$$

The average rate at which the microwave electric fields do work on the electrons in a unit volume of plasma is $P_{abs}/V = \mathrm{Re}(\mathbf{E}_{cc}\bullet\mathbf{j})/2$ [9], where $\mathbf{E}_{cc}$ is the complex conjugate of $\mathbf{E}$. If the resonant ECH mechanism dominates this energy transfer, the absorbed power density is mainly given by the right-hand circularly polarized component:

$$\begin{aligned}P_{abs}/V &= (\varepsilon_0|E_-|^2/2)\omega_{pe}^2\nu[(\omega-\Omega)^2+\nu^2]^{-1}\\ &= (\varepsilon_0|E_-|^2/2)\left(\omega_{pe}^2/\omega\right)\left\{(\nu/\omega)[(1-\Omega/\omega)^2+\nu^2/\omega^2]^{-1}\right\}.\end{aligned}$$

As $\nu/\omega \to 0$, the quantity in braces approaches $\pi\delta(1-\Omega/\omega)$, where $\delta(1-\Omega/\omega)$ is the Dirac delta function [10] so that in the limit of weak collisions,

$$P_{abs}/V = (\varepsilon_0|E_-|^2/2)\left(\omega_{pe}^2/\omega\right)\pi\delta(1-\Omega/\omega). \tag{4.16}$$

To determine the total power absorbed by the plasma, this expression must be integrated over the volume occupied by the plasma. As a concrete example, we will evaluate the power absorbed by the plasma confined in a simple magnetic mirror:

$$\begin{aligned}P_{abs} &= 2\pi\int\rho d\rho dz(\varepsilon_0|E_-|^2/2)\left(\omega_{pe}^2/\omega\right)\pi\delta(1-\Omega/\omega)\\ &= \pi^2\int\rho d\rho(dz/d\Omega)\varepsilon_0|E_-|^2\omega_{pe}^2\delta(1-\Omega/\omega)d\Omega/\omega\\ &= \pi^2\int\rho d\rho\varepsilon_0|E_-|^2\omega_{pe}^2(dz/d\Omega)|_{\Omega=\omega}.\end{aligned}$$

We can use the model magnetic-mirror field, Eq. (2.18), to evaluate $d\Omega/dz$ at the resonance surfaces and assume that the plasma density has a parabolic radial profile, $n(\rho) = n(0)(1-\rho^2/a^2)$. Then $(dz/d\Omega)_{|\Omega=\omega} = 2\pi(\Omega_o/L_c)[(M-M_{res})(M_{res}-1)]^{1/2}$, and taking into account both resonance surfaces we estimate the absorbed power to be

$$P_{abs} = (\pi a^2/4)L_c n(0)(e|E_-|^2/B_o)[(M-M_{res})(M_{res}-1)]^{-1/2}. \tag{4.17}$$

4.8.2
An Introduction to Collisionless Cyclotron Damping

In the second, "quasioptical" regime, it is possible to follow the path of an individual wave by solving the dispersion relation locally, as we will discuss later in this chapter. We can then evaluate the (spatial) damping rate at every point along that path by evaluating the imaginary component of the wave propagation vector, $\mathrm{Im}\,k = k_i$. The frequency, ω, is assumed to be real, i.e., we seek a steady-state solution to the dispersion relation. The dominant collisionless damping mechanism for ECH, often called cyclotron damping, is associated with kinetic effects arising from the thermal distribution of electron velocities and can be analyzed by solving the collisionless Boltzmann equation for the perturbed distribution function resulting from the presence of the wave. This analysis, first carried out by Vlasov [11], is quite general, but will be applied here to the specific cases of whistler waves propagating parallel to a uniform static magnetic field and O-modes propagating perpendicular to the magnetic field. We will return to more general applications of the Vlasov equation later. But first we consider a rudimentary analysis that will display the properties of the collisionless cyclotron damping mechanism in a relatively transparent form.

To do this, we examine a group of electrons flowing along a uniform, static magnetic field, $\mathbf{B}_o = B_o\mathbf{u}_z$, with velocity $\mathbf{v}_o = v_o\mathbf{u}_z$ and calculate their time-dependent perturbed velocity in the presence of a right-hand circularly polarized whistler wave with electric and magnetic fields given, respectively, by $\mathbf{E}_1 = \mathbf{E}_-(z,t) = \mathbf{E}_-\exp[i(kz - \omega t)]$ and $\mathbf{B}_1 = (k/\omega)\mathbf{u}_z \times \mathbf{E}_1$. We determine the perturbed electron velocity by solving the linearized collisionless Langevin equation:

$$m[\partial\mathbf{v}_1/\partial t + (\mathbf{v}_o \cdot \nabla)\mathbf{v}_1] = -e\{\mathbf{E}_1 + \mathbf{v}_1 \times [\mathbf{B}_o + (k/\omega)\mathbf{u}_z \times \mathbf{E}_1]\},$$

which in the present case becomes

$$\partial\mathbf{v}_1/\partial t + ikv_o\mathbf{v}_1 + \Omega\mathbf{v}_1 \times \mathbf{u}_z = -(e/m)[1-(kv_o/\omega)]\mathbf{E}_1.$$

Just as in the case of the unperturbed orbit equations of Chapter 3, the perpendicular components of the perturbed velocity are decoupled in rotating coordinates and we obtain the following equation for the time dependence of the right-hand component, v_-:

$$\partial v_-/\partial t + i(kv_o + \Omega)v_- = -(e/m)[1-(kv_o/\omega)]E_-\exp[i(kz-\omega t)].$$

The appropriate integrating factor is $\exp[i(kv_o + \Omega)t]$ and with it we obtain

$$\partial\{v_-\exp[i(kv_o+\Omega)t]\}/\partial t = -(e/m)[1-(kv_o/\omega)]\exp[i(kv_o+\Omega)t]E_-\exp[i(kz-\omega t)].$$

Integrating from $t = 0$ to $t = t$ and choosing $v_-(0) = 0$ we find $v_-(t)$ to be given by

$$v_-(t) = i(e/m)[1-(kv_o/\omega)]E_-\exp[i(kz-\omega t)]\times \{1-\exp[-i(kv_o+\Omega-\omega)t]\}/(kv_o+\Omega-\omega),$$

so that

$$v_-(t) = (e/m)[1-(kv_o/\omega)]E_-\exp[i(kz-\omega t) \times\{-\sin(kv_o+\Omega-\omega)t/(kv_o+\Omega-\omega) + i[1-\cos(kv_o+\Omega-\omega)t]/(kv_o+\Omega-\omega)\}. \quad (4.18)$$

This result neatly displays the essence of the cyclotron damping mechanism; namely, that if the Doppler-shifted resonance condition, $kv_o + \Omega - \omega = 0$, is satisfied, v_- increases linearly with time, according to the first term in braces, and the phase of v_- relative to that of the wave remains stationary according to the second term. If $\omega t \gg 1$, the first term in the braces has the properties of a delta function, since in the limit as $N \to \infty$, $(\sin xN)/x = \pi\delta(x)$.[10] Thus,

$$\sin(kv_o+\Omega-\omega)t/(kv_o+\Omega-\omega) \to \pi\delta(kv_o/\omega+\Omega/\omega-1)/\omega.$$

As we saw earlier, it is this kind of term that describes the damping process in which energy is transferred from the wave to the electrons. The average rate at which that transfer takes place can be obtained as before by evaluating $\mathrm{Re}(\mathbf{E}_{cc}\cdot\mathbf{j})/2$, where the current density is given by $j_- = -e\int dv_o f_o(v_o)v_-$ which in the present case is

$$\begin{aligned} j_- &= (e^2/m)E_-\exp[i(kz-\omega t)]\int dv_o f_o(v_o)[1-(kv_o/\omega)]\pi\delta(kv_o/\omega+\Omega/\omega-1)/\omega \\ &= (e^2/m)(\pi/k)E_-\exp[i(kz-\omega t)]\int d(kv_o/\omega)f_o(v_o)[1-(kv_o/\omega)]\delta(kv_o/\omega+\Omega/\omega-1) \\ &= (e^2/m)(\pi/k)E_-\exp[i(kz-\omega t)]f_o(v_o = v_{res})\Omega/\omega, \end{aligned}$$

where $v_{res} = (\omega - \Omega)/k$. If, for example, the electrons have a (one-dimensional) Gaussian distribution in the speed v_o, $f_o = n_e/(\alpha\sqrt{\pi})\exp(-v_o^2/\alpha^2)$, then

$$j_- = \sqrt{\pi}(e^2 n_e/m)(1/k\alpha)(\Omega/\omega)E_- \exp[i(kz-\omega t)\exp[-(\omega-\Omega)^2/k^2\alpha^2].$$

Note that in the limit in which the thermal speed is much less than the phase velocity of the wave, $k\alpha/\omega \ll 1$, the damping is negligible except very near the resonance surface. In fact, since $\exp(-x^2/\varepsilon)/\sqrt{\varepsilon} = \sqrt{\pi}\delta(x)$ in the limit as $\varepsilon \to 0$ [10], we recover exactly the earlier result, Eq. (4.16); namely, that the average power transferred from the wave to the electrons per unit volume is given by

$$P_{abs}/V = (\varepsilon_0|E_-|^2/2)(\omega_{pe}^2/\omega)(\Omega/\omega)\pi\delta(1-\Omega/\omega).$$

4.8.3
Cyclotron Damping of Whistler Waves

We next use the Vlasov equation to determine the perturbed electron distribution function resulting when a right-hand circularly polarized whistler wave with electric field $\mathbf{E} = E_-\exp[i(kz - \omega t)]$ propagates parallel to the static magnetic field through a more realistic model of the plasma equilibrium. The theoretical basis of the Vlasov equation is discussed at length by many authors [12] and we will proceed directly to its implementation for the present purpose by integrating in time along the unperturbed orbits of the electrons. For an infinite homogeneous plasma, the linearized Vlasov equation can be written compactly as

$$Df_1/Dt = -\mathbf{a}_1 \bullet \partial f_o/\partial \mathbf{v},$$

where Df_1/Dt is the convective derivative of the perturbed electron distribution and is to be integrated along the unperturbed electron orbits, and $\mathbf{a}_1$ is the electron acceleration due to the fields of the wave:

$$\mathbf{a}_1 = -(e/m)(\mathbf{E} + \mathbf{v} \times \mathbf{B}) = -(e/m)[(1-\mathbf{k}\cdot\mathbf{v}/\omega)\mathbf{E} + (\mathbf{v}\cdot\mathbf{E})\mathbf{k}/\omega].$$

We will choose an unperturbed electron distribution function, f_o, that describes an infinite, homogeneous plasma immersed in a uniform magnetic field, $\mathbf{B}_o = B_o\mathbf{u}_z$. Since the electrons generally thermalize rapidly along the magnetic field but can have various thermal as well as nonthermal distributions across the magnetic field, it is reasonable to assume a product form for $f_o(v_\perp, v_\parallel)$. Moreover, since we will usually deal with adiabatic electrons, for which the magnetic moment is a constant of the motion, we will assume that the equilibrium distribution function will have the following general form:

$$f_o(v_\perp v_\parallel) = \left[n_e/\left(\pi^{3/2}\alpha_\perp^2\alpha_\parallel\right)\right] g_o\left(v_\perp^2/\alpha_\perp^2\right)\exp\left(-v_\parallel^2/\alpha_\parallel^2\right). \qquad (4.19)$$

Here n_e is the electron density and thus we require that g_o is normalized so that $\int(v_\perp/\alpha_\perp)(dv_\perp/\alpha_\perp)g_o(v_\perp^2/\alpha_\perp^2) = 1$ where the integral is from zero to infinity.

For these equilibrium distribution functions,

$$\partial f_o/\partial \mathbf{v} = 2[n_e/(\pi^{3/2}\alpha_\perp^2\alpha_\parallel)]\exp\left(-v_\parallel^2/\alpha_\parallel^2\right) \times \left[(\mathbf{v}_\perp/\alpha_\perp^2)g'_o(v_\perp^2/\alpha_\perp^2) - (\mathbf{v}_\parallel/\alpha_\parallel^2)g_o(v_\perp^2/\alpha_\perp^2)\right],$$

where $g'_o(y) = dg_o/dy$. The linearized Vlasov equation now takes the following form:

$$Df_1/Dt = 2(e/m)\left[n_e/\left(\pi^{3/2}\alpha_\perp^2\alpha_\|\right)\right]\exp\left(-v_\|^2/\alpha_\|^2\right)[(1-\mathbf{k}\cdot\mathbf{v}/\omega)\mathbf{E}+(\mathbf{v}\cdot \mathrm{E})\mathbf{k}/\omega]$$
$$\bullet\left[(\mathbf{v}_\perp/\alpha_\perp^2)\mathbf{g}'_o(v_\perp^2/\alpha_\perp^2)-(\mathbf{v}_\|/\alpha_\|^2)\mathbf{g}_o(v_\perp^2/\alpha_\perp^2)\right]. \tag{4.20}$$

Since for this first example we are considering the effect of (transverse) whistler waves propagating parallel to the magnetic field, we set $\mathbf{k}\bullet\mathbf{v}_\perp = 0$ and $\mathbf{E}\bullet\mathbf{v}_\| = 0$ giving

$$Df_1/Dt = 2(e/m)\left[n_e/\left(\pi^{3/2}\alpha_\perp^2\alpha_\|\right)\right]\exp\left(-v_\|^2/\alpha_\|^2\right)$$
$$\times\,\mathbf{E}\cdot\mathbf{v}_\perp\left[(1-kv_\|/\omega)g'_o/\alpha_\perp^2-(kv_\|/\omega)g_o/\alpha_\|^2\right].$$

Note that $\mathbf{E}\bullet\mathbf{v}_\perp = E_xv_x + E_yv_y = E_+v_- + E_-v_+ = E_-v_+$ for our present case; namely, a right-hand circularly polarized wave. Also, recall that for the unperturbed orbits,

$$v_+ = v_\perp(0)\exp[i\phi(t)] \quad \text{where} \quad \phi(t) = \phi_o + \Omega t, \quad \text{and} \quad z = z_o + v_\| t.$$

Without loss of generality we can choose $z_o = 0$ so that

$$\mathbf{E}\cdot\mathbf{v}_\perp = |E_-|v_\perp(0)\exp[i\phi_o + i(kv_\| - \omega + \Omega)t].$$

This must be integrated in time from $t = -\infty$ to $t = 0$; and assuming that the perturbation vanishes at remote past times the result is the following expression for the perturbed distribution function at the time of observation:

$$f_1(0) = 2(e/m)\left[n_e/\left(\pi^{3/2}\alpha_\perp^2\alpha_\|\right)\right]\exp\left(-v_\|^2/\alpha_\|^2\right)[(1-kv_\|/\omega)g'_o/\alpha_\perp^2-(kv_\|/\omega)g_o/\alpha_\|^2]$$
$$\times\,|E_-|v_\perp(\exp i\phi)/[i(kv_\| - \omega + \Omega)].$$

For an isotropic Maxwell–Boltzmann equilibrium distribution function $\alpha_\perp^2 = \alpha_\|^2 = \alpha^2$, $g'_o = -g_o$, and thus $[(1-kv_\|/\omega)g'_o/\alpha_\perp^2-(kv_\|/\omega)g_o/\alpha_\|^2] = -g_o/\alpha^2$. In this case our expression for the perturbed distribution function becomes

$$f_1(0) = -2(e/m)\left[n_e/\left(\pi^{3/2}\alpha^5\right)\right]\exp\left[-\left(v_\perp^2+v_\|^2\right)/\alpha^2\right]|E_-|v_\perp(\exp i\phi)/[i(kv_\| - \omega + \Omega)].$$

The dispersion relation for the right-hand circularly polarized whistler wave results from using this perturbed distribution function to determine the current appearing in Maxwell's equations for the wave: we combine $\nabla\times\mathbf{E} = -\partial\mathbf{B}/\partial t$ and $\nabla\times\mathbf{H} = \mathbf{j} + \varepsilon_o\partial\mathbf{E}/\partial t$ to obtain the wave equation $\nabla\times(\nabla\times\mathbf{E}) = -\partial(\nabla\times\mathbf{B})/\partial t = -\partial(\mu_o\mathbf{j} + \mu_o\varepsilon_o\partial\mathbf{E}/\partial t)/\partial t$ which leads directly to $(n^2-1)\mathbf{E} = (i/\varepsilon_o\omega)\mathbf{j}$. The right-hand circular component of the perturbed current resulting from the whistler wave under consideration here is given by

$$j_- = -e\int d^3v v_- f_1$$
$$= -e\int v_\perp dv_\perp\int d\phi\int dv_\| v_\perp\exp(-i\phi)$$
$$\times 2(e/m)\left[n_e/\left(\pi^{3/2}\alpha^5\right)\right]\exp\left[-\left(v_\perp^2+v_\|^2\right)/\alpha^2\right]|E_-|v_\perp(\exp i\phi)/[i(kv_\| - \omega + \Omega)]$$

$$= -2i(e^2 n_e/m)|E_-| \int v_\perp^3/\alpha^3 dv_\perp/\alpha\left[\exp(-v_\perp^2/\alpha^2)\right]$$

$$\times(1/\sqrt{\pi}) \int dv_\parallel/\alpha\left[\exp(-v_\parallel^2/\alpha^2)\right]/[(kv_\parallel - \omega + \Omega)]$$

$$= -i(e^2 n_e/m)|E_-|Z(\zeta)/(k\alpha).$$

Here $Z(\zeta) = (1/\sqrt{\pi})\int dx \exp(-x^2)/(x-\zeta)$, where the integral is from $-\infty$ to ∞, is the plasma dispersion function [13] whose argument is $\zeta = (\omega - \Omega)/k\alpha$. Our dispersion relation for the right-hand circularly polarized whistler wave propagating parallel to the static magnetic field is thus given by the following equation for n, the (complex) index of refraction:

$$\begin{aligned}(n^2-1) &= (\omega_{pe}^2/\omega^2)(\omega/k\alpha)Z[(\omega-\Omega)/k\alpha] \\ &= (\omega_{pe}^2/\omega^2)(c/n\alpha)Z[(1-\Omega/\omega)(c/n\alpha)].\end{aligned} \tag{4.21}$$

Although this dispersion relation was derived for an infinite homogeneous plasma in a uniform magnetic field, it can be solved at closely spaced intervals along the magnetic field line to construct what amounts to a WKB solution for a weakly inhomogeneous equilibrium. The cumulative damping of an incident wave at any point z, along the path of the wave, is given by $\exp[-\int k_i(z)dz]$, where the integration ranges from the point at which the wave is first launched to the point of observation. The integral $2\int k_i(z)dz = \tau$, sometimes called the "optical depth" [14], characterizes the absorption of microwave power: $P_{abs} = P_{in}[1 - \exp(-\tau)]$. To evaluate τ, it is necessary to find the complex-n roots of the dispersion relation, a process which, in practice, employs computerized numerical root-finding techniques. We will use a more rudimentary and transparent albeit approximate approach here to illustrate the main features of propagation and absorption of whistler waves launched, for example, in the high-field region of a magnetic mirror. To this end, we write the dispersion relation Eq. (4.21) in the following form:

$$\mathrm{DIS} = n(n^2-1) - (\omega_{pe}^2/\omega^2)(c/\alpha)Z[(1-\Omega/\omega)(c/n\alpha)] = 0.$$

The real and imaginary parts are, respectively,

$$\mathrm{ReDIS} = n_r(n_r^2 - 3n_i^2 - 1) - (\omega_{pe}^2/\omega^2)(c/\alpha)Z_r(\zeta) = 0,$$

and

$$\mathrm{ImDIS} = n_i(3n_r^2 - n_i^2 - 1) - (\omega_{pe}^2/\omega^2)(c/\alpha)Z_i(\zeta) = 0.$$

Note that at the resonance surface where $\Omega/\omega = 1$, $Z_r(0) = 0$, and $Z_i(0) = \sqrt{\pi}$. The real part of the dispersion relation then reduces to $n_r^2 = 3n_i^2 + 1$ so that at the resonance the imaginary part is given simply by $2n_i(4n_i^2 + 1) = (\omega_{pe}^2/\omega^2)(c/\alpha)\sqrt{\pi}$.

We let $2n_i = \xi$ and $(\omega_{pe}^2/\omega^2)(c/\alpha)\sqrt{\pi} = \rho$. Since for most cases of interest $\rho \gg 1$, we can obtain an approximate solution for n_i by setting $\xi = \xi_o + \Delta\xi$ with $\xi_o^3 = \rho$. Then

$$\xi_o^3 + 3\xi_o^2\Delta\xi + 3\xi_o\Delta\xi^2 + \Delta\xi^3 + \xi_o + \Delta\xi = \rho,$$

and if we retain only the lowest order terms we find for the value of n_i at the resonance surface

$$n_i(0) = \xi/2 = 1.5\rho/(1 + \rho^{2/3}). \tag{4.22}$$

Very close to resonance where $|\zeta| < 1$, $Z(\zeta)$ can be approximated by a power series [13]:

$$Z(\zeta) = i\sqrt{\pi}\exp(-\zeta^2) - 2\zeta[1 - 2\zeta^2/3 + 4\zeta^4/15 - 8\zeta^6/105 + \cdots].$$

Far from resonance where $|\zeta| > 1$ and damping is negligible, $Z(\zeta)$ can be approximated by an asymptotic expansion [13]:

$$Z(\zeta) = i\sqrt{\pi}\exp(-\zeta^2) - \zeta^{-1}[1 + (1/2\zeta^2) + (3/4\zeta^4) + (15/8\zeta^6) + \cdots].$$

Note that in this limit we recover the cold-plasma dispersion relation for these whistler waves: $(n^2 - 1) = \omega_{pe}^2/\omega^2/[\omega(\Omega - \omega)]$.

For values of $|\zeta| \approx 1$ in lieu of sophisticated numerical techniques, we can interpolate among the tabulated values of $Z(\zeta)$ [13]. We can use these three approximations for Z to map the zeros of Re DIS and Im DIS in the complex ζ-plane and search for points (x_o, y_o) where the two contours cross and yield simultaneous solutions ReDIS = ImDIS = 0. Here we have set $\zeta = (1 - \Omega/\omega)(c/n\alpha) = x + iy$ so that n, the complex index of refraction, is given by

$$n_r + in_i = [(1 - \Omega/\omega)(c/\alpha)](x - iy)/(x^2 + y^2).$$

In this way we can obtain illustrative results such as those shown in Figure 4.10, where $T_e = 10$ eV, and Figure 4.11 for which $T_e = 100$ eV.

When the wave is far from the resonance surface ($\Omega/\omega \gg 1$), the real part of the index of refraction, n_r, is approximately equal to the value given by the cold-fluid model and damping of the wave is negligible. But when the wave reaches a point where the Doppler-shifted resonance condition can be satisfied by electrons in the tail of the thermal distribution, damping becomes nonvanishing and the imaginary part of the index of refraction, n_i, begins to increase and n_r begins to deviate significantly from the cold-fluid value. The Doppler-shifted resonance condition, $\omega - kv - \Omega = 0$, can be rewritten as

$$-(v/\alpha)_{res} = (\Omega/\omega - 1)(c/\alpha n_r),$$

and the relative density of electrons with speeds greater than this resonant value can be obtained from the error function: $\delta n_{res} = 1 - \text{erf}(v/\alpha)_{res}$. For example, the relative density of resonant electrons is less than 0.001 if $(v/\alpha)_{res}$ exceeds 2.325. We use this criterion to estimate the magnetic field strength at the onset of significant damping, $(\Omega/\omega - 1)_{onset} = \Delta_{onset}$, and thereby determine the thickness of the "resonance layer."

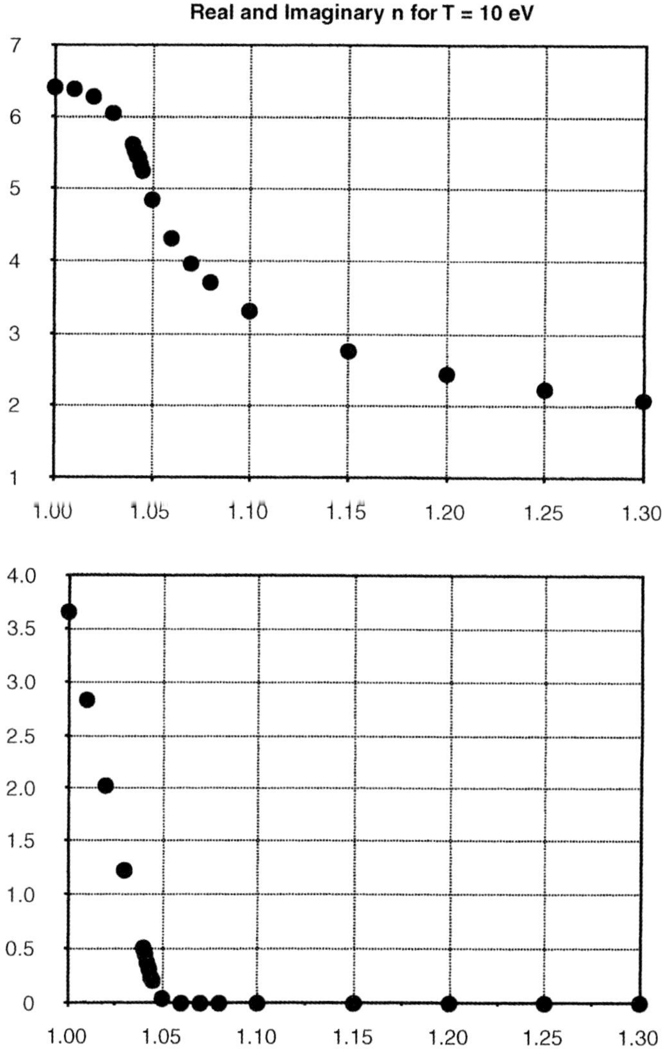

Figure 4.10 The complex index of refraction versus Ω/ω for whistler waves propagating parallel to a uniform magnetic field in a plasma whose electron temperature is 10 eV.

The cold-fluid model furnishes a reasonable estimate of n_r near the onset of damping, and combining this estimate with the Doppler-shifted resonance condition we obtain

$$\Delta_{\mathrm{onset}} \approx \left[-(v/\alpha)_{\mathrm{res}}(\alpha/c)(\omega_{\mathrm{pe}}/\omega)\right]^{2/3} \approx \left[2.325(\alpha/c)(\omega_{\mathrm{pe}}/\omega)\right]^{2/3}. \qquad (4.23)$$

Inside the resonance layer our numerical results show that the imaginary part of the index of refraction varies approximately as

$$n_i = n_i(0)(1-\Delta/\Delta_{\mathrm{onset}}),$$

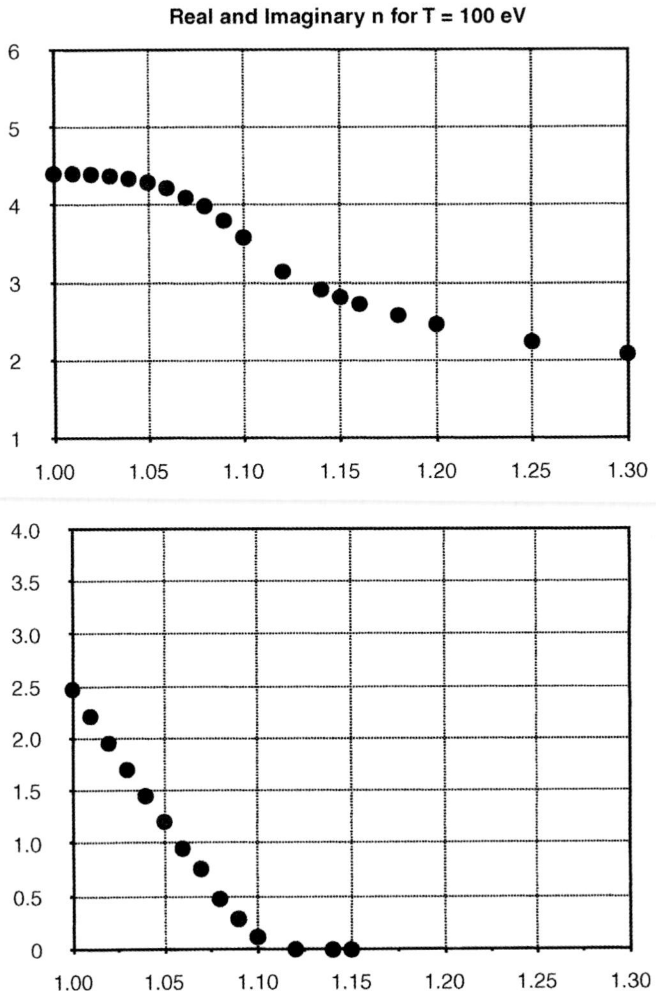

Figure 4.11 The complex index of refraction versus Ω/ω for whistler waves propagating parallel to a uniform magnetic field in a plasma whose electron temperature is 100 eV.

where an expression for $n_i(0)$ was given earlier in Eq. (4.22). The optical depth can now be evaluated for these whistler waves propagating toward the resonance surface from the point of high-field launch:

$$\tau = 2\int k_i(z)dz = [4\pi n_i(0)/\lambda]\int(1-\Delta/\Delta_{onset})(dz/d\Omega)d\Omega$$
$$= [4\pi n_i(0)/\lambda]\int(1-\Delta/\Delta_{onset})Ld\Delta \approx 2\pi n_i(0)\Delta_{onset}L/\lambda.$$

Here L is the magnetic scale length (assumed to be roughly constant in the resonance layer) and λ is the free-space wavelength $2\pi c/\omega$. For $T_e = 10$ eV we have found that the product $n_i(0)\Delta_{onset} = 0.17$, and it increases with electron temperature as $T_e^{1/6}$. Since the scale length is much greater than the wavelength, $L/\lambda \gg 1$ for quasioptical conditions, the optical thickness is very large for these whistler waves and the absorption of the wave is essentially complete within the resonance layer.

4.8.4 Cyclotron Damping of Waves Propagating as O-Modes

As a final example of wave absorption in the quasioptical regime we consider the damping of waves propagating in the O-mode perpendicular to a static and weakly inhomogeneous magnetic field. The damping of these waves has been evaluated by several different workers by incorporating several distinctly different kinetic effects, such as finite electron gyroradius [15] and weak relativistic effects [16], as well as mode conversion to cyclotron harmonic waves that only appear in the hot-plasma dispersion relation [17]. The results of the different approaches agree, and here we will adopt the finite electron gyroradius WKB approach taken by Antonsen and Manheimer [15]. From Eq. (4.20), the linearized Vlasov equation for waves with $\mathbf{E} = E\mathbf{u}_z$ and $\mathbf{k} = k\mathbf{u}_x$ propagating in an anisotropic bi-Maxwellian plasma is

$$Df_1/Dt = 2(e/m)\left[n_e/\left(\pi^{3/2}\alpha_\perp^2\alpha_\parallel\right)\right]\exp\left(-v_\parallel^2/\alpha_\parallel^2\right)[(1-kv_x/\omega)E\mathbf{u}_z + (\mathbf{v}_\parallel E)k\mathbf{u}_x/\omega]$$
$$\bullet\left[\left(\mathbf{v}_\perp/\alpha_\perp^2\right) + \mathbf{v}_\parallel/\alpha_\parallel^2\right]\left[-\mathbf{g}_o\left(v_\perp^2/\alpha_\perp^2\right)\right].$$

For the iostropic Maxwell–Boltzmann distribution, this reduces to

$$Df_1/Dt = -2(e/m)\left[n_e/\left(\pi^{3/2}\alpha^3\right)\right]\exp\left(-v_\parallel^2/\alpha^2\right)\exp\left(-v_\perp^2/\alpha^2\right)\left(v_\parallel/\alpha^2\right)E.$$

This is to be integrated in time using $E = |E|\exp[i(kx - \omega t)]$ with x given by the unperturbed orbit: $x = x_o + (v_\perp/\Omega)\sin(\phi_o + \Omega t) - (v_\perp/\Omega)\sin\phi_o$. We can set $x_o = 0$ and use the Bessel function relation [18] $\exp(\pm ia \sin b) = \Sigma J_n(a)\exp(\pm inb)$, where the index n ranges from $-\infty$ to ∞, to express E as the sum over m and n:

$$E = |E|\Sigma J_m(v_\perp/\Omega)\exp[im(\phi_o + \Omega t)]\Sigma J_n(v_\perp/\Omega)\exp(-in\phi_o)\exp(-i\omega t)$$
$$= |E|\Sigma J_m(v_\perp/\Omega)J_n(v_\perp/\Omega)\exp[i(m-n)\phi_o + (m\Omega-\omega)t)].$$

The linearized Vlasov equation for this case is then given by

$$Df_1/Dt = -2(e/m)[n_e/(\pi^{3/2}\alpha^3)]\exp(-v_\parallel^2/\alpha^2)\exp(-v_\perp^2/\alpha^2)(v_\perp/\alpha^2)$$
$$\times\ |E|\Sigma J_m(v_\perp/\Omega)J_n(v_\perp/\Omega)\exp[i(m-n)\phi_o + (m\Omega-\omega)t],$$

and the perturbed distribution function at $t = 0$ is

$$f_1 = -2(e/m)[n_e/(\pi^{3/2}\alpha^3)]\exp(-v_\parallel^2/\alpha^2)\exp(-v_\perp^2/\alpha^2)(v_\parallel/\alpha^2)$$
$$\times|E|\Sigma J_m(v_\perp/\Omega)J_n(v_\perp/\Omega)\exp[i(m-n)\phi]/[i(m\Omega-\omega)].$$

Note that this assumes that ω contains an infinitesimal positive imaginary part to ensure that the wave amplitude vanishes for remote (negative) times. The parallel current described by this perturbed distribution function is

$$j_{\|} = -e\int d^3v\, v_{\|} f_1 = -e\int d\phi \int v_{\perp} dv_{\perp} \int dv_{\|} v_{\|} f_1$$
$$= 2e|E| \int d\phi \int v_{\perp} dv_{\perp} \int dv_{\|} v_{\|} (e/m)[n_e/(\pi^{3/2}\alpha^3)] \exp(-v_{\|}^2/\alpha^2)$$
$$\times \exp(-v_{\perp}^2/\alpha^2)(v_{\|}/\alpha^2)\Sigma J_m(v_{\perp}/\Omega) J_n(v_{\perp}/\Omega) \exp[i(m-n)\phi]/[i(m\Omega-\omega)].$$

Since the integral over ϕ vanishes unless $m=n$ when it is equal to 2π, we have

$$j_{\|} = 4\pi e|E| \int v_{\perp} dv_{\perp} \int dv_{\|} v_{\|} (e/m)[n_e/(\pi^{3/2}\alpha^3)] \exp(-v_{\|}^2/\alpha^2)$$
$$\times \exp(-v_{\perp}^2/\alpha^2)(v_{\|}/\alpha^2)\Sigma J_n^2(v_{\perp}/\Omega)/[i(n\Omega-\omega)]$$
$$= 4(e^2 n_e/m)(1/\alpha^2)|E| \int v_{\perp} dv_{\perp} \exp(-v_{\perp}^2/\alpha^2)\Sigma J_n^2(v_{\perp}/\Omega)$$
$$\times 1/(\sqrt{\pi}\alpha) \int dv_{\|} \exp(-v_{\|}^2/\alpha^2)(v_{\|}^2/\alpha^2)/[i(n\Omega-\omega)]$$
$$= (e^2 n_e/m)|E|\Sigma \exp(-\lambda) I_n(\lambda)/[i(n\Omega-\omega)],$$

where the integration over perpendicular velocities is given by [19]

$$(2/\alpha^2)\int v_{\perp} dv_{\perp} \exp(-v_{\perp}^2/\alpha^2) J_n^2(v_{\perp}/\Omega) = \exp(-\lambda) I_n(\lambda) \quad \text{and} \quad \lambda = k^2\alpha^2/2\Omega^2.$$

This current perturbation is to be used in the O-mode wave equation

$$(n^2-1)|E| = (i/\varepsilon_o\omega) j_{\|} = (e^2 n_e/m\varepsilon_o)|E|\Sigma \exp(-\lambda) I_n(\lambda)/[\omega(n\Omega-\omega)],$$

so that the O-mode dispersion relation becomes

$$(kc/\omega)^2 = 1-\omega_{pe}^2 \Sigma \exp(-\lambda) I_n(\lambda)/[\omega(\omega-n\Omega)]. \tag{4.24}$$

For values of $\lambda \ll 1$, and if we retain only the lowest order terms,

$$\exp(-\lambda) I_o(\lambda) \approx (1-\lambda)[1+(\lambda/2)^2] \approx (1-\lambda)$$
$$\exp(-\lambda) I_1(\lambda) \approx (\lambda/2)(1-\lambda)[1+(1/2)(\lambda/2)^2] \approx (\lambda/2),$$

and the O-mode dispersion relation, Eq. (4.24), is given approximately by

$$(kc/\omega)^2 = 1-\omega_{pe}^2/\omega^2 - (\omega_{pe}^2/\omega^2)(k^2\alpha^2/4\Omega^2)[\omega/(\omega-\Omega)].$$

This is to be solved for $k = k_r + ik_i$ in a slowly varying magnetic field that, in the case of resonance at the fundamental gyrofrequency, we will model as $\Omega = \omega(1 - x/L)$. The first two terms on the right-hand side are the usual O-mode dispersion relation in the cold-plasma limit, from which we can estimate the real part of k:

$$(k_r c/\omega)^2 \approx 1-\omega_{pe}^2/\omega^2.$$

The imaginary part of k is then given by the remaining singular term

$$2k_ik_rc^2/\omega^2 \approx -\mathrm{Im}\{(\omega_{pe}^2/\omega^2)(k_r^2\alpha^2/4\omega^2)[\omega/(\omega-\Omega)]\},$$

$$2k_i \approx -\mathrm{Im}\{(\omega_{pe}^2/\omega^2)(\omega/c)(1-\omega_{pe}^2/\omega^2)^{1/2}(\alpha^2/4c^2)/(x/L+i\varepsilon)\}.$$

Here ε is a positive infinitesimal resulting from the causality condition, $\mathrm{Im}(\omega) > 0$.

The optical depth for these O-modes resonating at the fundamental of the electron gyrofrequency can then be obtained by using the Plemelj prescription to evaluate the singular integral over x and obtain,

$$\int 2k_i dx = -(\pi/2)(\omega_{pe}^2/\omega^2)(1-\omega_{pe}^2/\omega^2)^{1/2}(\alpha^2/2c^2)(\omega/c)L. \quad (4.25)$$

Although the factor $\alpha^2/2c^2 = T_e/mc^2 \ll 1$, the scale length, $L \gg c/\omega$, the free-space wavelength, and complete absorption is possible in large tokamaks, for example.

4.9 Electrostatic Plasma Waves

In addition to the electromagnetic waves described earlier in this chapter, plasmas also support electrostatic waves characterized by $\mathbf{E} = -\nabla\Phi = -i\mathbf{k}\Phi$. That is, the electric field of the wave is aligned parallel to the propagation vector and the waves are thus longitudinal rather than transverse. Since $\nabla \times \mathbf{E} = -\nabla \times \nabla\Phi = 0$, the wave has no fluctuating magnetic field and the linearized Vlasov equation reduces to

$$Df_1/Dt = -\mathbf{a}_1 \cdot \partial f_o/\partial \mathbf{v} = -i(e/m_e)\mathbf{k} \cdot \partial f_o/\partial \mathbf{v}\, \Phi.$$

The wave equation is Poisson's equation in which the charge density, ρ, is obtained from f_1:

$$\nabla \cdot \varepsilon_o(-\nabla\Phi) = \varepsilon_o k_2 \Phi = \rho = \int d^3v(-ef_1).$$

We first integrate the linearized Vlasov equation in time along the unperturbed orbits as before, assuming $\mathbf{k} = k_\perp \mathbf{u}_x + k_\| \mathbf{u}_z$ with the unperturbed orbits given by

$$x = x_o + (v_\perp/\Omega)\sin(\phi_o + \Omega t) - (v_\perp/\Omega)\sin\phi_o$$

$$z = z_o + v_\| t.$$

We choose $x_o = z_o = 0$ so that the phase of the wave along the unperturbed orbit is given by

$$\exp\{i[(k_\perp v_\perp/\Omega)\sin(\phi_o + \Omega t) - (k_\perp v_\perp/\Omega)\sin\phi_o + k_\| v_\| t - \omega t]\},$$

and using the now familiar Bessel expansion we have for the time-dependent wave amplitude

$$\Phi(t) = |\Phi| \Sigma J_m(k_\perp v_\perp/\Omega) J_n(k_\perp v_\perp/\Omega) \exp\{i[(m-n)\phi_o + (m\Omega + k_\| v_\| - \omega)t]\},$$

where the sum is over all values of m and n from $-\infty$ to ∞. The linearlized Vlasov equation can now be integrated in time from $t = -\infty$ to $t = 0$ giving

$$f_1(0) = -(en_e/m_e)\mathbf{k}\cdot\partial f_o/\partial\mathbf{v}|\Phi|\Sigma J_m(k_\perp v_\perp/\Omega)J_n(k_\perp v_\perp/\Omega)\exp[i(m-n)\phi_o] \times (m\Omega + k_\| v_\| - \omega)^{-1},$$

or,

$$f_1(0) = -(en_e/m_e)(k_\perp \cos\phi\partial f_o/\partial v_\perp + k_\|\partial f_o/\partial v_\|)|\Phi| \times\Sigma J_m(k_\perp v_\perp/\Omega)J_n(k_\perp v_\perp/\Omega)\exp[i(m-n)\phi](m\Omega + k_\| v_\| - \omega)^{-1}.$$

Note that we have normalized the equilibrium distribution function so that its integral over velocity space is unity. The term containing $\cos\phi$ can be manipulated using one of the Bessel function recursion relations [20], $J_{n-1}(z) + J_{n+1}(z) = 2nJ_n(z)/z$, and re-labeling the index n. The result is

$$f_1(0) = -(en_e/m_e)[(n\Omega/v_\perp)\partial f_o/\partial v_\perp + k_\|\partial f_o/\partial v_\|)]|\Phi| \times\Sigma J_m(k_\perp v_\perp/\Omega)J_n(k_\perp v_\perp/\Omega)\exp[i(m-n)\phi](m\Omega + k_\| v_\| - \omega)^{-1}.$$

The dispersion relation then follows from Poisson's equation is

$$1 = (\omega_{pe}^2/k^2)2\pi\int v_\perp dv_\perp \int dv_\| J_n^2(k_\perp v_\perp/\Omega)(n\Omega + k_\| v_\| - \omega)^{-1} \times [(n\Omega/v_\perp)\partial f_o/\partial v_\perp + k_\|\partial f_o/\partial v_\|)]. \quad (4.26)$$

This dispersion relation was first derived by Harris [21] and provides a more comprehensive description of electrostatic plasma waves than the earlier Bernstein analysis [22], which only considered waves that propagate perpendicular to the static magnetic field and thus are not damped. The Harris dispersion relation has been applied to broad classes of distribution functions that can be used to analyze the stability of anisotropic, mirror-confined plasmas [23]. The low-temperature limit can readily be obtained using the power series representation of the Bessel function and regarding as small parameters the quantities $k_\|v_\|/\omega$, $k_\|v_\|/(\omega - \Omega)$, and $k_\|v_\|/(\omega + \Omega)$. The result is

$$1 = (\omega_{pe}^2/\omega^2)(k_\|^2/k^2) + [\omega_{pe}^2/(\omega^2 - \Omega^2)](k_\perp^2/k^2).$$

We will return to more extended discussions of electrostatic waves later, particularly regarding their stability and role in absorbing incident microwave power through the conversion of electromagnetic waves into electrostatic waves at the upper hybrid resonance layer.

4.10
Estimates of the Electric Field Amplitude

Under some circumstances, the amplitude of the right-hand circularly polarized electric field near the resonance surface may be estimated from the level of microwave power coupled into the plasma. In the case of high-field launch whistler-wave heating, for example, the resonance layer absorbs virtually all the right-hand circularly polarized radiation incident on it, whereas the left-hand circularly polarized component passes through the resonance surface and is internally reflected at the LH cutoff. It can then propagate back into the high-field region where the microwave coupling structure is located. If this wave is subsequently reflected with a polarization reversal, it will ultimately be absorbed when it reaches the resonance surface. The situation is similar for X-modes launched in the high-field region of a tokamak and propagating at an oblique angle to the magnetic field. Under these circumstances, the microwave electric field strength can be estimated from the Poynting flux, $\mathbf{S} = \mathbf{E} \times \mathbf{H}$. From Maxwell's equations, one can easily show that

$$\nabla \cdot (\mathbf{E} \times \mathbf{H}) = -\varepsilon_o \mathbf{E} \cdot \partial \mathbf{E}/\partial t - (1/\mu)\mathbf{B} \cdot \partial \mathbf{B}/\partial t - \mathbf{E} \cdot \mathbf{j}.$$

Thus, in steady state and regions where there is no net transfer of energy between the wave and the plasma

$$P_\mu = \mathbf{S} \cdot \mathbf{A} = (\mathbf{E} \times \mathbf{H}) \cdot \mathbf{A}.$$

Here P_μ is the incident microwave power and $\mathbf{A}$ is the cross-sectional area of the microwave radiation pattern. Since for transverse waves $\mathbf{E} \times \mathbf{H} = c\varepsilon_o E^2 \mathbf{k} c/\omega$, and assuming that $\mathbf{k}$ is perpendicular to the surface, $\mathbf{A}$, $P_\mu/A = nc\varepsilon_o E^2$, whence

$$E^2 = P_\mu/(Anc\varepsilon_o).$$

This result can be used to estimate the electric-field strength along the trajectory of the microwave beam injected into the plasma, but clearly fails at cutoffs or resonances, where full-wave treatments are required.

The situation is very different in the case of weakly absorbed O-mode heating in plasmas confined in magnetic mirrors as well as low-temperature plasmas confined in smaller tokamaks. As demonstrated by Quon and Dandl [2], microwave power launched from the low-field region near the side wall of a cylindrical vacuum vessel with polarization chosen to couple to O-modes propagating in the plasma can be selectively absorbed by energetic electrons. These electrons can then be heated to relativistic energies ranging from hundreds of keV at lower microwave frequencies to multiple MeV at higher frequencies. The microwave electric-field strength can be significantly enhanced in this situation as the result of open-resonator effects produced by the vacuum vessel and the plasma wave optics. The degree of enhancement is determined by the effective Q of the plasma-loaded resonator and depends sensitively on the details of the particular experimental configuration. As mentioned earlier, in view of the complexity of the wave propagation in the low-field launch situation, it may be more realistic to regard the microwave electric fields as high-order cavity modes. We will discuss this topic later with particular experiments.

4.11
Ray Tracing in Inhomogeneous Plasmas

To this point, we have considered the damping only of waves propagating exactly parallel or perpendicular to the magnetic field. In general, the path taken by microwave power launched into an inhomogeneous plasma equilibrium at an arbitrary angle to the magnetic field is governed by refraction, reflection at cutoffs, and absorption at resonances. Determining the path taken by the power is often a complicated process, but it is greatly facilitated in situations where the variation in the plasma properties is negligible over distances comparable to the wavelength of the power. For many decades, it has been realized that this situation obtains in ionospheric plasmas, and it is becoming increasingly true in large plasma confinement experiments, where the plasma dimensions can be three orders of magnitude greater than the free-space wavelength of the microwave power used for ECH. Under these circumstances, one can apply the methodology of geometrical optics, implemented in suitable computer codes, to trace the path followed by microwave power launched at some initial point with a given initial trajectory. Codes developed for this purpose [24] have evolved over time and are now in widespread use in analyzing ECH applications to large plasmas. Here we will focus our attention on the theoretical basis [25] for the codes and interested readers can refer the sources cited in Refs. [24] for detailed descriptions of the codes.

We start with the scalar wave equation of optics: $\nabla^2\chi - (n^2/c^2)\partial^2\chi/\partial t^2 = 0$. If the medium through which the wave propagates is uniform so that n, the index of refraction, is constant, the wave equation has plane-wave solutions: $\chi = \chi_o \exp[i(\mathbf{k} \cdot \mathbf{r} - \omega t)]$ where $\mathbf{k}$ and ω must satisfy an appropriate dispersion relation, $n = kc/\omega$. For the moment we will choose $\mathbf{k} = k_o \mathbf{u}_x$, where $k_o = \omega/c$ is the value of the wave number in vacuum. Then $\chi = \chi_o \exp[ik_o(nx - ct)]$. We now consider a situation where n depends only on x and is a slowly varying function of x. We then look for solutions of the wave equation that are similar to plane waves; namely, $\chi = \chi_o \exp\{A(x) + ik_o[L(x) - ct]\}$. A(x) and L(x) will be assumed to be real. Differentiating χ twice in x and t and substituting the results into the wave equation yields

$$\{[d^2A/dx^2 + (dA/dx)^2 - k_o^2(dL/dx)^2 + n^2k_o^2] + ik_o[d^2L/dx^2 + 2(dA/dx)(dL/dx)]\}\chi = 0.$$

Since A and L are real, both expressions in brackets must vanish. Under the geometrical optics assumption that $\lambda_o = 2\pi/k_o$ is much smaller than the lengths characterizing the gradients of A and L, the dominant terms are

$$[-k_o^2(dL/dx)^2 + n^2k_o^2]\chi = 0,$$

giving a one-dimensional version of the eikonal equation of geometrical optics, $(dL/dx)^2\,\chi = n^2\,\chi$. The WKB solution for χ results from setting $L(x) = \pm\int n(x')dx'$:

$$\chi = \chi_o \exp\{A(x) + ik_o[(\pm)\int n(x')dx' - ct].$$

The eikonal equation is identical to the Hamilton–Jacobi equation; and indeed, Jacobi's form of the least action principle could be written as $\Delta\int n ds = \Delta\int ds/u = 0$, which are variations of Fermat's principle for the trajectories of light rays. We can, therefore, describe the trajectories of light rays in inhomogeneous plasmas by a Hamiltonian system in which the dispersion relation, $D(\mathbf{r},\mathbf{k},\omega,t) = 0$ plays the role of the Hamiltonian. The canonical equations for the ray trajectory are usually given in terms of a time-like parameter, τ:

$$\begin{aligned}
dr_j/d\tau &= \partial D/\partial k_j, \quad j = 1, 2, 3\\
dk_j/d\tau &= -\partial D/\partial r_j, \quad j = 1, 2, 3\\
d\omega/d\tau &= \partial D/\partial t, \quad \text{and}\\
dt/d\tau &= -\partial D/\partial \omega.
\end{aligned}$$

The path of the ray is advanced in time, t, according to the local value of the group velocity:

$$\mathbf{v}_g = d\mathbf{r}/dt = (d\mathbf{r}/d\tau)/(dt/d\tau) = (\partial D/\partial \mathbf{k})/(-\partial D/\partial \omega).$$

The wave vector is advanced in time along the ray trajectory using

$$d\mathbf{k}/dt = (d\mathbf{k}/d\tau)/(dt/d\tau) = (-\partial D/\partial \mathbf{r})/(-\partial D/\partial \omega).$$

Since we are usually concerned with time-independent equilibria, $\partial D/\partial t = 0$ and ω is constant. Also, if the equilibrium does not depend on one or more of the spatial coordinates, the corresponding components of $\mathbf{k}$ are constant. It is sometimes advantageous to introduce the distance along the ray trajectory, $ds = |\mathbf{v}_g|dt$ so that the ray trajectory and the wave vector are then advanced by the following equations:

$$\begin{aligned}
d\mathbf{r}/ds &= (1/|\mathbf{v}_g|)d\mathbf{r}/dt = -\text{sgn}(\partial D/\partial \omega)(\partial D/\partial \mathbf{k})/|\partial D/\partial \mathbf{k}| \quad \text{and}\\
d\mathbf{k}/ds &= (1/|\mathbf{v}_g|)d\mathbf{k}/dt = \text{sgn}(\partial D/\partial \omega)(\partial D/\partial \mathbf{r})/|\partial D/\partial \mathbf{k}|.
\end{aligned}$$

Most of the ray-tracing codes now in use evaluate the real part of the index of refraction at each point along the path of the ray using the cold-plasma dispersion relation. The damping or absorption of the wave energy can be calculated in different ways. From the wave point-of-view, the warm-plasma dispersion relation can be solved for the imaginary part of the index of refraction at each point along the pay. From the electron's point-of-view, the damping can be calculated using a suitable heating model, such as the Fokker–Planck model to be discussed in Chapter 6.

References

1 See, for example, John D. Kraus, *Antennas*, second edition, McGraw-Hill, New York (1980).

2 B.H. Quon and R.A. Dandl, Preferential electron–cyclotron heating of hot electrons and formation of overdense plasmas, *Phys. Fluids* **B1**, 2010 (1989); J.H. Booske, W.D. Getty, R.M. Gilgenbach, and R.A. Jong, Experiments on whistler mode electron–cyclotron resonance plasma startup and heating in an axisymmetric magnetic mirror, *Phys. Fluids* **28**, 3116 (1985).

3 R.A. Dandl, H.O. Eason, P.H. Edmonds, and A.C. England, *Nuclear Fusion* **11**, 411 (1971).

4 Thomas Howard Stix, *The Theory of Plasma Waves*, McGraw-Hill, New York (1962), Chapters 1 and 11; and many references cited in this work. Additional material has been added to *Waves in Plasmas* by Thomas Howard Stix, American Institute of Physics, New York (1992).

5 K.G. Budden, *The Propagation of Radio Waves*, Cambridge University Press, Cambridge (1985), Chapter 6.

6 P.C. Clemmow and R.F. Mullaly, *Physics of the Ionosphere*, 1955, pp. 340–350, The Physical Society of London; and W.P. Allis,"Waves in Plasma" MIT Research Laboratory of Electronics, Quarterly Progress Report 54.

7 H. Weitzner and D.B. Batchelor, *Phys. Fluids* **22**, 1355 (1979); D.B. Batchelor, *Plasma Phys.* **22**, 41 (1980) and references contained therein. See also J. Preinhalter and V. Kopecky, *J. Plasma Phys.* **10**, 1 (1973).

8 J.C. Sprott, *Phys. Fluids* **14**, 1795 (1971).

9 See Reference 4, p. 47.

10 Eugen Merzbacher, *Quantum Mechanics*, second edition, John Wiley & Sons New York, (1961) p. 84.

11 A.A. Vlasov, *J. Phys. (USSR)* **9**, 25 (1945).

12 See, for example, Bo Lehnert, *Dynamics of Charged Particles*, John Wiley & Sons, New York (1964) Chapter 5; and especially for this particular application, C.L. Olson, *Phys. Fluids* **15**, 160 (1972).

13 B.D. Fried and S.D. Conte, *The Plasma Dispersion Function*, Academic Press, New York (1961).

14 See, for example, George Bekefi, *Radiation Processes in Plasmas*, John Wiley & Sons, New York (1966) p. 38 ff.

15 T.M. Antonsen, Jr. and W.M. Manheimer, *Phys. Fluids* **21**, 2295 (1975).

16 M. Bornatici, F. Engelmann, S. Novak, and V. Petrillo, *Plasma Phys.* **23**, 1127 (1981); M. Bornatici *et al.*, *Nucl. Fusion* **23**, 1159 (1983).

17 R.A. Cairns and C.N. Lashmore-Davies, *Phys. Fluids* **25**, 1605 (1982).

18 C.D. Cantrell, *Modern Mathematical Methods for Physicists and Engineers*, Cambridge University Press, Cambridge (2000) pp. 690–693.

19 G.N. Watson, *A Treatise on the Theory of Bessel Functions*, Mathematical Library Edition, Cambridge University Press, Cambridge (1922) reprinted 1995, p. 395; see also, Yudell L. Luke, *Integrals of Bessel Functions*, McGraw-Hill, New York (1962), p. 314.

20 *Handbook of Mathematical Functions*, Edited by Milton Abramowitz and Irene A. Stegun, National Bureau of Standards Applied Mathematics Series. 55 (1965), reprinted by Dover Publications, New York (1970), p. 361, eq. 9.1.27.

21 E.G. Harris, *Phys. Rev. Lett.* **2**, 34 (1959); *J. Nucl. Energy* **C2**, 138 (1961) eq. 46.

22 I.B. Bernstein, *Phys. Rev.* **109**, 10 (1958).

23 G.E. Guest and R.A. Dory, *Phys. Fluids* **8**, 1853 (1965).

24 D.B. Batchelor, R.C. Goldfinger, and H. Weitzner, *IEEE Trans. Plasma Science*, **PS-8**, 78 (1980). For a recent description of several major codes, see R. Prater *et al.*, *Nucl. Fusion* **48**, 1 (2008).

25 Ira B. Bernstein, *Phys. Fluids* **18**, 320 (1975); Herbert Goldstein, *Classical Mechanics*, second edition, Addison-Westley, Reading, MA (1980) pp. 487–489; Thomas Howard Stix, *The Theory of Plasma Waves*, McGraw-Hill, New York (1962) Section 3.5.

Exercises

4.1. ***Determine the index of refraction and the polarization of the following waves:***

(a) ***an X-mode propagating perpendicular to the magnetic field in a cold plasma where*** $\omega_{pe}^2/\omega^2 = 0.8$ ***and*** $\Omega/\omega = 1.05$

(b) ***an X-mode propagating perpendicular to the magnetic field in a cold plasma where*** $\omega_{pe}^2/\omega^2 = 0.4$ ***and*** $\Omega/\omega = 0.52$

(c) *an X-mode propagating at an angle of 60° to the magnetic field in a cold plasma where* $\omega_{pe}^2/\omega^2 = 0.8$ *and* $\Omega/\omega = 1.05$

(d) *an X-mode propagating at an angle of 80° to the magnetic field in a cold plasma where* $\omega_{pe}^2/\omega^2 = 0.4$ *and* $\Omega/\omega = 0.52$.

4.2. *Determine the direction of the wave electric field at the upper hybrid resonance when an incident X-mode is propagating exactly perpendicular to the magnetic field.*

4.3. *Derive an expression for the polarization of X-modes propagating at an angle of 60° with respect to the magnetic field. Assume* $\omega = 1$, $\Omega = 1.5$, *and evaluate this polarization as a function of density for* $0 < \omega_{pe} < 1$.

4.4. *The Q ("Quality Factor") of a plasma-loaded microwave cavity is given by (see, for example, J. D. Jackson, Classical Electrodynamics, Wiley, New York (1962) p. 256)*

$$Q = \omega(\varepsilon_o E^2/2) V_{cavity}/P_{abs}$$

(a) *Using Eq. (4.17) for a simple magnetic mirror determine the value of Q for the following parameters:* $V_{cavity} = 30$ *l*, $M = 3.333$, $M_{res} = 1.905$, $L_c = 50$ *cm*, $a = 5$ *cm, and* $(\omega_{pe}^2/\omega^2)_{res} = 0.1$.

(b) *If 5 kW of 6.4 GHz microwave power is coupled into this cavity, what is the average value of the resulting microwave electric field strength?*

4.5. (a) *Evaluate the real and imaginary parts of the index of refraction for whistler waves propagating parallel to a uniform static magnetic field for the following parameters:* $T_e = 20$ *eV* $(\omega_{pe}^2/\omega^2) = 2$, *and* $\Omega/\omega = 1.02$.

(b) *For the same plasma parameters, evaluate the imaginary part of the index of refraction at the resonance surface, the magnetic field at the onset of damping, and the optical thickness of this plasma for these whistler waves.*

4.6. *Derive a formula for the optical depth of O-modes propagating perpendicular to the magnetic field and resonant at the second harmonic of the electron gyrofrequency in a Maxwellian plasma.*

5
Interaction of Electrons with Electromagnetic Fields at Resonance

In Chapter 4, we considered the wave processes that play major roles in ECH, whereas in this chapter we will consider the more microscopic processes by which electrons gain energy from the fields of those waves. Here the emphasis is on the interaction of individual electrons with the electric field at the resonance surface without particular regard for the waves that may be responsible for those fields. An essential element of cyclotron resonance heating is the occurrence of a temporal interval during which the phase of the electron gyration is nearly stationary relative to the phase of the RF wave. Several distinct mechanisms by which the electron and the wave can exchange energy will be discussed in this chapter, but none of them will have a significant effect except during this interval when the relative phase is stationary. Outside of this resonance interval, the relative phase changes rapidly in time, since the resonance results from the brief cancellation of two high-frequency oscillations; viz, the electron gyration and the fields of the electromagnetic wave. As a consequence of the high frequencies involved here, even relatively weak processes affecting the electron gyrophase can lead to randomization of the phases between resonances. Next we will first examine the kinematic processes that determine the duration of the resonance without considering the longer time behavior of the phase. Then in Section 5.2, we consider a more comprehensive dynamical description that can also account for the value of the phase at successive resonances. Some relativistic effects are discussed briefly in Section 5.3.

One possible model of the ECH process, the so-called stochastic model, treats the heating as a diffusion in velocity space brought about by many uncorrelated events; namely, successive transits of an electron through the resonance surfaces, where the electron's perpendicular energy and orientation in velocity space undergo abrupt changes. The implementation of such a model requires that we determine the properties of these resonant changes and the frequency with which they occur. A somewhat simplistic but hopefully transparent determination of these changes will be undertaken in Section 5.1. A more rigorous treatment of the heating process will be given in Section 5.2. There are, however, significant questions regarding the conditions under which the assumption of stochastic behavior is valid. These questions have been addressed by many researchers [1], and several circumstances involving correlations between successive transits through resonance have been identified. In Section 5.4, we describe two examples of such nonstochastic behavior

Electron Cyclotron Heating of Plasmas. Gareth Guest

ISBN: 978-3-527-40916-7

and the limit cycles to which they can give rise. Finally, in Section 5.5 we explore some nonlinear aspects of these limit cycles by employing a highly simplified mapping technique.

5.1
A Rudimentary Stochastic Model of ECH

As we have seen in Chapter 3, electrons confined in a magnetic mirror will bounce back and forth along magnetic lines of force. If the amplitude of this bounce motion is large enough or if the resonance surface is on the midplane, an electron can pass through the resonance surface where the (Doppler-shifted) wave frequency equals the local gyrofrequency. At resonance, the phase difference, ϕ, between the electron gyration and, for example, the right-hand circularly polarized component of the electric field, E_-, is stationary with some value, ϕ_{res}; but as the electron's motion along the magnetic field takes it beyond the resonance surface, ϕ will change. As we will see, relativistic increases in the electron mass can also cause the phase to change even in a uniform magnetic field. For this first heuristic picture of the stochastic model of ECH, we will concentrate on evaluating the effective duration of resonance, t_{eff}, and its dependence on the system parameters and the orientation of the electrons in velocity space.

Although t_{eff} has been defined in slightly different ways by different workers [2], in general the duration of resonance is a measure of the time interval throughout which ϕ stays within some maximum value relative to its value at the instant of exact resonance. During this time interval, the electron's (perpendicular) energy will change by an amount $\delta W_\perp$, which is given schematically by an expression of the form

$$\delta W_\perp = -e\int E_\perp v_\perp \cos\phi \, dt \approx -e|E_\perp|v_{\perp res} \cos\phi_{res}\, t_{eff}. \tag{5.1}$$

The rate at which the phase changes in time due to the electron's parallel motion or to the relativistic mass change is given by $\nu = d\phi/dt = \Omega + k_{||}v_{||} - \omega$, where $\nu = 0$ at resonance. For reasons that will become clear later, we can expand ν in a rapidly converging Taylor series about the instant of resonance, t_{res}. Recall the general form of the Taylor series given by

$$f(x) = f(a) + (x-a)f'(a) + (x-a)^2 f''(a)/2! + (x-a)^3 f'''(a)/3! + \cdots$$

Applying this to $\nu = d\phi/dt$ and keeping in mind that ν vanishes at resonance, we have for times near the instant of resonance, t_{res}

$$\nu(t) = (t-t_{res})\nu'(t_{res}) + (t-t_{res})^2\nu''(t_{res})/2! + (t-t_{res})^3\nu'''(t_{res})/3! + \cdots$$

Thus, the phase shift $\delta\phi$ after some time interval $\delta t_\pm$ is given by

$$\begin{aligned}\delta\phi = \phi-\phi_{res} &= \int d(t-t_{res})[(t-t_{res})\nu'(t_{res}) + (t-t_{res})^2\nu''(t_{res})/2! \\ &\quad + (t-t_{res})^3\nu'''(t_{res})/3! + \cdots] \\ &= \delta t_\pm^2\nu'(t_{res})/2 \pm \delta t_\pm^3\nu''(t_{res})/3! + \delta t_\pm^4\nu'''(t_{res})/4! + \cdots\end{aligned} \tag{5.2}$$

Here $\delta t_+ = (t - t_{res}) > 0$ and $\delta t_- = (t - t_{res}) < 0$ are the limiting time intervals corresponding to the periods after the electron passes through resonance and prior to resonance, respectively. If for the moment we arbitrarily set the maximum phase difference as $\delta\phi_{max} = \pm\pi/4$ and retain only the first two terms in the expansion, we can use simple numerical techniques to estimate an effective duration of resonance, t_{eff}, given by $\delta t_+ + \delta t_-$, from the lowest order terms of Eq. (5.2):

$$\delta t_\pm^2 \nu'(t_{res})/2 \pm \delta t_\pm^3 \nu''(t_{res})/3! = \pm\pi/4. \tag{5.3}$$

Allowing for the relativistic change in electron mass as well as the electron's parallel motion, we can express the rate of change of ν at resonance as

$$\nu'(t_{res}) = \{\Omega[v_{||}\mathrm{dln}B/dz - dW/dt/(\gamma mc^2) + k_{||}(dv_{||}/dt)/\Omega]\}_{res} \tag{5.4}$$

The first term, $v_{||}\mathrm{dln}B/dz$, is usually the determining factor in limiting the duration of resonance unless the magnetic field is locally spatially uniform or the electron turns at or just beyond resonance. In the former case, the relativistic increase in electron mass may limit the duration of resonance. In the latter case, the next order term in the Taylor series will determine the duration of resonance. The magnitude of the second nonrelativistic contribution, $k_{||}(dv_{||}/dt)/\Omega$, can be estimated from Newton's second law using the $\mu \cdot \nabla B$ force: $m dv_{||}/dt = -\mu dB/dz$ so that

$$k_{||}(dv_{||}/dt)/\Omega = -(k_{||}/k)(kc/\omega)(\omega/\Omega)(v_\perp/c)(v_\perp/2)\mathrm{dln}B/dz.$$

Unless the resonance occurs very near the electron turning point (where $v_{||} = 0$), this term will be smaller by roughly $v_\perp/c$ than the $v_{||}\mathrm{dln}B/dz$ term. If the electron turns well beyond the resonance, the duration of resonance is given approximately by the quadratic term of the Taylor series:

$$t_{eff} = \delta t_+ + \delta t_- \approx 2\left[\pi/\left|2\nu'(t_{res})\right|\right]^{1/2} \quad \text{if } \nu'(t_{res}) \neq 0. \tag{5.5}$$

If $\nu'(t_{res}) = 0$, we can use the next order term in the series to obtain

$$t_{eff} = \delta t_+ + \delta t_- \approx 2\left[3\pi/\left|2\nu''(t_{res})\right|\right]^{1/3} \quad \text{if } \nu'(t_{res}) = 0. \tag{5.6}$$

In either case, under the assumption that the successive resonance encounters are uncorrelated we can estimate the resulting energy diffusion coefficient, D_W to be given by

$$D_W = \langle \delta W_\perp^2 \rangle \nu_{trs} = (1/2)\left(e|E_\perp|v_\perp t_{eff}\right)^2 \nu_{trs} \tag{5.7}$$

The frequency with which an electron trapped in a magnetic mirror passes through a resonance surface, ν_{trs}, is four times the bounce frequency if the electron turns well beyond the resonance and twice the bounce frequency if the electron turns at or just beyond the resonance or if the resonance surface is at the midplane. We estimate the heating rate in this stochastic model as

$$dW_\perp/dt = D_W/2W_\perp = (1/2)(e^2/m)|E_\perp|^2\, t_{eff}^2\, \nu_{trs}. \tag{5.8a}$$

We now illustrate some of the properties of this stochastic heating model by applying it to the simple magnetic mirror configuration described in Chapter 2. For more clarity, we will neglect the Doppler shift and the relativistic effects and approximate the rate of phase change simply by $\nu = \Omega(z) - \omega$ so that.

$$\nu'(t_{res}) = (d\nu/dt)_{res} = \left[v_{||} d\Omega/dz\right]_{res}$$

and

$$\nu''(t_{res}) = \left[v_{||}^2 d^2\Omega/dz^2 + (dv_{||}/dt) d\Omega/dz\right]_{res}.$$

For this illustrative case, we approximate the variation of magnetic intensity along the magnetic lines of force by our earlier expression for the simple magnetic mirror,

$$\begin{aligned} \Omega(z) &= (\Omega_o/2)[(M+1)-(M-1)\cos k_o z], \\ \text{so that} \quad d\Omega/dz &= (k_o\Omega_o/2)(M-1)\sin k_o z \\ \text{and} \quad d^2\Omega/dz^2 &= (k_o^2\Omega_o/2)(M-1)\cos k_o z. \end{aligned}$$

At the resonance surface,

$$\begin{aligned} 2B_{res}/B_o \equiv 2M_{res} &= (M+1)-(M-1)\cos k_o z_{res} \\ \text{whence} \quad \cos k_o z_{res} &= (M+1-2M_{res})/(M-1) \\ \text{and} \quad \sin k_o z_{res} &= \pm[2/(M-1)][(M-M_{res})(M_{res}-1)]^{1/2}. \end{aligned} \tag{5.9}$$

If the electron motion between successive transits of the resonance surfaces is adiabatic, the parallel velocity at resonance will be given by

$$v_{||\,res}^2 = (2/m)(\varepsilon-\mu B_{res}) = (2\mu B_o/m)(\varepsilon/\mu B_o - B_{res}/B_o),$$

where ε is the electron's total kinetic energy so that

$$v_{||\,res} = v_{\perp o}(M_t - M_{res})^{1/2}.$$

We can evaluate $dv_{||}/dt$ at resonance as before using Newton's second law:

$$\begin{aligned} m dv_{||}/dt &= -\mu dB/dz, \\ \text{giving} \quad (dv_{||}/dt)_{res} &= -(1/4)k_o v_{\perp o}^2 (M-1)\sin k_o z_{res}. \end{aligned}$$

When applied to this model of the simple magnetic mirror configuration, the parameters entering our stochastic heating model thus become

$$\nu'(t_{res}) = k_o v_{\perp o}\Omega_o[(M_t-M_{res})(M-M_{res})(M_{res}-1)]^{1/2} \tag{5.10}$$

and

$$\nu''(t_{res}) = (1/2)(k_o v_{\perp o})^2\Omega_o \times [(M_t-M_{res})(M+1-2M_{res})-(M-M_{res})(M_{res}-1)] \tag{5.11}$$

These expressions can then be used in our equation for the time-dependent phase difference to follow the electron's phase in time as it passes through resonance:

$$\phi - \phi_{res} = (t-t_{res})^2\nu'(t_{res})/2 + (t-t_{res})^3\nu''(t_{res})/3! + (t-t_{res})^4\nu'''(t_{res})/4! + \cdots$$

It is convenient to use the bounce frequency, ω_b, to define a dimensionless time variable, $\tau \equiv \omega_b t$ where the bounce frequency in the simple magnetic mirror was given by Eq. (3.11):

$$2\omega_b = k_o v_{\perp o}(M-1)^{1/2}\{(\pi/2)K^{-1}[(M_t-1)/(M-1)]\} \approx k_o v_{\perp o}(M-1)^{1/2}.$$

Retaining only the first two orders in the Taylor series, our equation for the time-dependent phase difference becomes,

$$\phi-\phi_{res} = (\tau-\tau_{res})^2\, \nu'(t_{res})/(2\omega_b^2) + (\tau-\tau_{res})^3\, \nu''(t_{res})/(6\omega_b^3).$$

From Eqs. (5.10) and (5.11),

$$\nu'(t_{res})/(2\omega_b^2) = (1/2)(k_o v_{\perp o}/\omega_b)^2(\Omega_o/k_o v_{\perp o}) \times[(M_t-M_{res})(M-M_{res})(M_{res}-1)]^{1/2}$$

and

$$\nu''(t_{res})/(6\omega_b^3) = (1/12)(k_o v_{\perp o}/\omega_b)^3(\Omega_o/k_o v_{\perp o}) \times[(M_t-M_{res})(M+1-2M_{res})-(M-M_{res})(M_{res}-1)],$$

where

$$k_o v_{\perp o}/\omega_b = (4/\pi)(M-1)^{-1/2}K[(M_t-1)/(M-1)] \quad \text{and}$$
$$\Omega_o/k_o v_{\perp o} = L_c/(2\pi\rho_o).$$

Recall that $L_c = 2\pi/k_o$ is the separation of the two coils forming the simple magnetic mirror configuration and ρ_o is the electron gyroradius at the midplane. We choose typical values for the mirror ratio, $M = 2$, the mirror ratio at resonance, $M_{res} = 1.4$, and the size parameter, $L_c/(2\pi\rho_o) = 150$ and numerically evaluate the phase difference, $\phi - \phi_{res}$, as a function of the dimensionless time, $(\tau - \tau_{res})$, for values of the mirror ratio at the electron turning point, M_t, ranging from M_{res} to M. The first of these illustrative results for $M_t = M_{res}$ is shown in Figure 5.1.

Since $\nu'(t_{res}) = 0$ for $M_t = M_{res}$, the phase difference is determined solely by the cubic term; and since $\nu''(t_{res}) < 0$ the phase difference is positive as the electron approaches the resonance, $(\tau - \tau_{res}) < 0$, and negative after the electron has turned at resonance and is moving back toward the midplane, $(\tau - \tau_{res}) > 0$. The effective duration of resonance, τ_{eff}, is indicated on the figure.

For values of M_t slightly greater than M_{res}, the quadratic term is nonzero and with increasing values of M_t it soon becomes dominant, leading to a positive extremum in the phase difference for positive times as shown in Figure 5.2, where $M_t = 1.41$.

With further increases in M_t, this is followed by an abrupt decrease in $\delta\tau_+$, the time after resonance when the phase shift reaches $\pi/4$, since the limiting value is now $+\pi/4$ rather than $-\pi/4$, as was the case in Figure 5.1. The effective duration of resonance decreases further as the cubic term becomes increasingly weaker relative to the quadratic term, as suggested by Figure 5.3, for which $M_t = 1.45$, where the parabolic form of the phase shift associated with the quadratic term is almost symmetrical about the point of resonance.

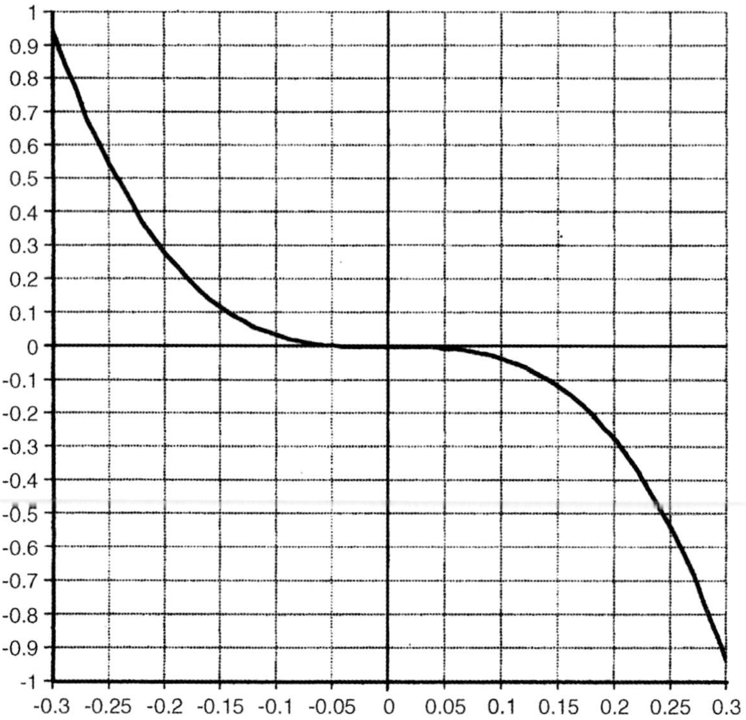

Figure 5.1 The phase difference, $\delta\phi = \phi - \phi_{res}$, versus $\tau = \omega_b t$ for electrons turning at the resonance surface, $M_t = M_{res} = 1.40$. Note that τ_{eff} is the interval between the times when $\delta\phi = \pm\pi/4 = \pm 0.7854$.

The effective duration of resonance for these illustrative cases is summarized in Figure 5.4 for $1.40 < M_t < 1.41$ ("electrons turning at or just beyond resonance") and in Figure 5.5 for $1.41 < M_t < 2$ ("electrons turning well beyond resonance").

The stochastic heating rate in this heuristic model is given by Eq. (5.8a), which we rewrite as follows:

$$dW_\perp/dt = \left(e|E_\perp|^2/B_o\right)\mathfrak{G}\left(M, M_{res}, M_t\right). \tag{5.8b}$$

For the simple magnetic mirror model considered here, and for electrons turning well beyond resonance, the electron pitch-angle dependence is contained in the function

$$\mathfrak{G}(M, M_{res}, M_t) = \{(M-1)/[(M_t - M_{res})(M - M_{res})(M_{res} - 1)]\}^{1/2} \times\{(\pi/2)K^{-1}[(M_t - 1)/(M-1)]\}.$$

With the same values of M, M_{res}, and M_t used earlier to illustrate the pitch-angle dependence of the duration of resonance, τ_{eff}, we can use this expression for $\mathfrak{G}(M, M_{res}, M_t)$ to estimate heating rates for electrons turning far enough beyond resonance that the quadratic behavior dominates. In this way, we find that the heating rate for electrons with $M_t = 1.43$, for example, is five times greater than the heating rate of electrons with $M_t = 1.9$.

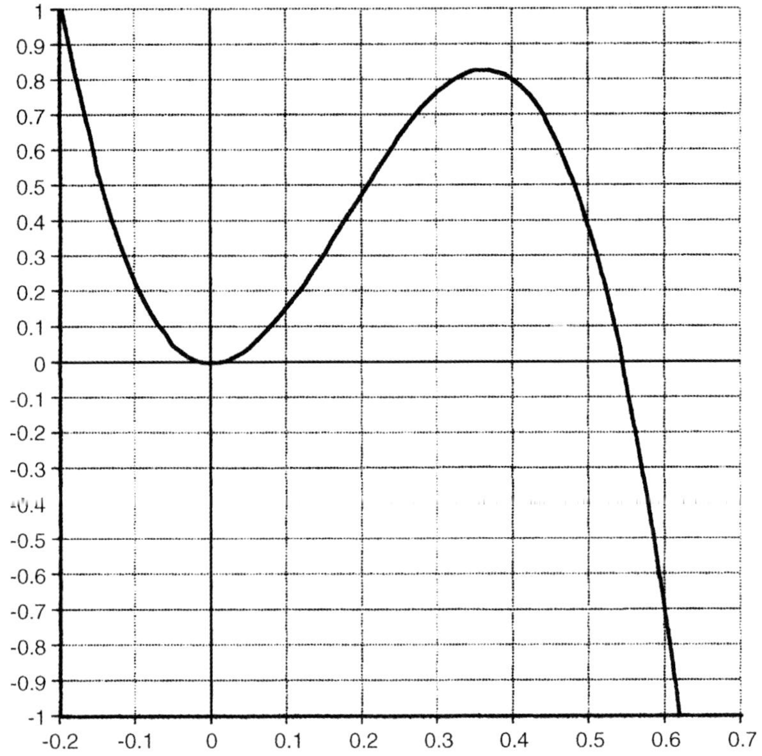

Figure 5.2 The phase difference, $\phi - \phi_{res}$, versus $\tau = \omega_b t$ for electrons turning just beyond the resonance surface where $M_t = 1.41$.

The point to be emphasized here is that electrons that turn just beyond resonance experience an effective duration of resonance that is four to five times longer than electrons that turn well beyond resonance and a heating rate that can be an order of magnitude larger. Furthermore, as we shall see, the turning points of heated electrons are being moved toward the resonance surface by the heating process itself. We, therefore, anticipate a significant dependence of the heating rate on the equilibrium distribution of electrons in turning points, with higher rates for electrons turning near resonance. Indeed, in magnetic mirror experiments where the equilibrium is strongly affected by the heating, we can reasonably expect the equilibrium distribution function of heated electrons to be peaked for electrons turning near resonance, an expectation supported by experimental data.

The locations of the electron turning points, as specified by the magnetic intensity at the turning point, B_t, are changed by the heating in such a way that the turning points of nonrelativistic electrons tend to accumulate near the resonance surface. Since $B_t = \varepsilon/\mu$, the change in the turning point due to the heating at the resonance surface is given by

$$\begin{aligned} \delta B_t &= \left[(\partial\varepsilon/\partial W_\perp)/\mu - (\varepsilon/\mu^2)\partial\mu/\partial W_\perp\right]_{res}\delta W_\perp \\ &= (1 - B_t/B_{res})\delta W_\perp/\mu. \end{aligned} \tag{5.12}$$

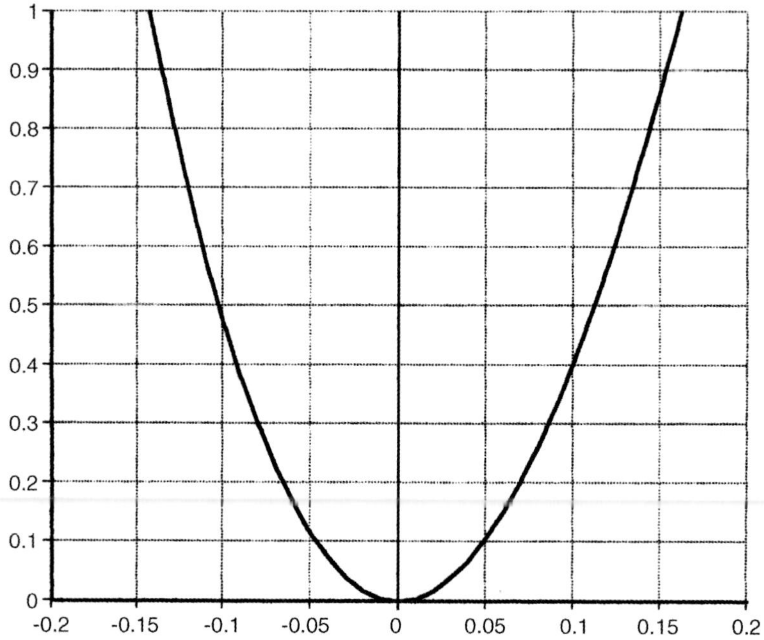

Figure 5.3 The phase difference, $\phi - \phi_{res}$, versus $\tau = \omega_b t$ for electrons turning beyond the resonance surface where $M_t = 1.45$.

Since the electrons must turn at or beyond the resonance surface if they are to pass through resonance, $B_t/B_{res} > 1$. The magnetic intensity at the turning point thus decreases with heating until electrons turn near resonance, the bracket vanishes, and δB_t itself vanishes.

5.2
Dynamics of the Fundamental Resonance Interaction

We now consider in a more rigorous way the changes in the energy and velocity-space orientation of an individual electron resulting from a single transit through a resonance surface that is illuminated by microwave power. The formalism used here will also lead to the identification of limit cycles that can affect the long-term behavior of heated electrons. We start with the nonrelativistic equation of motion of electrons in our model simple magnetic mirror field. To include relativistic effects, we will find later that it is only necessary to replace the rest mass, m, by the relativistic mass, γm.

The equation of motion for an electron moving in a magnetostatic field, $\mathbf{B}_o$, and the fields of an electromagnetic wave, $\mathbf{E}_1$ and $\mathbf{B}_1$, follows directly from Eq. (3.1):

$$\mathbf{F} = m d\mathbf{v}/dt = -e(\mathbf{E} + \mathbf{v} \times \mathbf{B}) = -e\mathbf{E}_1 - e\,\mathbf{v} \times (\mathbf{B}_o + \mathbf{B}_1).$$

If $\mathbf{B}_1$, the magnetic field of the wave, is mainly parallel to $\mathbf{B}_o$, the static magnetic field, its contribution to the total Lorentz force on the electron will usually be

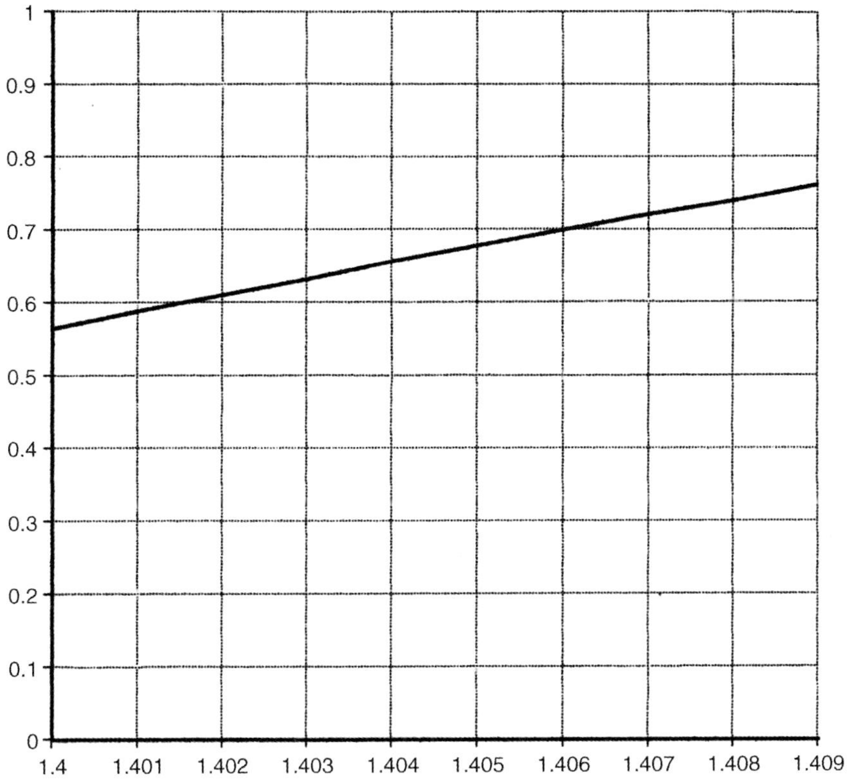

Figure 5.4 The duration of resonance versus M_t for $1.40 \leq M_t < 1.41$.

negligible. This is the case, for instance, if the wave is an X-mode propagating nearly perpendicular to the magnetostatic field. If, however, the magnetic field of the wave is largely perpendicular to the static magnetic field, it can play an essential role in the exchange of energy between the wave and the plasma electrons. This is the case for an O-mode propagating perpendicular to the magnetostatic field with the electric field of the wave aligned parallel to the magnetostatic field. We consider each of these circumstances separately, starting with the case exemplified by X-mode heating.

5.2.1 Dynamics of the Electron Interaction With X-Mode Waves

Assuming that the wave varies in space and time as $\exp[i(\mathbf{k} \cdot \mathbf{r} - \omega t)]$, we have from Maxwell's equations $\mathbf{k} \cdot \mathbf{E}_1 = \omega \mathbf{B}_1$; and if we set $e\mathbf{B}_o/m = \Omega_o \mathbf{u}_z$, we can write the equation of motion as

$$dv/dt - \Omega_o \mathbf{u}_z \cdot \mathbf{v} = -(e/m)[\mathbf{E}_1(1 - \mathbf{k} \cdot \mathbf{v}/\omega) + (\mathbf{v} \cdot \mathbf{E}_1)\mathbf{k}/\omega].$$

For the fields of X-modes, we can neglect the terms proportional to kv/ω that arise from the contribution of the wave magnetic field to the Lorentz force. Then in circular

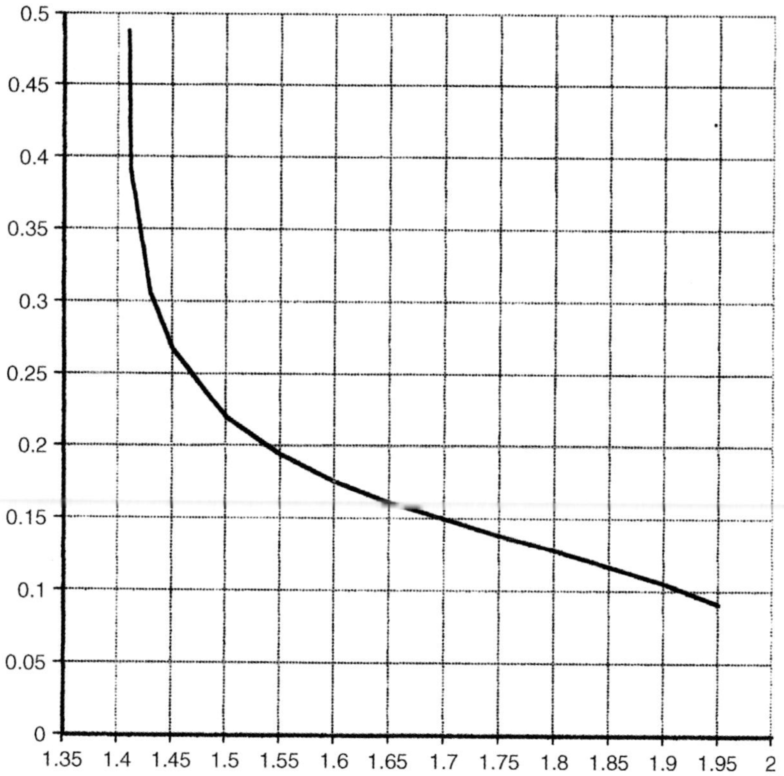

Figure 5.5 The duration of resonance versus M_t for $1.41 \leq M_t < 2$.

coordinates, the perpendicular components of the resulting equation of motion were given by Eq. (3.3):

$$\begin{aligned} dv_+/dt - i\Omega v_+ &= -eE_+/m \\ dv_-/dt + i\Omega v_- &= -eE_-/m. \end{aligned}$$

We focus our attention on the right-hand circularly polarized component that can resonate with electron gyration. Note that we omit the subscripts "o", "e", and "1" in these dynamical equations for the electron motion. Ion dynamics are generally negligible at electron gyrofrequencies and will not be included in the following discussion. An integrating factor for the right-hand circularly polarized component is $\exp(i \int \Omega dt)$ so that

$$v_-(t)\exp\left(i\int \Omega dt\right) = v_-(0) - (e/m)\int dt\, E_-(\mathbf{r},t)\exp\left(i\int \Omega dt\right).$$

We linearize this equation by evaluating the field along the unperturbed orbit of the electron, assuming the field to be a superposition of plane waves propagating at an arbitrary angle with respect to the static magnetic field, $E_-(\mathbf{r},t) = |E_-| \exp[i(\mathbf{k}\cdot\mathbf{r} - \omega t)]$ where $\mathbf{r}(t)$ is the unperturbed electron orbit. Without loss of generality, we can once again choose the propagation vector to lie in the $x-z$ plane and arbitrarily choose the initial position of the electron at the origin. Then from Eq. (3.5)

$$\mathbf{k}\cdot\mathbf{r} = k_\perp \rho \sin\left(\phi_0 + \int \Omega dt\right) - k_\perp \rho \sin\phi_0 + k_{||}\int v_{||} dt.$$

In order to separate $\int\Omega dt$ from the argument of $\sin(\phi_0 + \int\Omega dt)$, we use the Bessel function identity [3], $\exp(\pm ib\sin\theta) = \Sigma J_n(b)\exp(\pm in\theta)$ and obtain the following expression for the right-hand circularly polarized component of the electron velocity for times $t > 0$:

$$v_-(t)\exp\left(i\int\Omega dt\right) = v_-(0) - (e/m)|E_-|\exp(-ik_\perp\rho\sin\phi_0)\Sigma J_n(k_\perp\rho)\exp(in\phi_0)$$
$$\times\int dt \exp\left\{i\int dt[(n+1)\Omega + k_{||}v_{||} - \omega]\right\}.$$

The summation is over all (integer) values of the index n from $-\infty$ to ∞. The condition for resonance at the fundamental electron gyrofrequency, $\nu = \Omega + k||v|| - \omega = 0$, is given by the $n = 0$ term; and the corresponding change in electron velocity due to the resonance will then be described for times greater than that at which the electron passes through fundamental resonance by

$$v_-(t) = \exp\left(-i\int\Omega dt\right)$$
$$\times\left[v_-(0) - (e/m)|E_-|\exp(-ik_\perp\rho\sin\phi_0)J_0(k_\perp\rho)\int dt\exp\left(i\int dt\nu\right)\right].$$

Note that the condition for X-mode resonance at the *n*th harmonic of the electron gyrofrequency, $\nu = n\Omega + k_{||}v_{||} - \omega = 0$, is given by the $n-1$ term; and the corresponding change in electron velocity due to the nth harmonic resonance will then be described by

$$v_-(t) = \exp\left(-i\int\Omega dt\right)$$
$$\times\left[v_-(0) - (e/m)|E_-|\exp(-ik_\perp\rho\sin\phi_0)J_{n-1}(k_\perp\rho)\int dt\exp\left(i\int dt\nu\right)\right].$$

We now introduce the time-history integral, $H(t)$, defined by

$$H(t) = \int dt\exp\left(i\int dt\nu\right) = \int dt\exp\{i[\phi(t) - \phi_o]\},$$

where, as before, $\phi(t) - \phi_o = \int dt\ \nu = \int dt(\Omega + k_{||}v_{||} - \omega)$ so that the fundamental X-mode resonance leads to the following change in v_-:

$$v_-(t) = \exp\left(-i\int\Omega dt\right)[v_-(0) - (e/m)|E_-|\exp(-ik_\perp\rho\sin\phi_0)J_0(k_\perp\rho)H(t)].$$

The method of stationary phase [4] provides an asymptotic estimate of the time-history integral in the limit $\Omega_o/\omega_b \gg 1$. We write

$$H = H(t) = \int dt\exp\left(i\int dt\,\nu\right) = \int dt\exp(i\,x\,\psi),$$

where $x = \Omega/\omega_b$ and $\psi = \int(v/\Omega)\omega_b dt$. At resonance $\psi' = 0$; and if $\psi'' \neq 0$ the method of stationary phase gives for the asymptotic value of H

$$H = \delta t_{\pm} \exp[ix\,\psi(t_{res}) \pm i\pi/4],$$

where the + sign applies if $\psi''(t_{res}) > 0$ and the – sign applies if $\psi''(t_{res}) < 0$ so that $\delta t_{\pm}$ corresponds to the same quantities introduced in Section 5.1. In the limit in which the two time intervals are approximately equal, we have for electrons turning well beyond the resonance surface

$$H = t_{eff} \exp[i(\phi_{res} - \phi_o)] \quad \text{with} \quad t_{eff} = \{2!/[x|\psi''(t_{res})|]\}^{1/2}\Gamma(1/2).$$

Since $\Gamma(1/2) = \sqrt{\pi}$, this result is identical to Eq. (5.5). If the electron turns at resonance, the stationary phase technique yields

$$t_{eff} = 2\{3!/[x|\psi'''(t_{res})|]\}^{1/2}\,\Gamma(1/3)/3,$$

which agrees closely with Eq. (5.6). Note that $\phi_{res} - \phi_o = x\,\psi(t_{res}) \pm \pi/4$. In the limit of $x \gg 1$, the change in the right-hand circular component of the electron velocity resulting from a single transit of the fundamental resonance surface is

$$v_-(t) = \exp\left(-i\int \Omega\, dt\right)$$
$$\times\{v_-(0) - (\varepsilon/m)|E_-|J_0(k_\perp\rho)\, t_{eff} \exp[i(\phi_{res} - \phi_o)]\exp(-ik_\perp\rho \sin\phi_o)\}.$$

Recall that the electron's perpendicular kinetic energy is given by $W_\perp = mv_-^*(t)v_-(t)$ and $v-(0) = v_\perp(0)\exp(-i\phi_o)/\sqrt{2}$. Thus we obtain for the electron's energy after a single transit through the fundamental resonance surface with X-mode illumination

$$W_\perp = W_\perp(0) - e[v_\perp(0)/\sqrt{2}]|E_-|J_0(k_\perp\rho)t_{eff}\{\exp[i(k_\perp\rho \sin\phi_o - \phi_{res})]$$
$$+ \exp[-i(k_\perp\rho \sin\phi_o - \phi_{res})]\} + (e^2/m)|E_-|^2 J_0^2(k_\perp\rho)t_{eff}^2.$$

Note that $\{\exp[i(k_\perp\,\rho \sin\phi_o - \phi_{res})] + \exp[-i(k_\perp\,\rho\sin\phi_o - \phi_{res})]\}/\sqrt{2}$

$$= \sqrt{2}\cos(k_\perp\rho \sin\phi_o - \phi_{res})$$
$$= \sqrt{2}[\cos(k_\perp\rho \sin\phi_o)\cos\phi_{res} + \sin(k_\perp\,\rho\sin\phi_o)\sin\phi_{res}]$$
$$= \sqrt{2}\{\cos\phi_{res}[J_0(k_\perp\rho) + 2\Sigma J_{2n}(k_\perp\rho)\cos 2n\,\phi_o]$$
$$+ \sin\phi_{res}[2\Sigma J_{2n+1}(k_\perp\rho)\sin(2n+1)\phi_o]\},$$

where we have made use of the two Bessel function relations [5]

$$\cos(a \sin b) = J_0(a) + 2\Sigma J_{2n}(a)\cos 2nb$$
$$\sin(a \sin b) = 2\Sigma J_{2n+1}(a)\sin(2n+1)b.$$

Both of the summations are from $n = 1$ to $n = \infty$. To lowest order in $k_\perp\rho$, the change in $W_\perp$ after a single transit of the fundamental X-mode resonance surface is, therefore, given by

$$\Delta W_\perp = -e\sqrt{2}|E_-|v_\perp(0)J_0^2(k_\perp\rho)t_{eff}\cos\phi_{res} + (e^2/m)|E_-|^2 J_0^2(k_\perp\rho)t_{eff}^2. \tag{5.13}$$

Since $\sqrt{2}|E_-| = |E_\perp|$, apart from the factor $J_0^2(k_\perp\rho)$, the first term on the right-hand side is similar to Eq. (5.1). As we noted earlier, this term depends on ϕ_{res}, the phase of the microwave electric field relative to the electron gyration at resonance. The resulting change in energy can be positive or negative and will vanish when averaged over a population of electrons with random gyrophases. By contrast, the final term is positive definite and describes a phase-independent increase in $W_\perp$. This increase is proportional to t_{eff}^2 and closely resembles the result of the stochastic heating model of Section 5.1. Note that if we average Eq. (5.13) over the gyrophase angle and multiply the result by $\nu_{trs,}$ we recover our expression for the heating rate derived from the stochastic model. The positive definite term arises because electrons that are accelerating when they cross the resonance surface have greater average velocities during the resonance interval than those that are decelerating when they cross the resonance surface.

5.2.2 Dynamics of the Electron Interaction With Parallel RF Electric Fields

First consider the direct transfer of wave energy into electron motion along the magnetostatic field as described by the parallel component of the equation of motion:

$$m dv_{||}/dt = -eE_{||} = -e|E_{||}|\exp[i(\mathbf{k}\cdot\mathbf{r}-\omega t)].$$

In the absence of finite gyroradius phenomena, the average of $v_{||}$ over a wave period would be constant, but finite gyroradius effects introduce the possibility of cyclotron resonance effects. Just as was done earlier, we linearize this equation by evaluating the electron position at any time using the unperturbed orbits with the same assumptions as before and obtain the following expression for the electron's parallel velocity after passing through resonance:

$$v_{||}(t) = v_{||}(0)-(e/m)|E_{||}|\exp(-ik_\perp\rho\sin\phi_o)\Sigma J_n(k_\perp\rho)\exp(in\phi_o) \times \int dt\exp\left[\int dt(n\Omega + k_{||}v_{||}-\omega)\right].$$

For fundamental resonance, $n = 1$ and we have

$$\begin{aligned} v_{||}(t) &= v_{||}(0)-(e/m)|E_{||}|J_1(k_\perp\rho)\exp(-ik_\perp\rho\sin\phi_o)\exp(i\phi_o)H_1(t) \\ &= v_{||}(0)-(e/m)|E_{||}|J_1(k_\perp\rho)t_{eff}\exp(i\phi_{res}-ik_\perp\rho\sin\phi_o) \\ &\approx v_{||}(0)-(e/m)|E_{||}|J_1(k_\perp\rho)J_0(k_\perp\rho)t_{eff}\cos\phi_{res}. \end{aligned}$$

The corresponding change in the "parallel" kinetic energy is then

$$\Delta W_{||}(t) = -v_{||}(0)e|E_{||}|J_1(k_\perp\rho)J_0(k_\perp\rho)t_{eff}\cos\phi_{res} + (m/2)\left[(e/m)|E_{||}|J_1(k_\perp\rho)J_0(k_\perp\rho)t_{eff}\cos\phi_{res}\right]^2.$$

Averaging over random gyrophases (at resonance) yields

$$\langle\Delta W_{||}\rangle = (e^2/4m)\left[|E_{||}|J_1(k_\perp\rho)J_0(k_\perp\rho)t_{eff}\right]^2.$$

It is instructive to compare this with our earlier result for the phase-independent change in the "perpendicular" kinetic energy: $\Delta W_\perp = (e^2/2m)|E_\perp|^2 J_0^2(k_\perp \rho) t_{eff}^2$. Evidentially the direct transfer of wave energy into "parallel" kinetic energy is smaller by a factor of $[J_1(k_\perp\rho)|E_{||}|/|E_\perp|]^2$, which is of the order of magnitude of $(v_\perp/c)^2$ and thus considered negligible except for relativistic-electron plasmas.

5.2.3
Dynamics of the Electron Interaction with O-Mode Waves

To this point, we have considered the work done on gyrating electrons by RF electric fields, including (1) the perpendicular electric field of a right-hand circularly polarized electromagnetic wave, and (2) the parallel electric field of the RF wave. Next we will describe a different mechanism by which waves can exchange energy with electrons through the combined actions of the electric and magnetic fields of the waves. An important practical example of this mechanism occurs for O-modes propagating perpendicular to the magnetostatic field. Hamiltonian mechanics provide a convenient description of the electron dynamics [6]. We model the magnetostatic and wave fields in the resonance zone using the following vector potential:

$$\mathbf{A} = \mathbf{u}_y B_o x + \mathbf{u}_z A_1 \exp[i(kx-\omega t)].$$

The magnetic and electric fields for this vector potential are given by the real parts of

$$\begin{aligned}\mathbf{B} &= \nabla \times \mathbf{A} = \mathbf{u}_z B_o - \mathbf{u}_y ikA_1 \exp[i(kx-\omega t)]\\ \mathbf{E} &= -\partial\mathbf{A}/\partial t = \mathbf{u}_z i\omega A_1 \exp[i(kx-\omega t)].\end{aligned}$$

The Lagrangian,

$$\begin{aligned}\mathcal{L} &= mv^2/2 - e\mathrm{v}\cdot \mathrm{A}\\ &= m(v_x^2+v_y^2+v_z^2)/2 - e\{v_y B_o x + v_z A_1 \exp[i(kx-\omega t)]\}\end{aligned}$$

and the generalized momenta, $p_i = \partial\mathcal{L}/\partial v_i$ are therefore

$$\begin{aligned}p_x &= mv_x\\ p_y &= mv_y - eB_o x\\ p_z &= mv_z - eA_1 \exp[i(kx-\omega t)].\end{aligned}$$

If these expressions are solved for the components of the velocity and used to express the Lagrangian in terms of the generalized momenta and coordinates, we obtain

$$\mathcal{L} = (p_x^2+p_y^2+p_z^2)/2m - m\Omega_o^2 x^2/2 - (e^2A_1^2/2m)\exp[2i(kx-\omega t)].$$

The Hamiltonian is then given by $\mathfrak{H} = \mathbf{p}\cdot\mathbf{v} - \mathcal{L}$ and we have

$$\begin{aligned}\mathfrak{H} = &(p_x^2+p_y^2+p_z^2)/2m + p_y\Omega_o x + m\Omega_o^2x^2/2 + p_z(eA_1/m)\exp[i(kx-\omega t)]\\ &+ (e^2A_1^2/2m)\exp[2i(kx-\omega t)].\end{aligned}$$

Since the Hamiltonian is independent of y and z, their respective conjugate momenta are constants of the motion:

$$p_y = mv_y - eB_o x = p_{yo} = \text{constant}$$
$$p_z = mv_z - eA_1\exp[i(kx-\omega t)] = p_{zo} = \text{constant}.$$

The rate of change of p_x (which equals F_x, the x-component of the force on the electron) is given by the canonical equation of motion:

$$dp_x/dt = mdv_x/dt = F_x = -\partial\mathfrak{H}/\partial x$$
$$= -p_y\Omega_o - m\Omega_o^2 x - ikp_z(eA_1/m)\exp[i(kx-\omega t)] - ik(e^2A_1^2/m)\exp[2i(kx-\omega t)].$$

Substituting from our expressions for p_y and p_z we obtain for the force in the x-direction

$$F_x = -m\Omega_o v_y - ikv_z(eA_1)\exp[i(kx-\omega t)].$$

The instantaneous rate at which this force changes the (perpendicular) kinetic energy is then given by

$$dW_\perp/dt = mv_y dv_y/dt + mv_x dv_x/dt$$
$$= v_y d(p_{yo}/m + m\Omega_o x)/dt - v_x\{m\Omega_o v_y + ikv_z(eA_1)\exp[i(kx-\omega t)]\}$$
$$= -ikv_x v_z(eA_1)\exp[i(kx-\omega t)].$$

Taking the real part then gives the instantaneous heating rate as

$$dW_\perp/dt = kv_x v_z(eA_1)\sin(kx-\omega t).$$

The results obtained in this Hamiltonian formulation can also be obtained directly from Newton's second law including the Lorentz force due to $\mathbf{B}_1$ in a transparent way that facilitates interpretation. Again ignoring the weak spatial gradients in the magnetic field within the resonance zone we have

$$md\mathbf{v}/dt = -e(\mathbf{E} + \mathbf{v}\times\mathbf{B}) = -e[E_1\mathbf{u}_z + \mathbf{v}\times(B_o\mathbf{u}_z + B_1\mathbf{u}_y)]$$

giving three equations for the components of $md\mathbf{v}/dt$:

$$mdv_x/dt = -ev_y B_o + ev_z B_1$$
$$mdv_y/dt = ev_x B_o$$
$$mdv_z/dt = -eE_1 - ev_x B_1$$

so that the rate of change of the perpendicular energy is

$$dW_\perp/dt = m(v_x dv_x/dt + v_y dv_y/dt) = ev_x v_z B_1 = ev_x v_z(kA_1)\sin(kx-\omega t).$$

The rate of change of parallel momentum is given by

$$mdv_z/dt = -eE_1 - ev_x B_1 = e\omega A_1\sin(kx-\omega t) - ev_x kA_1\sin(kx-\omega t)$$
$$= eA_1 d\cos(kx-\omega t)/dt$$

whence

$$mv_z = eA_1\cos(kx-\omega t) + p_{zo}.$$

We can use this last expression to distinguish between two classes of electrons depending on their value of p_{zo}. The first class is comprised of electrons that have nonvanishing values of parallel momentum at the resonance surface, $p_{zo} \neq 0$. If the expression for mv_z is averaged over a wave period (which equals a gyroperiod at resonance) we have $\langle mv_z \rangle = p_{zo}$ and for these electrons the instantaneous rate of change of the perpendicular energy is given by $dW_\perp/dt = ev_x v_z(kA_1)\sin(kx - \omega t)$. A case of particular interest is the heating of mirror-trapped electrons where the unperturbed orbits are described by adiabatic kinematics so that

$$v_x = v_\perp \cos(\phi_o + \Omega t) = (2W_\perp/m)^{1/2}\cos(\phi_o + \Omega t) = (2\mu B/m)^{1/2}\cos(\phi_o + \Omega t)$$

and

$$v_z = \sigma[2W_{||}/m]^{1/2} = \sigma[2(\varepsilon - \mu B)/m]^{1/2}, \text{ where } \sigma = \pm 1.$$

With these substitutions, we can easily show that the instantaneous rate at which electrons exchange energy with the O-mode wave at resonance is

$$dW_\perp/dt = \sigma(ekA_1/m)(W_{\perp res}W_{||res})^{1/2}\{\sin(\phi_o + kx) - \sin[(\Omega + \omega)t + \phi_o - kx]\}.$$

In the limit $k\rho \ll 1$, the change in energy at the fundamental resonance is given approximately by

$$\delta W_\perp = \sigma \sin\phi_o(ekA_1/m)(W_{\perp res}W_{||res})^{1/2}t_{eff}.$$

Electrons for which $\sigma \sin\phi_o > 0$ at the resonance surface will gain energy, while electrons with $\sigma \sin\phi_o < 0$ at the resonance surface will loose energy. If the phase of gyration relative to the wave varies randomly from one resonance to the next stochastic, heating will result, as described earlier in this chapter. The resulting heating rate is then

$$dW_\perp/dt = D_W/2W_\perp = (1/2)(e^2/m)E_1^2 t_{eff}^2 \nu_{trs}(n/2)^2 W_{||res}/mc^2 \tag{5.14}$$

where we have set

$$D_W = \langle \delta W_\perp^2 \rangle \nu_{trs} = (1/2)[(ekA_1/m)(W_{\perp res}W_{||res})^{1/2} t_{eff}]^2 \nu_{trs}.$$

As in the earlier discussion of stochastic heating, ν_{trs} is the frequency with which the electrons pass through resonance surfaces and $n = kc/\omega$ is the index of refraction.

The second class of electrons is characterized by $p_{zo} = 0$ and is comprised of electrons that are trapped in the resonance zone by the electric field of the O-mode wave. In practice, this may occur if the resonance is at a local minimum in the magnetostatic field or if the electron is turning at the resonance surface. The resulting nonlinear orbits are more complex and additional resonances are possible. This situation has been discussed at length by Carter *et al.* [7] and will be dealt with only briefly and qualitatively here to suggest how the additional resonances may come about. Although not justified clearly, the expression for the instantaneous rate of exchange of energy between the electron and the wave will be linearized for the present heuristic purposes by integrating along the unperturbed orbits using

$$v_x = v_\perp \cos\left(\phi_o + \int \Omega dt\right) \text{ and } v_z = (eA_1/m)\cos(kx - \omega t),$$

where the expression for v_z is given by the conserved momentum. Then,

$$
\begin{aligned}
W_\perp(t)-W_\perp(0) &= -ikv_\perp(e^2A_1^2/m)\int dt\cos\left(\phi_o+\int \Omega dt\right)\exp[2i(kx-\omega t)] \\
&= -ikv_\perp(e^2A_1^2/m)\int dt\cos\left(\phi_o+\int \Omega\, dt\right)\exp(-2ik\,\rho\,\sin\phi_o) \\
&\quad \times \Sigma J_n(2k\rho)\exp(in\phi_o)\times\int dt\exp\left[i\int dt\,(\nu\Omega+k_{||}v_{||}-2\omega)\right] \\
&= -i\,kv_\perp(e^2A_1^2/2m)\exp(-2ik\rho\,\sin\phi_o) \\
&\quad \times\left\{\Sigma J_n(2k\rho)\exp[i(n+1)\phi_o]\int dt\exp i\int dt[(n+1)\Omega+k_{||}v_{||}-2\omega]\right. \\
&\quad \left.+\Sigma J_n(2k\rho)\exp[i(n-1)\phi_o]\int dt\exp i\int dt[(n-1)\Omega+k_{||}v_{||}-2\omega]\right\}.
\end{aligned}
$$

This nonlinear mechanism exhibits resonances at $2\omega=(n\pm1)\Omega$. For the lowest-order resonance, $\omega=\Omega/2$, we have for the change in perpendicular energy

$$
\begin{aligned}
W_\perp(t)-W_\perp(0) &= kv_\perp(e^2A_1^2/2m)J_o^2(2k\rho)t_{\mathrm{eff}}\sin\phi_{\mathrm{res}} \\
&= (kc/\omega)(v_\perp/c)(\Omega/2\omega)(eE_1^2/B_o)J_o^2(2k\rho)t_{\mathrm{eff}}\sin\phi_{\mathrm{res}}.
\end{aligned}
$$

The distinctive $\Omega/2$ resonance can readily be understood from the expression for the rate of change of the electron's perpendicular kinetic energy:

$$dW_\perp/dt = kv_xv_z(eA_1)\sin(kx-\omega t)$$

with

$$mv_z = eA_1\cos(kx-\omega t)\quad \text{and}\quad v_x = v_\perp\cos(\phi_o+\Omega t)$$

Note the $2\cos x\sin x=\sin 2x$ and $2\cos x\sin y=\sin(x+y)-\sin(x-y)$ so that $dW_\perp/dt = (kv_\perp/m)(eA_1/2)^2[\sin(\phi_o+\Omega t+2kx-2\omega t)-\sin(\phi_o+\Omega t-2kx+2\omega t)]$. We can make use of $\sin(x+y)=\sin x\cos y+\cos x\sin y$ and the Bessel function relations[5]

$$\cos(a\sin b) = J_o(a)+2\Sigma J_{2n}(a)\cos 2nb$$

and

$$\sin(a\sin b) = 2\Sigma J_{2n+1}(a)\sin(2n+1)b,$$

where both summations are from $n=1$ to $n=\infty$. With these substitutions we obtain

$$
\begin{aligned}
dW_\perp/dt = {}& (kv_\perp/m)(eA_1/2)^2\{[\sin(\phi_o+\Omega t-2\omega t)\cos 2kx \\
&+\cos(\phi_o+\Omega t-2\omega t)\sin 2kx]-[\sin(\phi_o+\Omega t+2\omega t)\cos 2kx \\
&-\cos(\phi_o+\Omega t+2\omega t)\sin 2kx]\} \\
= {}& (kv_\perp/m)(eA_1/2)^2\{\sin(\phi_o+\Omega t-2\omega t)[J_o(k\rho)
\end{aligned}
$$

$$+2\Sigma J_{2n}(k\rho)\cos 2n(\phi_o + \Omega t)] + \cos(\phi_o + \Omega t - 2\omega t)$$
$$[2\Sigma J_{2n+1}(k\rho)\sin(2n+1)(\phi_o + \Omega t)] - \sin(\phi_o + \Omega t + 2\omega t)$$
$$[J_o(k\rho) + 2\Sigma J_{2n}(k\rho)\cos 2n(\phi_o + \Omega t)]$$
$$-\cos(\phi_o + \Omega t + 2\omega t)[2\Sigma J_{2n+1}(k\rho)\sin(2n+1)(\phi_o + \Omega t)]\}.$$

When this expression is integrated in time the rapidly oscillating terms vanish, and for $\omega = \Omega/2$ the change in $W_\perp$ during the electron's transit of the resonance zone is given by

$$\delta W_\perp = (kv_\perp/m)(eA_1/2)^2 \sin\phi_o J_o(k\rho)t_{eff.}$$

5.3
Heating of Relativistic Electrons

Several relativistic effects were noted in Chapter 3, including most notably the energy-dependent electron gyrofrequency, $\Omega = eB/(\gamma m)$. If for the moment we neglect the Doppler effect, which is of considerable importance in tokamak applications, the condition for resonance at the *n*th harmonic of the gyrofrequency becomes

$$\nu = n\Omega + k_{||}v_{||} - \omega \approx neB/(\gamma m) - \omega = 0.$$

We see that the fundamental resonance surface, for example, will move toward higher magnetic fields as the electron energy increases and will ultimately disappear when $B_{res}(n=1)$ exceeds B_{max}. However, if the mirror ratio is greater than or roughly equal to 2, a second-harmonic resonance surface will appear near the midplane and resonant heating can continue for these relativistic electrons. The change in $W_\perp$ for these electrons after each transit of the (second harmonic) resonance surface with X-mode illumination is

$$\Delta W_\perp = (e^2/m)|E_-|^2 J_1^2(k_\perp\rho)t_{eff}^2,$$

where we have once again omitted the phase-dependent term. The argument of the Bessel function is given in order of magnitude by

$$k_\perp\rho = (k_\perp/k)(kc/\omega)(\omega/\Omega_{res})(v_\perp/v)(v/c)$$
$$= O(1)O(1)O(n)O(1)[(\gamma^2-1)/\gamma^2]^{1/2}.$$

The energy-dependent factor, $[(\gamma^2 - 1)/\gamma^2]^{1/2}$, increases from roughly 0.2 at an energy of 10 keV to almost 0.9 for energies above 500 keV. Since, for comparable values of t_{eff} the ratio of the heating rate for second-harmonic resonance relative to that for fundamental resonance is $[J_1(k_\perp\rho)/J_0(k_\perp\rho)]^2$, the relative heating rate for second-harmonic resonance will increase from roughly 1% to 25% over this same energy range. As we shall see in later chapters, the electron confinement time increases with energy, and relativistic electrons can continue to gain energy provided their heating rate exceeds the rate at which they loose energy, for example, by synchrotron radiation, until they reach the maximum energy for adiabatic invariance of the magnetic moment. If the mirror ratio of the region illuminated by microwave power is less than 2, as is frequently the case in tokamak applications of ECH, the heating will be limited by the maximum energy for which the resonance condition can be satisfied, a situation sometimes referred to as "relativistic heating gaps." We will see later how such gaps can be bridged by employing multiple-frequency ECH.

5.4 Limit Cycles

For electrons that have gained a substantial amount of energy from their first few transits through resonance (or from other forms of heating), the energy increment in the next succeeding transit through resonance may be much less than the initial energy just prior to resonance. In this case, the dynamics of the resonant interaction can be analyzed using the unperturbed bounce orbit, $z(t) = z_t \sin \omega_b t$, where z_t is the axial location of the turning point and ω_b is the bounce frequency. The time-dependent phase factor, $\phi = \int dt\nu = \int dt(\Omega + k_{||}v_{||} - \omega)$, which varies rapidly except in the neighborhood of resonance, can be expressed conveniently for the model magnetic-mirror field in such a circumstance. If we neglect the Doppler shift, the rate of change of the difference in phase between the electron gyration and the phase of the wave is given for the simple magnetic mirror field by the following expression:

$$\begin{aligned}\nu = \Omega[z(t)] - \omega &= (\Omega_o/2)\{[(M+1)-(M-1)\cos(k_o z_t \sin \omega_b t)]\\ &\quad -[(M+1)-(M-1)\cos(k_o z_{res})]\}\\ &= (\Omega_o/2)(M-1)[\cos(k_o z_{res}) - \cos(k_o z_t \sin \omega_b t)],\end{aligned}$$

where z_{res} is the axial location of the resonance. We can again make use of the Bessel function generating function the [5] $\cos(b \sin\theta) = J_0(b) + 2\Sigma J_{2n}(b)\cos(2n\theta)$, where the sum over n runs from $n = 1$ to ∞. With this substitution, we obtain for the phase factor,

$$\begin{aligned}\phi(t) &= \int dt(\Omega-\omega) = \int dt\{(\Omega_o/2)(M-1)\\ &\quad \times [\cos(k_o z_{res}) - J_0(k_o z_t) - 2\Sigma J_{2n}(k_o z_t)\cos(2n\omega_b t)]\}\\ &= \phi(0) + (\Omega_o/\omega_b)[(M-1)/2]\{[\cos(k_o z_{res}) - J_0(k_o z_t)]\omega_b t\\ &\quad - \Sigma n^{-1} J_{2n}(k_o z_t)\sin(2n\omega_b t)\}. \end{aligned} \quad (5.15)$$

Since $(\Omega_o/\omega_b)(M-1)/2 \gg 1$, small changes in the bracketed function in Eq. (5.15) result in very large changes in $\phi(t)$. It is, therefore, reasonable to expand this function to evaluate small changes in the phase near the instant of resonance, just as was done in Section 5.1. We define dimensionless variables $\zeta_{res} = k_o z_{res}$, $\zeta_t = k_o z_t$ and $\tau = \omega_b t$. Then the function in braces in Eq. (5.15) is

$$F(\zeta_{res}, \zeta_t, \tau) = [\cos \zeta_{res} - J_0(\zeta_t)]\tau - \Sigma n^{-1} J_{2n}(\zeta_t)\sin(2n\tau).$$

To evaluate the duration of resonance, $\delta\tau$, we expand $F(\zeta_{res}, \zeta_t, \tau)$ to form

$$\delta F = F(\tau_{res} + \delta\tau) - F(\tau_{res}) = (\partial F/\partial\tau)_{res}\delta\tau + (\partial^2 F/\partial\tau^2)_{res}(\delta\tau)^2/2 + \cdots$$

The indicated derivatives are

$$(\partial F/\partial \tau)_{res} = \cos\zeta_{res} - [J_0(\zeta_t) + 2\Sigma J_{2n}(\zeta_t)\cos(2n\tau_{res})] = 0,$$
$$(\partial^2 F/\partial \tau^2)_{res} = 4\Sigma n J_{2n}(\zeta_t)\sin(2n\tau_{res}),$$
$$(\partial^3 F/\partial \tau^3)_{res} = 8\Sigma n^2 J_{2n}(\zeta_t)\cos(2n\tau_{res}),$$

etc.

If the electron turns at resonance, $\zeta_t = \zeta_{res}$, then $\sin\tau_{res} = 1$, $\tau_{res} = \pi/2$, and thus $\sin(2n\tau_{res}) = 0$ and $\cos(2n\tau_{res}) = (-1)^n$. For this case, $(\partial^2 F/\partial \tau^2)_{res} = 0$ and $(\partial^3 F/\partial \tau^3)_{res} = 8\,\Sigma(-1)^n n^2 J_{2n}(\zeta_t)$. The duration of resonance corresponding to the expression from Section 5.1 is then given by

$$\delta\tau = \{6\delta\phi(\omega_b/\Omega_o)[(M-1)/2]/[8\Sigma n^2 J_{2n}(\zeta_t)\cos(2n\tau_{res})]\}^{1/3},$$

with $\delta\phi = \pm\pi/4$. More generally, the duration of resonance can be found by solving the cubic equation,

$$(\omega_b/\Omega_o)[(M-1)/2]\delta\phi = 2\Sigma n J_{2n}(\zeta_t)\sin(2n\tau_{res})(\delta\tau)^2 + (4/3)\Sigma n^2 J_{2n}(\zeta_t)\cos(2n\tau_{res})(\delta\tau)^3.$$

In this way, we can recover results very similar to those displayed in Section 5.1. The closed form for the phase factor derived in this section, however, indicates the possibility that for certain classes of electrons coherent limit cycles can exist in which the electron energy oscillates around a fixed value. If, for example, an electron turning at the resonance surface experiences a phase change of $\pm(2N+1)\pi$ between successive resonances, the energy increment at the first resonance will be exactly cancelled at the second resonance. This situation has been termed "superadiabaticity" [8]. Similarly, if an electron turns just beyond the resonance surface, it will experience two closely spaced transits through resonance. If the heating fields are coherent over the spatial extent that includes this portion of the electron's orbit, heating at the two successive resonances can add or cancel depending on the phase change. Cancellation of the two successive resonant interactions leads to a null in the heating, and hence the designation of these conditions as "null-heating surfaces".

In the case of superadiabaticity, the electrons turn at the resonance surface so that $\zeta_t = \zeta_{res}$. If the first resonance occurs at $\tau_{res1} = \pi/2$, the second will occur at $\tau_{res2} = 3\pi/2$, and the phase change between the two successive resonances is

$$\phi(\tau_{res2}) - \phi(\tau_{res1}) = (\Omega_o/2\omega_b)(M-1)[\cos(k_o z_{res}) - J_0(k_o z_{res})]\pi. \quad (5.16)$$

Recall that the bounce frequency for this model magnetic-mirror field depends on energy and the location of the turning points as given by Eq. (3.11):

$$2\omega_b = k_o v_{\perp o}(M-1)^{1/2}\{(\pi/2)K^{-1}[(M_t-1)/(M-1)]\} \approx k_o v_{\perp o}(M-1)^{1/2},$$

where K is the elliptic integral. Fixed points require $\phi(\tau_{res2}) - \phi(\tau_{res1}) = \pm(2N+1)\pi$, which then determines critical values of energy, ε_N, for given values of the turning points which, in the case of superadiabaticity, coincide with the resonance surfaces.

Our expression for the phase change does not include the effect of heating at the first resonance, and if this heating results in a large change in the phase at the second resonance, the fixed points will be unstable. In this way, the limit cycle can be broken by strong enough heating. To estimate this effect, we note that heating changes only the bounce frequency so that for the fixed points to be stable we require

$$\Delta(\delta\phi) = (\partial\delta\phi/\partial W_\perp)\Delta W_\perp = -(\delta\phi/2\varepsilon)\Delta W_\perp \leq \pm\pi/2.$$

We therefore anticipate the possibility of superadiabatic behavior mainly at higher energies and with small energy increments per transit of resonance.

In the case of null-heating surfaces, the criterion for cancellation between two successive transits through resonance is again $\phi(\tau_{res2}) - \phi(\tau_{res1}) = \pm(2N + 1)\pi$, but in this case the two resonance times are related by $\tau_{res2} = \pi - \tau_{res1}$ and we now find for the phase difference between the two closely spaced resonances

$$\phi(\tau_{res2}) - \phi(\tau_{res1}) = (\Omega_o/\omega_b)(M-1)\{[\cos\zeta_{res} - J_0(\zeta_t)](\pi/2 - \tau_{res1}) + \Sigma n^{-1} J_{2n}(\zeta_t)\sin(2n\tau_{res1})\}. \quad (5.17)$$

For a given resonance, position the null-heating conditions determine critical values of the turning point as functions of energy. The stability of these fixed points is determined as in the case of superadiabaticity by the conditions under which $\phi(\tau_{res2})$ is changed by $\pm\pi/2$ as a result of the heating that occurs at the first resonance.

5.5 Nonlinear Effects: Mapping Approaches

The use of mapping techniques to study chaotic systems has been described extensively by Lichtenberg and Liebermann [9], and their work is an invaluable resource. In this section, we construct a rudimentary two-step mapping to explore in an approximate way the properties of the superadiabatic limit cycles in simple magnetic-mirror configurations. In this approach, the electron gyrophase relative to the electric field is advanced from one resonance to the next using Eq. (5.16), while the electron energy is similarly advanced using Eq. (5.1). From the mth resonance to the m + 1th resonance, the mapping takes the following form:

$$\phi_{m+1} = \phi_m + \delta\phi_m$$

and

$$W_{\perp m+1} = W_{\perp m} + \delta W_{\perp m+1}$$

where

$$\delta\phi_m = (\Omega_o/2\omega_{bm})(M-1)[\cos(k_o z_{res}) - J_0(k_o z_{res})]\pi$$

and

$$\delta W_{\perp m+1} = -e\int E_{-} v_{\perp} \cos\phi \, dt \approx -e|E_{-}| v_{\perp m+1} \cos\phi_{m+1} \, t_{\text{eff } m+1}.$$

Since superadiabatic electrons turn at the resonance surfaces, the effective duration of resonance in the simple magnetic-mirror configuration is given by Eq. (5.6) by setting $M_t = M_{res}$:

$$t_{eff} \approx 2\left[3\pi/|(2v''(t_{res})|\right]^{1/3}$$

and, from Eq. (5.11),

$$v''(t_{res}) = -(1/2)(k_o v_{\perp o})^2 \Omega_o (M - M_{res})(M_{res} - 1).$$

Between the resonance surfaces, the electron's magnetic moment is invariant and $v_{\perp o}^2 = v_{\perp res}^2 / M_{res}$. Recall that the bounce period for the model magnetic-mirror field depends on energy and the location of the turning points as given by Eq. (3.11); we therefore substitute the following expression into the equation for the phase step:

$$\pi/2\omega_{bm} = (2/k_o v_{\perp m})[M_{res}/(M-1)]^{1/2} K[(M_{res}-1)/(M-1)]$$

so that our mapping equations become

$$\phi_{m+1} = \phi_m + (2\Omega_o/k_o v_{\perp m})[M_{res}(M-1)]^{1/2} K[(M_{res}-1)/(M-1)] \times [\cos(k_o z_{res}) - J_0(k_o z_{res})]$$

and

$$W_{\perp m+1} = W_{\perp m} - 2(e|E_{-}|/k_o)(k_o v_{\perp m+1}/\Omega_o)^{1/3} \times \{3\pi M_{res}/[(M - M_{res})/(M_{res}-1)]\}^{1/3} \cos\phi_{m+1}.$$

If we define $U_m = k_o v_{\perp m}/\Omega_o$ and substitute for $W_{\perp} = (m/2)(\Omega_o U/k_o)^2$ to rewrite the mapping equations, we obtain the following somewhat more compact expressions:

$$\phi_{m+1} = \phi_m + C_1/U_m$$

and

$$U_{m+1}^2 = U_m^2 - C_2 U_{m+1}^{1/3} \cos\phi_{m+1}$$

where

$$C_1 = 2[M_{res}(M-1)]^{1/2} K[(M_{res}-1)/(M-1)][\cos(k_o z_{res}) - J_0(k_o z_{res})]$$

and

$$C_2 = 4(e/m)\left(k_o|E_-|/\Omega_o^2\right)\left\{3\pi M_{res}/[(M-M_{res})/(M_{res}-1)]\right\}^{1/3}.$$

We can iterate the equation that advances U starting with $U_{m+1}^{1/3}(0) = U_m^{1/3}$ on the right-hand side and using the resulting values of U_{m+1} for the subsequent iterations. The condition for superadiabatic response at the resonance is

$$|C_1|/U_{res}^{(N)} = (2N+1)\pi$$

and the corresponding resonant energies are

$$W_\perp^{(N)} = (m/2)\left(\Omega_o U_{res}^{(N)}/k_o\right)^2 = (m/2)[(\Omega_o/k_o)|C_1|/(2N+1)\pi]^2.$$

Some of the features of this mapping can be illustrated using experimental parameters from studies reported recently [10] which we will discuss in a later chapter. The simple magnetic-mirror configuration in these experiments had an overall mirror ratio of $M = 1.43$, the effective separation of the mirrors was $L_c = 21$ cm, and the frequency of the microwave power was $f_\mu = 2.45$ GHz. To illustrate the mapping approach, we apply it to the experimental case in which the magnetic intensity on the midplane was set at the value $B_o = 820$ G so that the mirror ratio at resonance was $M_{res} = 875/820 = 1.067$. In this case, the superadiabatic resonances are at (perpendicular) energies given by

$$W_{\perp res} = 7800\ \text{eV}/(2N+1)^2 = 7800\ \text{eV},\ 867\ \text{eV},\ 312\ \text{eV},\ 159\ \text{eV}, \ldots$$

We select for our illustrative mapping the $N = 2$ resonance and arbitrarily set the initial gyrophase at $\phi_0 = \pi/4$ and the initial energy at its resonant value. The resulting energies at the subsequent resonances are then evaluated for values of the right-hand circularly polarized electric field strength, $|E_-|$ increasing from 0.5 to 6 V cm^{-1}. The results are displayed in the series of plots shown in Figures 5.6 (a)–(e).

In the first plot, the energy remains near the resonant value but varies slowly as the shift in gyrophase between resonances slips away from -5π. This slow variation in $W_{\perp m}$ is more pronounced when $|E_-|$ is doubled, as is seen in the second plot, where nulls in the energy increment, corresponding to values of the gyrophase that are odd integral multiples of $\pi/2$, occur at $m = 5$ and $m = 35$. The next doubling of $|E_-|$ results in a higher frequency quasiperiodic pattern, but the pattern remains visible. The next doubling to a value of $|E_-| = 4$ V cm^{-1} leads to the breakup of the quasiperiodic pattern, but the energy excursions remain bounded. If $|E_-|$ is increased to 6 V cm^{-1}, the magnitude of the energy excursions away from the resonant value increases sharply, indicating that the step-to-step phase increment has changed enough to destroy the limit cycle. We can estimate the critical conditions for this breakup to occur from the first of the mapping relations:

$$\phi_{m+1} = \phi_m + C_1/U_m.$$

The shift in the step-to-step change in the gyrophase, $\Delta(\delta\phi_m)$, brought about by the change in the perpendicular energy, $\Delta W_\perp$, is given by

$$\Delta(\delta\phi_m) = -(C_1/U_m)(\Delta U_m/U_m) = -(C_1/U_m)(\Delta W_\perp/W_\perp)/2.$$

As was indicated earlier, we anticipate that the limit cycles will be broken up if $\Delta(\delta\phi_m)$ exceeds $\pi/2$. Since $\delta\phi_m = -(C_1/U_m) = (2N + 1)\pi$, our condition for disrupting the Nth limit cycle is $\Delta W_\perp/W_\perp \geq 1/(2N + 1)$.

For our illustrative case with $N = 2$, we anticipate the breakup of the associated limit cycle for $\Delta W_\perp \geq 62$ eV. The rms step size in $W_\perp$ for $|E_-| = 6\,\mathrm{V\,cm^{-1}}$, the value at which the limit cycle is disrupted in the case shown earlier, is around 59 eV. We will discuss this issue of limit cycles later in connection with control of the temperature anisotropy and the generation of high-energy density plasmas.

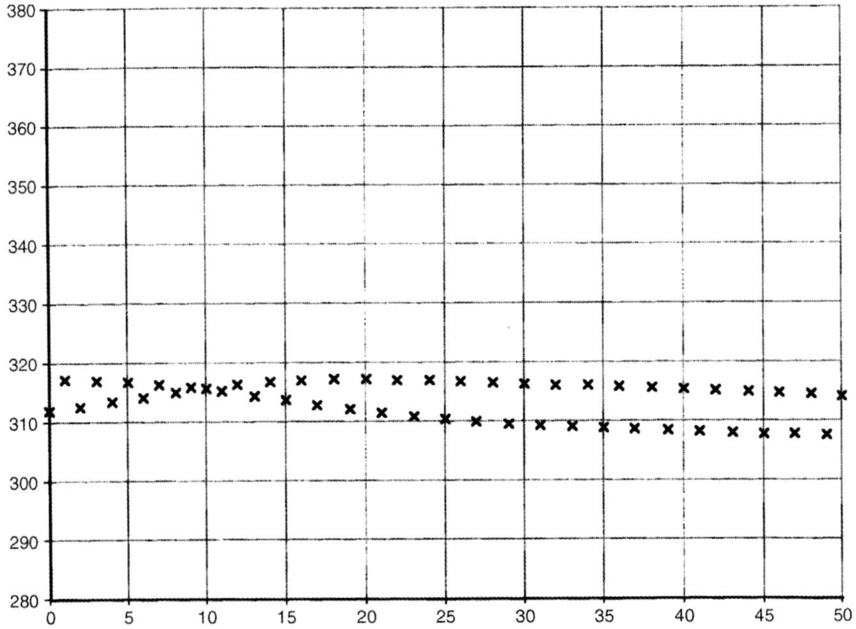

(a)

Figure 5.6 (a) Results of the rudimentary mapping algorithm for $|E_-| = 0.5\,\mathrm{V\,cm^{-1}}$. (b) Results of the rudimentary mapping algorithm for $|E_-| = 1\,\mathrm{V\,cm^{-1}}$. (c) Results of the rudimentary mapping algorithm for $|E_-| = 2\,\mathrm{V\,cm^{-1}}$. (d) Results of the rudimentary mapping algorithm for $|E_-| = 4\,\mathrm{V\,cm^{-1}}$. (e) Results of the rudimentary mapping algorithm for $|E_-| = 6\,\mathrm{V\,cm^{-1}}$.

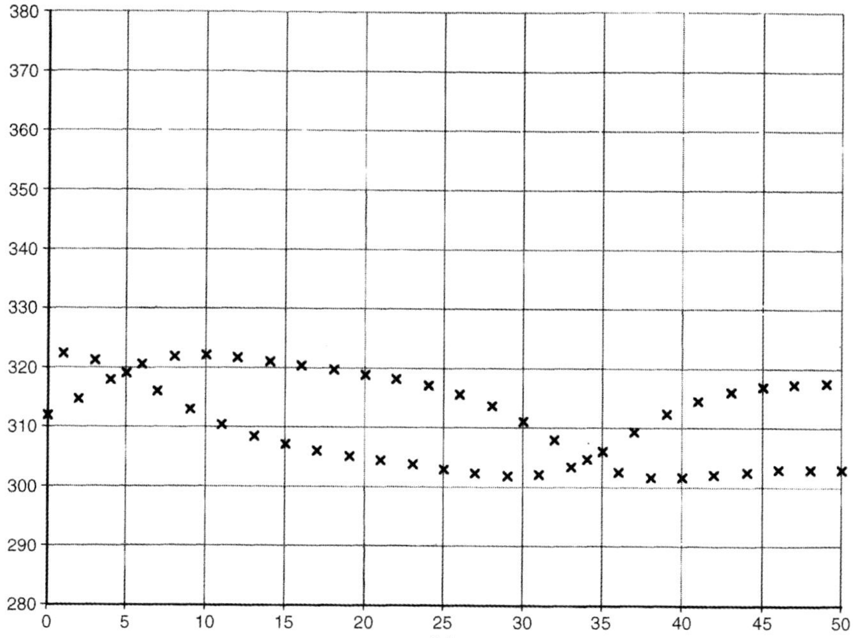

(b)

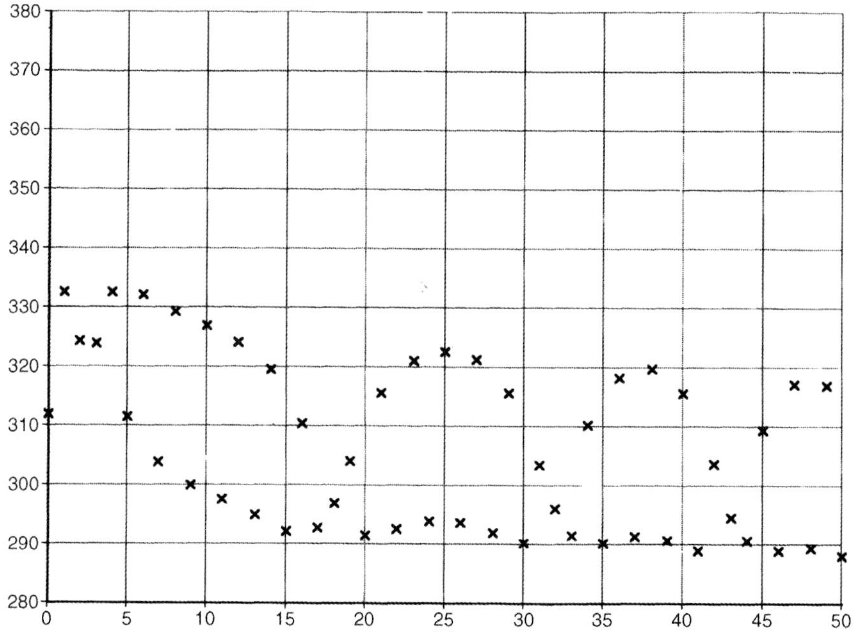

(c)

Figure 5.6 *(Continued)*

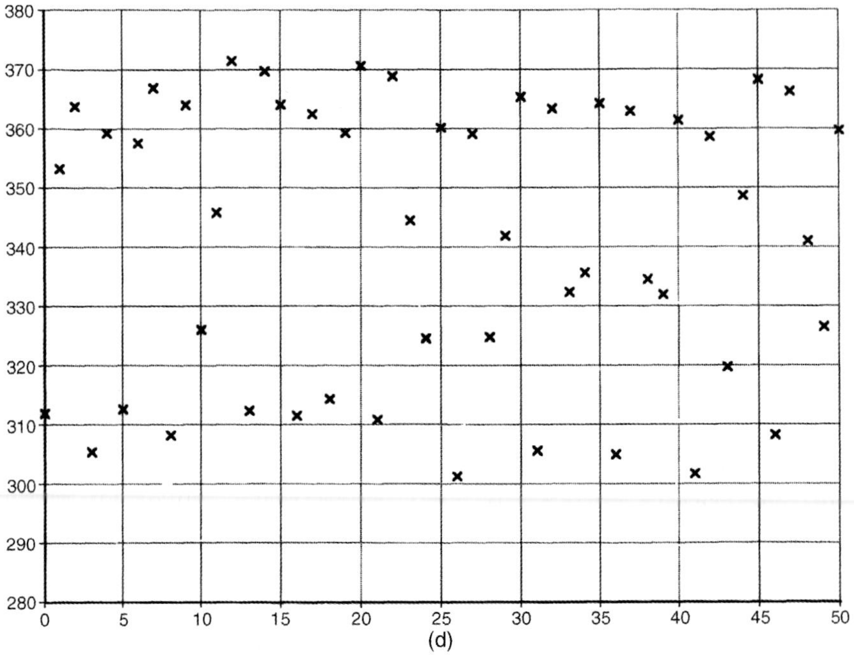

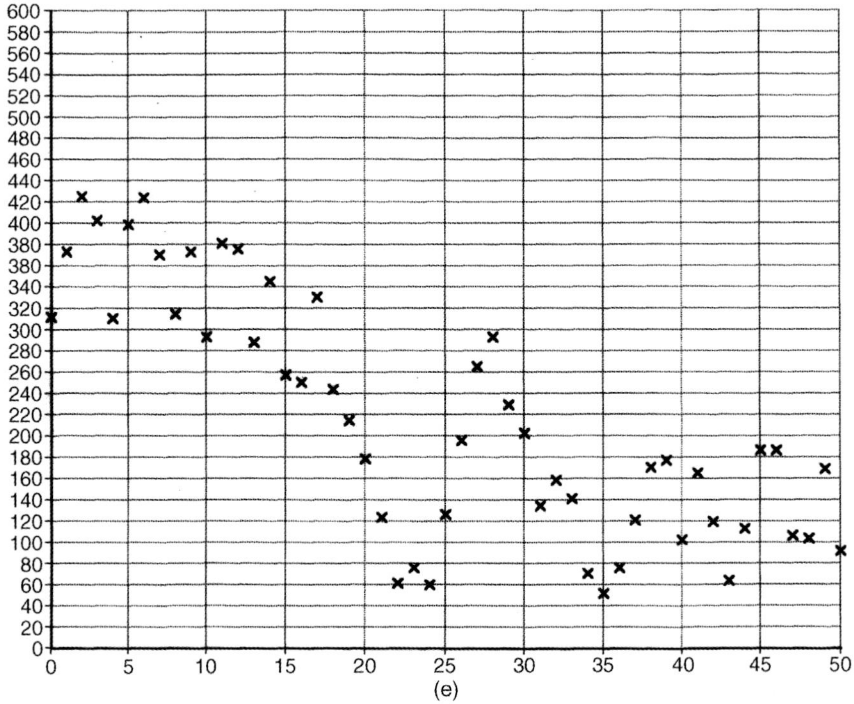

Figure 5.6 *(Continued)*

References

1 See, for example, T.H. Stix, "The Role of Stochasticity in Radiofrequency Plasma Heating", Joint Varenna-Grenoble International Symposium on Heating in Toroidal Plasmas, Grenoble, 3–7 July, 1978 and the extensive list of references contained therein, as well as Thomas Howard Stix, *Waves in Plasmas*, pp. 465–477, American Institute of Physics, New York (1992). See also, F. Jaeger, A.J. Lichtenberg, and M.A. Lieberman, *Plasma Phys.* **14**, pp. 1073–1100 (1972) and references cited therein.

2 James E. Howard, *Plasma Phys.* **23**, 597 (1981) and references cited therein.

3 C.D. Cantrell, *Modern Mathematical Methods for Physicists and Engineers*, Cambridge University Press, Cambridge (2000), pp. 690–693.

4 Carl M. Bender and Steven A. Orsag, *Advanced Mathematical Methods for Scientists and Engineers*, pp. 276–280, McGraw-Hill, New York (1978).

5 *Handbook of Mathematical Functions*, edited by Milton Abramowitz and Irene A. Stegun, National Bureau of Standards (1964), reprinted by Dover Publications, New York, (1970) p. 361.

6 See, for example, E.V. Suvorov and M.D. Tokman, *Plasma Phys.* **25**, 723 (1983).

7 M.D. Carter, J.D. Callen, D.B. Batchelor, and R.C. Goldfinger, *Phys. Fluids* **29**, 100 (1986) and references cited therein.

8 M. Seidl, *Plasma Phys. (J. Nucl. Energy Part C)* **6**, 597 (1964); see also M.N. Rosenbluth, *Phys. Rev. Lett.* **29**, 408 (1972).

9 A.J. Lichtenberg and M.A. Liebermann, *Regular and Stochastic Motion*, Springer, New York (1983) and second edition (1992).

10 K.S. Gulyaev *et al.*, *J. Phys. D: Appl. Phys.* **27**, 2349 (1994).

Exercises

5.1 ***Verify that Eq. (5.8b) follows from Eq. (5.8a).***

5.2 ***A simple magnetic mirror with a mirror ratio on axis $M = 2.2$ consists of two coils separated by 71 cm. Microwave power at a frequency of 9.5 GHz is coupled into a plasma confined in this configuration and the magnetic intensity is adjusted so that fundamental resonance occurs for $M_{res} = B_{res}/B_o = 1.6$. Estimate the following quantities for a group of 100 eV electrons turning at a location where $M_t = B_t/B_o = 1.7$: (a) $(v_{||} d\Omega/dz)_{res}$, (b) $(k_{||} dv_{||}/dt)_{res}$, (c) t_{eff}, (d) ω_b, (e) $\Delta W_\perp$ in each transit through the resonance (f) $dW_\perp/dt$ if $|E_-| = 10\ \mathrm{V\,cm^{-1}}$, and (g) $(d\Omega/d\gamma)(d\gamma/dt)$.***

6
Equilibrium

This chapter and Chapter 7 address some of the fundamental collective plasma phenomena that affect the outcome of electron cyclotron heating (ECH) experiments, particularly the issues of equilibrium and stability. A full theoretical treatment of plasma equilibrium would necessitate rather elaborate transport models of plasmas specific to each particular magnetic field and heating configuration under discussion. Instead of such an ambitious program, we shall limit ourselves here to an analysis of the spatially averaged properties of somewhat generic plasmas and return later to more detailed discussions of the equilibria of some specific configurations. In this chapter, we emphasize plasmas in which the equilibrium is effectively determined by the processes associated with ECH, as contrasted with those cases, such as tokamak applications of ECH, where the basic equilibrium is determined by other processes and is only modified by ECH.

The transport processes governing the steady-state equilibria of electron cyclotron heated plasmas naturally separate into two distinct groups differentiated by characteristic equilibration times. The more rapid of these, quasineutrality, or the condition for charge balance, is established in the time required for electrons to traverse the plasma region. In contrast, the conditions under which the charged particle density and temperature become stationary are achieved only after the much longer times that characterize the ionization and heating processes responsible for creating new ion–electron pairs and heating them to the equilibrium temperature. Together with the particle and energy loss rates, they determine the final equilibrium. We now address these detailed balance issues as they apply to the spatially averaged properties of radially bounded but open-ended plasmas exemplified by magnetic-mirror configurations. In tokamaks, stellarators and high-energy density mirror-confined plasmas, the equilibria must also satisfy pressure balance conditions. We will consider some of these issues later as they affect particular experiments.

6.1
Charge Balance

A defining characteristic of the plasma medium is that electrical neutrality is maintained to a high degree throughout the body of the plasma, a characteristic

Electron Cyclotron Heating of Plasmas. Gareth Guest

ISBN: 978-3-527-40916-7

usually referred to as quasineutrality [1]. This electrical neutrality is maintained despite the great disparity in mass between the plasma electrons and ions. The far more mobile electrons must be prevented from escaping from the plasma more rapidly than the massive ions. This is usually accomplished by electrostatic fields which arise spontaneously to reduce the loss rate of electrons (and simultaneously increase the loss rate of ions). This electrostatic field, the so-called ambipolar field, maintains quasineutrality by ensuring equal loss rates of ions and electrons from the body of the plasma. As is customary, we will describe this ambipolar electric field through the associated electrostatic potential, $\Phi(x,y,z)$.

In open-ended magnetized plasmas, particularly plasmas confined in the simple magnetic-mirror configuration discussed in Chapter 2, the rate of transport of plasma across the magnetic field is generally much slower than the rate of transport along the static magnetic field [2]. The conditions for charge balance must therefore be satisfied along each magnetic line of force, and we shall neglect, for the time being, the transverse directions and examine only the variation of the ambipolar potential along the magnetic field direction, $\Phi(z)$. This potential takes the form of a "positive well" which traps electrons and thereby retards their escape from the plasma. Electrons trapped in this ambipolar potential well rapidly thermalize into a Maxwell–Boltzmann distribution for which the density, $n(z)$, varies along the magnetic line of force according to

$$n(z) = n(z_o)\exp\{e[\Phi(z)-\Phi(z_o)]/T_e\}$$

In what follows, we choose z_o to be the position at which the potential is maximum and set $\Phi(z_o) = \Phi_o$ and $\Phi_{surface} = 0$. Here T_e is the electron temperature in eV. Since electrons will escape from the plasma if they are heated to an energy that is greater than the depth of the ambipolar potential well, their loss rate is determined by their heating rate, dW_e/dt. The ion loss rate is determined by the rate at which they flow along the magnetic field and the resulting flux of ions through the plasma surface. For quasineutrality, the two loss rates must be equal. Consider first the electrons, whose loss rates are governed by their heating rate but whose lifetimes must in any event equal the common "ambipolar" lifetime, τ_{amb}:

$$|d\ln n_e/dt|_{loss} = (dW_e/dt)/e\Phi_o = \tau_{amb}^{-1}.$$

Since $(dW_e/dt)\tau_{amb} \approx T_e$, we require an ambipolar potential well whose depth is roughly equal to the electron temperature: $e\Phi_o \approx T_e$.

The rate at which ions flow along the magnetic field to the surface of the plasma will be increased as a result of the acceleration by the ambipolar electric fields. Conservation of energy dictates that in the absence of collisions

$$M_i v_{io}^2/2 + q_i\Phi_o = M_i v_{i,surface}^2/2 + q_i\Phi_{surface}$$

and since $\Phi_{surface} = 0$ and $e\Phi_o \approx T_e$, the ion speed at the surface is given by

$$\begin{aligned} v_{i,surface} &= \{2T_{io}/M_i + 2q_i\Phi_o/M_i\}^{1/2} \approx [(2/M_i)(T_{io} + q_i T_e/e]^{1/2} \\ &= c_s \end{aligned} \tag{6.1}$$

where c_s is the ion sound speed (for a more rigorous and comprehensive derivation of the ion sound speed, see Ref. [3]). Thus, we anticipate that ions will escape at the ion sound speed and accordingly set $\tau_{amb} = L_{esc}/c_s$. The "escape" distance, L_{esc}, depends on the particular configuration being modeled. For the simple magnetic-mirror configuration, we have seen that the heated electrons tend to accumulate around the resonance surfaces. It is therefore reasonable in this case to approximate L_{esc} by half of the axial separation of the resonance surfaces. In some toroidal plasmas, one can also define an escape distance that is determined by the geometry of the outermost "scrape-off layer" and the limiters. We will return to this issue in discussing particular applications.

6.2 Particle and Power Balance

In the interior of the magnetic-mirror-confined plasma, the conditions governing the self-consistent values of the plasma density and temperature are particle balance (i.e., the balance between the rates of creation of ion–electron pairs and their loss) and power balance (i.e., the balance between the rates of heating and energy loss) for each species. We begin our discussion of these conditions by separating the plasma electrons somewhat arbitrarily into three possible groups, not all of which will be populated in any given situation. The groups are separated according to their energy and the dominant particle- and power-balance mechanisms operative in each group. We will take steps to refine this arbitrary separation later by employing kinetic models that vary continuously in energy.

1. The electrons with the lowest average energies form the first of the three groups; their density will be denoted by n_{e1}. They are created by ionization of the background gas and are electrostatically confined by the ambipolar potential discussed earlier. Typically, their average energies will be in the 10–100 eV range. Group 1 electrons loose energy mainly through inelastic collisions with neutral atoms of the background gas.

2. Electrons in the second group, with density n_{e2}, have energies between 100 eV and 10 keV and are sufficiently energetic to be magnetically confined by the magnetic-mirror effect. They result if some of the colder, Group 1 electrons are heated rapidly enough to become mirror trapped before they can escape at the ambipolar loss rate. Group 2 electrons loose energy mainly by slowing down on the Group 1 electrons as well as through inelastic collisions with gas atoms. Their confinement time in Group 2 is determined by scattering into the loss cone and the time required to heat them into Group 3.

3. Electrons in the third group, with density n_{e3}, typically have average energies in the 100–1000 keV range and therefore have low rates of Coulomb scattering and correspondingly long mirror confinement times. They result if some of the electrons in the second group are heated more rapidly than they are lost by scattering into the mirror loss cone. They loose energy mainly by synchrotron radiation as well as by cooling on lower energy electrons through Coulomb collisions. If their energies reach the limit beyond which adiabatic invariance breaks down, they can escape and take their full energy with them.

We can express the conditions for steady-state particle balance for electrons in each of these three groups in the following three very schematic equations:

$$\begin{aligned} dn_{e1}/dt &= n_{e1}n_o\langle\sigma_{ion}v_e\rangle_1 + n_{e2}n_o\langle\sigma_{ion}v_e\rangle_2 + n_{e3}n_o\langle\sigma_{ion}v_e\rangle_3 \\ &\quad - n_{e1}/\tau_1 - n_{e1}(dW/dt)_1/\Delta W_{2,1} = 0 \\ dn_{e2}/dt &= n_{e1}(dW/dt)_1/\Delta W_{2,1} - n_{e2}/\tau_2 - n_{e2}(dW/dt)_2/\Delta W_{3,2} = 0 \\ dn_{e3}/dt &= n_{e2}(dW/dt)_2/\Delta W_{3,2} - n_{e3}/\tau_3 - n_{e3}(dW/dt)_3/\Delta W_{escape} = 0 \end{aligned} \tag{6.2}$$

The ionization rates averaged over the electron energy distributions for electrons in each of the three groups are $\langle\sigma_{ion}v_e\rangle_1$, $\langle\sigma_{ion}v_e\rangle_2$, and $\langle\sigma_{ion}v_e\rangle_3$. In what follows we shall usually adopt the formulas for these ionization rates summarized by Barnett [4] or their empirical values as given by Freeman and Jones [4]. The average electron confinement times for each of the three groups are τ_1, τ_2, and τ_3; and the average heating rates for each group are $(dW/dt)_1$, $(dW/dt)_2$, and $(dW/dt)_3$. The energy at the boundary separating the cold- and warm-electron groups is denoted by $\Delta W_{2,1}$ and is typically of the order of 100 eV. The energy at the boundary separating the warm- and hot-electron groups, $\Delta W_{3,2}$, is typically of several orders of magnitude larger, on the order of 100 keV. ΔW_{escape} is the energy increase required for a hot electron to become nonadiabatic and can be several mega electron volts in relatively large, high-field experiments.

The conditions under which the rates of heating and energy loss balance for each group of electrons depend on the heating and cooling mechanisms dominant in each group, here indicated by the subscript $s = 1, 2, 3$ and will generally have the following, again, schematic form:

$$d(1.5n_eT_e)_s/dt = \mathcal{P}_{in,s} - \mathcal{P}_{out,s} - (1.5n_eT_e)_s/\tau_{Es} = 0. \tag{6.3}$$

Here $\mathcal{P}_{in,s}$ and $\mathcal{P}_{out,s}$ are the power (per unit volume) flowing into and out of the electrons in group s, respectively, and τ_{Es} is the overall energy confinement time for the group. In what follows we explore the consequences of these schematic equations.

6.2.1 Particle and Energy Balance for Group 1

We start with the simplest case of interest, namely, a plasma in which there are no hot electrons. Mirror-confined plasmas heated using high-field launch are good examples of this case. Since in this case, all plasma electrons are in Group 1, we can drop the subscript and write for Eqs. (6.2) and (6.3):

$$dn_e/dt = n_en_o\langle\sigma v_e\rangle_{ioniz} - n_e/\tau_e = 0,$$

and

$$d(1.5n_eT_e)/dt = \mathcal{P}_\mu - \mathcal{P}_x - \mathcal{P}_i - \mathcal{P}_{ei} - (1.5n_eT_e)/\tau_E = 0, \tag{6.4}$$

where n_e and n_o are the local values of the electron and neutral gas densities, respectively, $\langle\sigma v_e\rangle_{ioniz}$ is the ionization rate coefficient averaged over the electron distribution, τ_e and τ_E are the electron (particle) and energy confinement times, respectively, and $\mathcal{P}_{\mu}$, $\mathcal{P}_{x}$ and $\mathcal{P}_{i}$ are the power densities associated with the absorption of microwave power and excitation and ionization of gas atoms, respectively:

$$\mathcal{P}_x = n_e n_o \langle \sigma v_e \rangle_x E_x,$$

and

$$\mathcal{P}_i = n_e n_o \langle \sigma v_e \rangle_{ioniz} E_{ioniz},$$

where $\mathcal{P}_{ei}$ is the power density transferred to plasma ions by plasma electrons through Coulomb collisions; it will usually be negligible. In keeping with our earlier discussion, we shall assume that the average lifetime of plasma electrons is given by the ambipolar confinement time: $\tau_e = L_{esc}/c_s$. Later, we will return to a discussion of the effect on the equilibrium of populations of energetic mirror-confined electrons in Groups 2 and 3. For the moment, we restrict our consideration to a plasma containing only low temperature, Group 1 electrons.

The energy confinement time is generally less than the particle confinement time by virtue of the additional processes such as thermal conduction and radiation, by which energy can be lost from the interior of the plasma. We shall somewhat arbitrarily set $\tau_E = \tau_e/2$. In steady state, our particle and power balance conditions then become

$$n_o \langle \sigma v_e \rangle_{ioniz} \tau_e = 1 \qquad (6.5a)$$

and

$$\mathcal{P}_{\mu}/n_e = 1.5T_e/\tau_E + n_o(\langle \sigma v_e \rangle_x E_x + \langle \sigma v_e \rangle_{ioniz} E_{ioniz}). \qquad (6.5b)$$

If the particle lifetime is governed by ambipolar flow out of the plasma, the balance of particle creation and loss rates, Eq. (6.5a), will be satisfied in a plasma containing only singly charged ions if

$$n_o L_{esc} = p_o L_{esc}(3.537 \times 10^{16}\ \mathrm{cm^{-3}\ Torr^{-1}}) = (2T_e/M_i)^{1/2}/\langle \sigma v_e \rangle_{ioniz}.$$

The right-hand side of this equation depends only on the electron temperature and the type of gas; the left-hand side depends only on the local gas density (pressure) and the effective dimensions of the plasma. Evidently, the steady-state electron temperature is governed primarily by the local gas density or gas pressure, p_o, through the temperature-dependent ionization rate constants and more specifically, the temperature- and gas species-dependent function $(2T_e/M_i)^{1/2}/\langle \sigma v_e \rangle_{ioniz}$.

We can illustrate some of the properties of this Group 1 particle balance condition by employing the Thompson formula to provide a closed-form estimate of the electron impact ionization cross section [4]. From Barnett's equation (31) of Ref. [4], it is given by

$$\sigma_{ioniz} = 4n(E_H/E_i)^2(E_i/E)[1-(E_i/E)]\pi a_o^2,$$

where n is the number of electrons in the outer shell of the target atom, $E_H = 13.60$ eV is the ionization potential of the hydrogen atom, E_i is the ionization potential of the target atom, and E is the energy of the incident electron. $\pi a_o^2 = 8.797 \times 10^{-17}\ \mathrm{cm}^2$. If argon is used for our illustrative case, $n = 6$ and $E_i = 15.76$ eV. The resulting cross section is then multiplied by the electron speed and their product is averaged over a

Maxwell–Boltzmann electron energy distribution function:

$$f(E)dE = 2n_e(E/\pi)^{1/2}(kT_e)^{-3/2}\exp(-E/kT_e)dE,$$

where n_e is the electron density, k is Boltzmann's constant, and T_e is the electron temperature. For convenience, let $x = E/E_i$ and $y = E_i/kT_e$ so that

$$\begin{aligned}\sigma_{ioniz}v &= 4n(E_H/E_i)^2\pi a_o^2(2E_i/m)^{1/2}x^{-1/2}[1-(1/x)], \quad \text{if } x \geq 1\\ &= 0, \quad \text{if } x < 1\end{aligned}$$

and the Maxwell–Boltzmann distribution function is

$$f(x)dx = 2n_e(1/\pi)^{1/2}y^{3/2}x^{1/2}\exp(-xy)dx.$$

Integrating over the energy parameter x, we obtain

$$\langle\sigma_{ioniz}v\rangle = 8n(E_H/E_i)^2\pi a_o^2(2E_i/\pi m)^{1/2}y^{1/2}[\exp(-y)-yE_1(y)], \tag{6.6}$$

where $E_1(y)$ is the exponential integral [5]. The Group 1 particle balance condition for argon gas can then be displayed as shown in Figure 6.1.

Here p_oL_{esc} (in mTorr cm) is plotted on the horizontal axis, and the associated values of kT_e (in eV) for steady-state particle balance on the vertical axis. At the highest gas

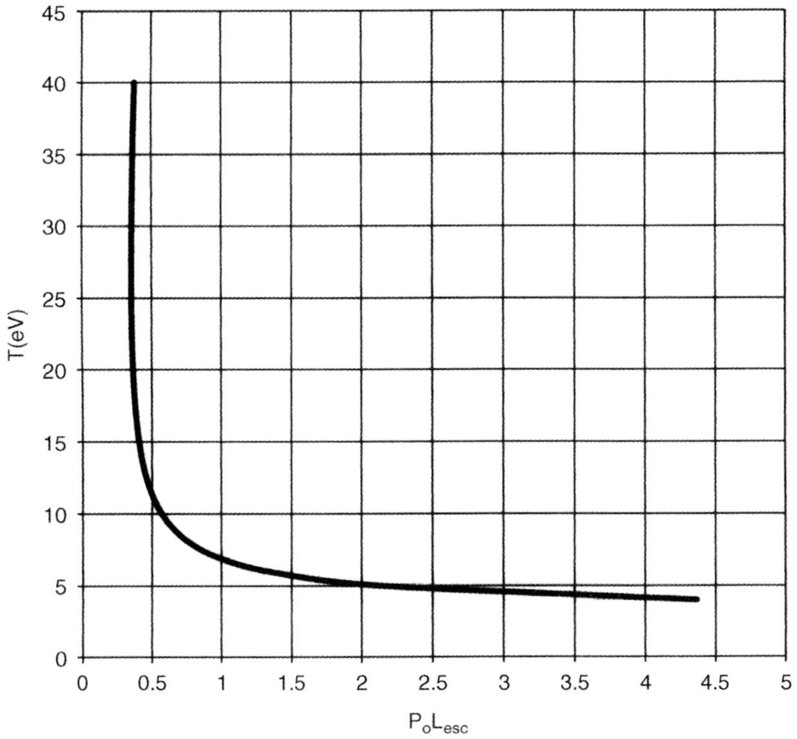

Figure 6.1 The electron temperature in eV versus the product of the neutral gas pressure, p_o, in mTorr, and the characteristic length L_{esc} in cm.

pressures shown in this plot, the electron temperature is well below the ionization potential, and ionization depends on electrons in the tail of the distribution function. It is clear from Figure 6.1 that for gas pressures less than a critical value, there are no steady-state equilibria [6]. This critical value is determined solely by the characteristic length, L_{esc}, and the particular type of gas used. In the case of argon, this critical value is estimated to be approximately 0.35 mTorr cm. If, for example, $L_{esc} = 10\,\text{cm}$, then the critical pressure is 3.5×10^{-5} Torr.

For values of n_o and T_e that satisfy the steady-state particle balance condition, the particle and power balance conditions, Eqs. (6.5a) and (6.5b), can be combined to give

$$\mathcal{P}_\mu/n_e = 1.5T_e/\tau_E + [E_x\langle\sigma v_e\rangle_x/\langle\sigma v_e\rangle_{ioniz} + E_{ioniz}]/\tau_e. \tag{6.7}$$

The power-balance condition then determines the ratio of microwave power density to the plasma electron density, $\mathcal{P}_\mu/n_e$. Note that, just as the electron temperature is determined entirely by the gas density, the plasma electron density will be determined, for this temperature, by the absorbed microwave power density. Thus, the plasma density is predicted to be proportional to microwave power if the neutral gas pressure is fixed. The inelastic rate constants for argon, for example, can again be estimated from the Thompson formula; but more accurate values of the total energy expended per ionization event are available [7] and we use them to display $n_e/(\mathcal{P}_\mu L_{esc})$ in Figure 6.2.

As the gas pressure approaches the critical value the electron density abruptly decreases, whereas the electron temperature abruptly rises. Our model suggests that the ambipolar potential will also increase rapidly with further reductions in the gas pressure, and for pressures below a critical value, steady-state equilibria will not exist. The nature of this critical pressure is discussed at length in Ref. [6]. The present model suggests that the plasma parameters will exhibit relaxation oscillations as the increasing electron temperature leads to loss rates that exceed the rate at which ionization can maintain the plasma density.

6.3 Breakdown and Start-up

In considering the conditions for particle and power balance for electrons in Group 1, we assumed that the electrons were in local thermodynamic equilibrium, i.e., distributed in energy with a Maxwell–Boltzmann distribution function. This is not necessarily correct at breakdown, when ionization may be produced in large part by electrons that have acquired energies greater than E_i but are not yet thermalized. For breakdown to occur in magnetic-mirror traps, these electrons must reach energies greater than or comparable to E_i and remain in the mirror longer than an ionization time, $\tau_{ioniz} = (n_o\sigma_{ioniz}v_e)^{-1}$. We can anticipate that under the optimal conditions of pressure and power for breakdown to occur, the time to reach the energy at which σ_{ioniz} approaches its maximum, typically 60–100 eV, is roughly equal to the ionization time. In what follows we consider three ECH breakdown

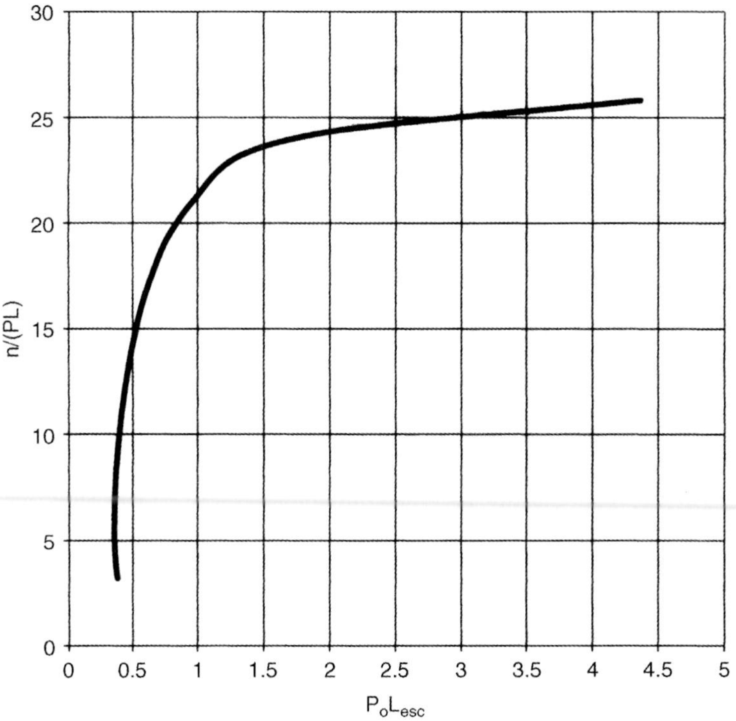

Figure 6.2 The ratio of the electron density, in electrons per cm^3, to the product of the absorbed microwave power density, in Watts per cm^3, and the characteristic "escape" length in cm.

situations that differ only in the location of the resonance surface and consequently, the heating dynamics as discussed earlier in Chapter 5.

6.3.1 Breakdown by Heating on the Midplane of a Magnetic Mirror

In the first case to be considered, the resonance surface is on the midplane of a magnetic mirror, equivalent in some respects to heating in a uniform magnetic field. Unlike that case, however, the heated electrons will remain in the trap until they are scattered into the loss cone through collisions with the gas atoms, whereas in the uniform magnetic field case, they will escape if scattered through even a very small angle. In both cases, the duration of resonance is governed by the relativistic mass increase. As in earlier discussions of the duration of resonance, we set the maximum phase shift from the value at resonance at $\pm\pi/4$, although this is clearly somewhat arbitrary in this case. Thus,

$$\delta\phi_{max} = \pm\pi/4 = \int dt\{[eB/\gamma(t)m] - \omega_{\mu}\}, \tag{6.8}$$

where the integral is from 0 to t_{max} and m is the electron rest mass. The Lorentz factor, $\gamma(t)$, is given by $\gamma(t) = 1 + W(t)/mc^2$ and since $W \ll mc^2$ for breakdown, we have $1/\gamma(t) \approx 1 - W(t)/mc^2$. At resonance, $eB/m = \omega_\mu$ and thus Eq. (6.8) requires that $\delta\phi_{max} = \pm\pi/4 = -\omega_\mu \int dtW(t)/mc^2$, or, taking the lower sign,

$$1/8 = f_\mu \int dtW(t)/mc^2,$$

where f_μ is the microwave frequency, $f_\mu = \omega_\mu/2\pi$. The heating is coherent rather than stochastic in this situation, so we set

$$dW_\perp/dt = mv_\perp dv_\perp/dt = -eE_\perp v_\perp \cos\phi_{res},$$

and obtain

$$v_\perp(t) = v_\perp(0) - (eE_\perp \cos\phi_{res}/m)t,$$

and thus

$$v_\perp^2(t) = v_\perp^2(0) - 2v_\perp(0)(eE_\perp \cos\phi_{res}/m)t + (eE_\perp \cos\phi_{res}/m)^2 t^2.$$

If we assume a random distribution of gyrophase angles and average this expression over ϕ_{res}, the value at the instant of resonance, we have for the energy at time t

$$W_\perp(t)/mc^2 = W_\perp(0)/mc^2 + (eE_\perp\lambda_\mu/2mc^2)^2 f_\mu^2 t^2,$$

where $\lambda_\mu = c/f_\mu$ is the (vacuum) wavelength of the microwave power. Our estimate of the maximum duration of resonance is given by the cubic equation,

$$1/8 = [W_\perp(0)/mc^2] f_\mu t_{max} + (1/3)(eE_\perp\lambda_\mu/2mc^2)^2 (f_\mu t_{max})^3.$$

If we neglect $[W_\perp(0)/mc^2]f_\mu t_{max}$, we estimate the duration of resonance to be

$$f_\mu t_{max} = (3/2)^{1/3}(eE_\perp\lambda_\mu/mc^2)^{-2/3}. \tag{6.9}$$

The corresponding value of the energy at this time is

$$W_\perp(t_{max})/mc^2 = (1/4)[(3/2)(eE_\perp\lambda_\mu/mc^2)]^{2/3}. \tag{6.10}$$

For optimum breakdown conditions, $W_\perp(t_{max})$ should be near the energy at which the ionization cross section reaches its maximum value, roughly 100 eV for most gases, and the time to reach this energy, t_{max}, should be near the ionization time. The first condition and Eq. (6.10) determine the microwave power under optimum breakdown conditions through the value of $E_\perp$. Using this value of $E_\perp$ together with the second condition and Eq. (6.9), we can determine the optimum gas pressure. For example, for a microwave frequency of $f_\mu = 2.45$ GHz, and with $W_\perp(t_{max}) = 76$ eV, we estimate $E_\perp = 0.4\ \mathrm{V\,cm^{-1}}$ and the duration of resonance $t_{max} = 1\ \mu s$. The resulting estimate of the optimal gas pressure is then about 1.5×10^{-4} Torr.

6.3.2
Breakdown with Heating Well Off the Midplane

We next consider the situation when the resonance surfaces are far enough off the midplane so that we can assume the heating to be stochastic. From Chapter 5, Eq. (5.8), the stochastic heating rate is approximately given by

$$dW_{\perp}/dt = (e^2/m)|E_{-}|^2 t_{eff}^2 \nu_{trs} \approx E_{max}/\tau_i, \tag{6.11}$$

where for the optimum breakdown conditions, we again require $W_{\perp}(\tau_i) = E_{max}$, the energy at which the ionization rate constant is maximum. τ_i is the same (neutral gas pressure dependent) ionization time as before. Since we are dealing with the initiation of breakdown, we can reasonably assume that the heated electrons are turning at the resonance surface, so that from Eqs. (5.6) and (5.11), the duration of resonance is given by $t_{eff} = 2[(3\pi)/|2\nu''(t_{res})|]^{1/3}$. For the simple magnetic mirror, $\nu''(t_{res}) = -(k_o v_{\perp o})^2 \Omega_o (M - M_{res})(M_{res} - 1)/2$ and the frequency with which electrons encounter the resonance surfaces is $\nu_{trs} = 2/\tau_b$. Here, as in earlier discussions, $k_o = 2\pi/L_c$, where L_c is the distance between the magnetic mirrors. Using our approximate expression for the bounce time, we have

$$1/\tau_b \approx k_o v_{\perp o}(M-1)^{1/2}/\pi.$$

With these substitutions, we obtain the following expression for the microwave electric field strength under optimum breakdown conditions in the simple magnetic mirror:

$$|E_{-}|^2 \approx 0.088[E_{max}/(e\tau_i)]B_o(k_o v_{\perp o}/\Omega_o)^{1/3}[(M-M_{res})(M_{res}-1)]^{2/3}/(M-1)^{1/2}. \tag{6.12}$$

Clearly, this approaches a minimum as M_{res} approaches M.

6.3.3
Breakdown with Heating near the Midplane

If the resonance surfaces are near but not on the midplane, it is likely that limit cycles will play a major role is determining the minimum microwave electric field strength for breakdown, since the field strength will have to be large enough to destabilize any limit cycles whose characteristic energies are below E_{max} in order for the electrons to reach E_{max}. We first evaluate the *N*th-order limit cycle energy, ε_N, for a given magnetic-mirror configuration and then calculate the corresponding values of the microwave electric field, $|E_{-}|$, for which the limit cycle will be unstable and allow the electrons to be heated to the optimum energy for ionization. If, as we assumed earlier, the heated electrons are turning at the resonance surfaces, the applicable limit cycles are due to "superadiabaticity." That is, the gyrophase relative to the microwave electric field changes between two successive resonances by an amount given by $\pm(2N + 1)\pi$, where $N = 0, 1, 2, 3, \ldots$. From Eq. (5.14), the circumstances under which this can occur are obtained from the condition that

$$\delta\phi = \phi(\tau_{res2}) - \phi(\tau_{res1}) = (\Omega_o/2\omega_b)(M-1)[\cos(k_o z_{res}) - J_0(k_o z_{res})]\pi$$
$$= \pm(2N+1)\pi.$$

Recall that the bounce frequency for this model magnetic-mirror field depends on energy and the location of the turning points as given by Eq. (3.11). If the resonance surfaces are very near the midplane, the bounce frequency is closely approximated by the right-hand version of that expression:

$$2\omega_b = k_o v_{\perp o}(M-1)^{1/2}\{(\pi/2)K^{-1}[(M_t-1)/(M-1)]\} \approx k_o v_{\perp o}(M-1)^{1/2}.$$

The condition for superadiabaticity then becomes

$$k_o v_{\perp o}/\Omega_o = \pm(M-1)^{1/2}[\cos(k_o z_{res}) - J_0(k_o z_{res})]/(2N+1).$$

From this equation, we can readily obtain a corresponding condition for the energy:

$$\varepsilon_N = (e^2/2m)M_{res}(M-1)\{B_0 L_c[\cos(k_0 z_{res}) - J_0(k_0 z_{res})]/(2N+1)2\pi]^2.$$

The *N*th-order limit cycle will be destabilized if during the initial transit of resonance, the perpendicular energy changes by an amount given by

$$\Delta(\delta\phi) = (\partial\delta\phi/\partial W_\perp)\Delta W_\perp = -(\delta\phi/2\varepsilon)\Delta W_\perp > \pm\pi,$$

which, since $\delta\phi = \pm(2N+1)\pi$, becomes

$$\Delta W_\perp > 2\varepsilon_N/(2N+1).$$

Substituting from $\Delta W_\perp = -eE_\perp v_{\perp res}\cos\phi_{res} t_{eff}$, we obtain the following expression for the minimum electric field strength needed to destabilize the *N*th-order limit cycle,

$$e|E_-|(k_o v_{\perp o}/\Omega_o)^{1/3} M_{res}^{1/2}\cos\phi_{res}[3\pi/(M-M_{res})(M_{res}-1)]^{1/3} > \varepsilon_N k_o/(2N+1) \quad (6.13)$$

The minimum electric field strength to destabilize the limit cycle corresponds to $\cos\phi_{res} = \pm 1$. On average, this value will be increased by $\sqrt{2}$.

These results reproduce at least qualitatively some of the features of the experimental results reported by Gulyaev *et al.* [8] for breakdown in a simple magnetic-mirror configuration. For this configuration, the magnetic field was reduced in small increments and the minimum microwave power for breakdown was recorded for several fixed pressures. Two minima in the power for breakdown were observed: one when the resonance surface is just inside the magnetic mirror and another when the resonance is on the midplane. Although the relation between ECH power and the electric field strength at the resonance is not given in their paper, we can estimate the values of the following parameters for their experimental apparatus:

$$M = 1.43, k_o = 2\pi/21\ \text{cm, and } B_{res} = 875\ \text{GHz}.$$

Using results from Section 6.3.1, for heating at the midplane, $B_o = 875$ G, we estimated the optimum electric field strength to be 0.4 V cm^{-1} for an argon pressure of 1.5×10^{-4} Torr. From Section 6.3.2, for heating far enough off the midplane to give stochastic heating, we find that the optimum electric field strength is around $1\,\mathrm{V\,cm^{-1}}$ for heating just inside the peak magnetic field ($B_o \approx 700$ Gauss) rising to values around $1.5\,\mathrm{V\,cm^{-1}}$ for $B_o \approx 725$ Gauss. In Section 6.3.3, we find that the electric field strength required to destabilize the first five limit cycles peaks near $3\,\mathrm{V\,cm^{-1}}$ for $N = 0$ at $B_o \approx 870\,\mathrm{G}$, decreasing to roughly $1\,\mathrm{V\,cm^{-1}}$ for $N = 4$ at $B_o \approx 830\,\mathrm{G}$. If the electric field strength varies as $P^{1/2}$, we can expect the breakdown power to increase by almost two orders of magnitude as the magnetic field at the center is lowered from 875 G to somewhat less than 870 G. For lower values of B_o, the breakdown power will be comparable to the values estimated for stochastic heating.

6.4
ECH Runaway: Groups 2 and 3

If the ambient gas pressure is reduced so as to approach the critical lower limit discussed earlier, the plasma electron temperature increases from values of a few electron volts to values of several tens of electron volts. A thermal distribution of electrons whose temperature is in this range may contain a significant number of electrons with energies greater than that for which the inelastic collision rate is maximum. These electrons may run away to high energies if they are adequately confined and subjected to further heating. The relative density and average energy of such ECH runaway electrons depend on the details of the heating and confinement processes. By employing magnetic-mirror fields for confinement and novel ECH techniques for preferentially heating the more energetic electrons, Dandl *et al.* [9] achieved stable, steady-state, relativistic-electron plasmas with plasma kinetic pressures nearly equal to the magnetostatic pressure and with average energies of several million electron volts. In this section, we consider the processes that affect the creation of relativistic-electron plasmas such as these and the corresponding requirements on the heating parameters.

6.4.1
Particle Balance for Electrons in Group 2

In the energy range we have labeled Group 2, $100 \leq W_e \leq 10\,\mathrm{keV}$, the dominant dynamical processes entering the condition for power balance are RF heating and pitch-angle diffusion, cooling of the heated electrons by Coulomb collisions with the cooler electrons, and inelastic collisions with neutral atoms. In highly ionized plasmas, inelastic collisions with neutral atoms will usually be negligible; but in plasmas having substantial densities of neutral atoms, Group 2 electrons will loose energy at a significant rate through inelastic collisions. The total cooling rate is given approximately by Book [10].

$$-dW_e/dt \approx n_o \sigma_{\mathrm{inelastic}} v_e \Delta W_x + 7 \times 10^{-6} n_e \ln\Lambda W_e^{-1/2}\,\mathrm{cm^3\,eV^{3/2}\,sec^{-1}}, \quad (6.14)$$

where n_o is the neutral density, ΔW_x is the characteristic energy transfer per collision, and $\sigma_{inelastic}$ is the cross section for inelastic collisions at the speed v_e. The Coulomb cooling rate, the second term on the right, decreases monotonically with electron energy, whereas the rate for inelastic impact with neutral atoms has a maximum value for electron energies around 100–200 eV, depending on the particular type of gas. In atomic hydrogen, for example, the product $\sigma_{inelastic}v_e$ for electron impact ionization reaches a maximum value of $4\times10^{-8}\,cm^3/s$ at energy of 100 eV. If $\Delta W_x \approx 50$ eV for this particular inelastic process, we can estimate the peak cooling rate due to this process as a function of gas pressure. For example, at a gas pressure of 3×10^{-5} Torr, corresponding to a neutral atomic density of $10^{12}\,cm^{-3}$, the peak cooling rate due to inelastic collisions will be around 2×10^6 eV/s. At an energy of 100 eV, the rate at which the Group 2 electrons transfer energy to the cooler Group 1 electrons through dynamical friction, the second term above, is usually somewhat larger. For example, in a plasma where the Group 1 electrons have a density of $5\times10^{11}\,cm^{-3}$, this cooling rate may exceed 7×10^6 eV/s. If the RF heating rate exceeds the cooling rate in this energy range, the net heating rate will remain positive and will in fact increase with increasing fast-electron energy until additional cooling or loss mechanisms become important.

By way of illustration, we consider a simple magnetic-mirror configuration for which the fundamental heating rate of electrons turning well beyond the resonance surface is given by an expression of the form

$$dW_\perp/dt = (e|E_\perp|^2/B_o)\mathfrak{G}(M, M_{res}, M_t), \tag{6.15}$$

with $\mathfrak{G}(M,M_{res},M_t) = \{(M-1)/[(M_t - M_{res})(M - M_{res})(M_{res} - 1)]\}^{1/2}\{(\pi/2)K^{-1}[(M_t - 1)/(M-1)]\}$. This function, specific to the simple magnetic mirror, describes the dependence of the heating rate on the electron turning point. As we saw in Chapter 5, $\mathfrak{G}(M, M_{res}, M_t)$ decreases by roughly a factor of 5 as M_t increases from values just greater than M_{res} to values just less than M. This pitch-angle dependence of the heating rate quantifies the degree to which electrons turning just beyond the resonance surface are heated more rapidly than electrons turning nearer the mirror throat. The microwave electric field strength required to heat electrons from Group 1 into Group 2 can be estimated for a given situation by evaluating the terms in the heating and cooling rates. The condition for the type of runaway envisioned here is the following:

$$\begin{aligned}(e|E_\perp|^2/B_o)\mathfrak{G}(M, M_{res}, M_t) &> n_o\sigma_{inelastic}v_e\Delta W_x \\ &+ 7\times10^{-6} n_e \ln\Lambda W_e^{-1/2}\,cm^3\,eV^{3/2}\,sec^{-1}.\end{aligned} \tag{6.16}$$

For a conservative estimate, we take the maximum value of the first term on the right-hand side and intermediate values of M_t for $\mathfrak{G}(M,M_{res},M_t)$. As a typical magnetic-mirror example, we choose $B_o = 3$ kG, $M = 2$, $M_{res} = 1.4$, $M_t = 1.7$, $n_o = 10^{12}\,cm^{-3}$, $W_e = 100$ eV, $\sigma_{inelastic}v_e = 4\times10^{-8}\,cm^3\,s^{-1}$, $\Delta W_x = 50$ eV, and $n_{e1} = 5\times10^{11}\,cm^{-3}$. We then find that if the RF electric field strength, $|E_\perp|$, is between 10 and 20 $V\,cm^{-1}$, the net heating rate will be positive and a significant fraction of the electrons will be heated from Group 1 into Group 2.

In addition to the drag on heated electrons from dynamical friction, Coulomb scattering also leads to pitch-angle diffusion. In a magnetic mirror, this diffusion will cause some of the heated electrons to reach the loss cone and escape from the mirror trap. This diffusion process has been described by Rose and Clark [11] in terms of the mean-square pitch-angle displacement after an elapsed time, t:

$$\langle\Delta\theta^2\rangle = (1/\tau_\theta)t, \tag{6.17}$$

where for an electron of energy W_e, we have

$$(1/\tau_\theta) = 3\times10^{-6}\, n_e \ln\Lambda W_e^{-3/2}\, eV^{3/2}\, cm^3\, sec^{-1}. \tag{6.18}$$

In an infinitesimal time interval, Δt, the pitch angle and the electron energy change by an amount given by

$$2\theta_i\Delta\theta = (1/\tau_\theta)\Delta t = (1/\tau_\theta)\Delta W_e/(dW_e/dt),$$

so that the fractional change in the electron pitch angle is related to the change in energy due to RF heating by

$$d\theta/\theta_i = [3\times10^{-6} n_e \ln\Lambda W_e^{-3/2} cm^3 eV^{3/2}\, s^{-1}][dW_e/(dW_e/dt)]/2\theta_i^2. \tag{6.19}$$

Since the heating rate, dW_e/dt, is approximately independent of energy, we can integrate the expression in Eq. (6.19) over the energy range spanned by Group 2 to obtain

$$(\theta_f-\theta_i)/\theta_i = (3\times10^{-6} n_e \ln\Lambda\, W_e^{-3/2} cm^3 eV^{3/2}\, s^{-1})/[\theta_i^2(dW_e/dt)] \times(W_{e,i}^{-1/2} - W_{e,f}^{-1/2}). \tag{6.20}$$

For heating rates large enough to offset the cooling mechanisms described above, this increase in the electron's pitch angle will be very small. Only those electrons with pitch angles very near the loss cone will be lost as the electrons are heated to the upper boundary of Group 2.

More generally, if the electron heating is regarded as a diffusion in energy, then there is an associated diffusion in the electron pitch angle, as measured, for example, by the mirror ratio at the electron turning point, $M_t = \varepsilon/\mu B_o$. As discussed in Chapter 5, M_t undergoes a change, ΔM_t, when the electron crosses the resonance surface:

$$\Delta M_t = (1-M_t/M_{res})(\Delta W_\perp/W_\perp),$$

where $W_\perp$ is the perpendicular energy evaluated at the midplane. Assuming that the electron gyrophase is randomized between successive transits of the resonance surfaces, we can derive a diffusion coefficient for M_t and show that it is proportional to the energy diffusion coefficient:

$$M_t dM_t/dt = (1-M_t/M_{res})^2 W_\perp^{-1} dW_\perp/dt,$$

so that

$$M_t dM_t/(1-M_t/M_{res})^2 = dW_\perp/W_\perp.$$

For notational convenience, we set $\eta = M_t/M_{res}$ and integrate to find the relation between the change in the turning point and the increase in electron energy:

$$\ln[(\eta_f - 1)/(\eta_i - 1)] + (\eta_i - 1)^{-1} - (\eta_f - 1)^{-1} = (1/M_{res}^2)\ln(W_{\perp f}/W_{\perp i}). \quad (6.21)$$

If we take the "final" value of M_t to be the mirror ratio, M, $\eta_f = M/M_{res}$, we can evaluate η_i as a function of the increase in electron energy, $W_{\perp f}/W_{\perp i}$. In this way, we can determine the width of a depletion zone that widens as the electron energy increases. Heated electrons within this depletion zone will diffuse into the loss cone and escape. As an illustrative example, we choose $M = 2$ and $M_{res} = 1.5$ and solve for $W_{\perp f}/W_{\perp i}$ as a function of η_i. At the upper limit of the electron energy that we have labeled Group 2, the depletion zone in this illustrative case extends from the loss cone, $\varepsilon/\mu B_o = M$, to an effective mirror ratio, $M_{effective} = M_{res}\eta_i = 1.825$. Effectively, the loss cone is gradually widened as the electrons gain energy.

6.4.2
Particle and Power Balance for Electrons in Group 3

In the energy range we have labeled Group 3, the dominant dynamical processes are RF heating at overtones of the relativistic electron gyrofrequency and energy loss by synchrotron radiation and Coulomb scattering. In addition, magnetic-mirror trapped electrons can be lost as a result of the breakdown of the invariance of the electron magnetic moment. The rate at which individual electrons loose energy by synchrotron radiation is [12]:

$$-dW/dt = [e^2/(6\pi\varepsilon_o c)](eB_o/m)(\gamma^2 - 1). \quad (6.22)$$

where m is the electron rest mass and γ is the relativistic factor, $\gamma = 1 + W/mc^2$. In typical cases, this radiative energy loss rate for individual electrons in Group 3 can increase from 10^4 eV s^{-1} at the lower energies to more than 10^5 eV s^{-1} at the higher energies.

We have made extensive use of the adiabatic motion of electrons and indicated that this invariance will fail if the magnetic intensity varies significantly over the electron gyration path. In fact, the magnetic moment of an energetic mirror-confined electron can undergo a substantial change when the electron passes through the local minimum in the magnetic intensity as it moves along the magnetic line of force. The resulting change in the electron pitch angle can lead to loss of the electron in a single time of flight through the system. The rate at which these changes in the magnetic moment occur has been analyzed by Cohen et al. [13]. According to their analysis, the fractional change in the magnetic moment each time an electron passes the magnetic minimum is given by the following expression

$$\Delta\mu/\mu \cong A\exp(-KL_{||}/\rho). \quad (6.23)$$

In this expression, A is a constant $3 \le A \le 5$, $L_{||}$ is obtained from a quadratic fit to the variation of the magnetic intensity in the neighborhood of the local minimum, and ρ is the electron gyroradius. The function K depends on the electron pitch angle and can be conveniently written in terms of the mirror ratio at the electron turning point, $M_t = \varepsilon/\mu B_o$:

$$K = (M_t/2)\{[(M_t - 1)/(2\sqrt{M_t})]\ln[(\sqrt{M_t}+1)/(\sqrt{M_t}-1)] - 1\}. \tag{6.24}$$

The fractional change in the mirror ratio at the turning point, M_t, is proportional to the fractional change in the magnetic moment: $\Delta M_t/M_t = -\Delta\mu/\mu$. A succession of random changes in the magnetic moment will therefore lead to diffusion of the turning points and we can roughly estimate the time, t_{na}, required for an electron's turning point to reach the loss cone after starting from the midplane:

$$t_{na} \approx (M-1)^2/D_{\eta\eta}.$$

The diffusion coefficient is given by

$$D_{\eta\eta} \cong \langle(\Delta M_t)^2\rangle\nu_b = \langle M_t^2(\Delta\mu/\mu)^2\rangle\nu_b,$$

and if we let $\nu_{na} = 1/t_{na}$, we have

$$\nu_{na} = [1/(M-1)^2]\langle Mt_t^2(\Delta\mu/\mu)^2\rangle\nu_b.$$

The average value of M_t^2 during the diffusion into the loss cone is $(M^2 - 1)/2$ and we obtain

$$\nu_{na} = (1/2)[(M+1)/(M-1)]\langle(\Delta\mu/\mu)^2\rangle\nu_b. \tag{6.25}$$

For the field of a simple magnetic mirror, $L_{||} = k_o(M - 1)^{1/2}/2$, where $k_o = 2\pi/L_c$, and the relativistic electron gyroradius is given approximately by $\rho = (mc/eB)(\gamma^2 - 1)^{1/2}$. We can substitute these simple magnetic-mirror results into the expression for ν_{na} to find an approximate value for the nonadiabatic scattering rate relative to the bounce frequency:

$$\nu_{na}/\nu_b = (1/2)[(M+1)/(M-1)]A^2\exp(-2KL_{||}/\rho). \tag{6.26}$$

6.5
Fokker–Planck Models of Hot-Electron Equilibria

In the earlier chapters, we have used two different kinetic equations to describe the electron dynamics that play a role in electron cyclotron heating. The first was a simple Langevin equation that described the response of individual "cold" electrons to RF electric and magnetic fields. This equation included electron collisions in a rudimentary way and provided a useful description of the propagation of electromagnetic waves in cold plasmas. We then used the (collisionless) linearized Vlasov equation to describe the response of a specified equilibrium electron distribution function to the electromagnetic fields of waves propagating in the plasma in order to calculate the damping of these waves. In view of the dynamical processes described in the earlier sections of this chapter, it is clear that we need a more comprehensive kinetic equation that can describe the equilibrium form of the electron distribution function resulting from the interplay of all of the competing dynamical processes: RF heating, Coulomb collisions, synchrotron radiation, and nonadiabatic changes in the magnetic moment of the electrons that result in scattering high-energy electrons into the loss cone of magnetic mirrors. Considerable progress has been made in developing a suitable kinetic equation based on the Fokker–Planck equation, in which a Fokker–Planck collision operator is added to the Boltzmann equation that formed the basis of the Vlasov model. This collision operator, originally developed for binary interactions obeying an inverse-square force law [14], describes the results of a large number of collisions, each of which produces only a very small change in the velocity of the particles. The results of these collisions can be described by a probability distribution function, $P(\mathbf{v},\Delta\mathbf{v})$, such that the particle distribution function, $f(\mathbf{r},\mathbf{v},t)$, evolves from preceding distribution functions according to

$$f(\mathbf{r},\mathbf{v},t) = \int f(\mathbf{r},\mathbf{v}-\Delta\mathbf{v},t-\Delta t)P(\mathbf{v}-\Delta\mathbf{v},\Delta\mathbf{v})d^3\Delta v, \tag{6.27}$$

where the probability distribution function is normalized to unity:

$$1 = \int P(\mathbf{v},\Delta\mathbf{v})d^3\Delta v.$$

Since we have assumed that the changes $\Delta\mathbf{v}$ are very small, we can expand the integrand in Eq. (6.27); and if only the lowest order terms are retained and the limit of $\Delta t \to 0$ is taken, we obtain the following formal expression for the Fokker–Planck collision operator:

$$\partial f/\partial t = -\nabla_v \cdot [\langle\Delta\mathbf{v}/\Delta t\rangle f(\mathbf{r},\mathbf{v},t)] + (1/2)\nabla_v\nabla_v : [\langle\Delta\mathbf{v}\Delta\mathbf{v}/\Delta t\rangle f(\mathbf{r},\mathbf{v},t)]. \tag{6.28}$$

The quantity in brackets in the first term on the right-hand side of this expression is called the dynamical friction:

$$\langle\Delta\mathbf{v}/\Delta t\rangle = \int(\Delta\mathbf{v}/\Delta t)P(\mathbf{v},\Delta\mathbf{v})d^3\Delta v. \tag{6.29}$$

The quantity in brackets in the second term on the right-hand side of this expression is called the diffusion tensor:

$$\langle\Delta\mathbf{v}\Delta\mathbf{v}/\Delta t\rangle = \int(\Delta\mathbf{v}\Delta\mathbf{v}/\Delta t)P(\mathbf{v},\Delta\mathbf{v})d^3\Delta v. \tag{6.30}$$

Explicit formulas for the dynamical friction and the diffusion tensor have been derived for plasmas that include the case of Coulomb scattering of fast electrons by a Maxwell–Boltzmann distribution of lower temperature electrons [15]. The dynamical friction force exerted on energetic electrons by a population of lower temperature electrons is given by

$$\begin{aligned}\mathbf{F}_{hc} &= -(1/4\pi)(e^2/\varepsilon_o)^2(2/m)n_c\ln\Lambda\mathbf{v}/v^3[\phi(y)-y\phi'(y)]\\ &= -(1/2)\nu_{Coul}m\mathbf{v}[\phi(y)-y\phi'(y)],\end{aligned} \tag{6.31}$$

where we have defined a Coulomb scattering rate, ν_{Coul}, using the Rutherford cross section, σ_{Coul}:

$$\nu_{Coul} = n\sigma_{Coul}v = n[(1/4\pi)(e^2/\varepsilon_o)^2\ln\Lambda/(m_r^2v^4)]v, \tag{6.32}$$

where m_r is the reduced mass and $\phi(y)$ is the error function of the argument $y = v/v_c$, where v_c is the average speed of the lower-temperature electrons. The rate at which the dynamical friction force transfers energy from the energetic electrons to the lower temperature electrons is then given by

$$dW/dt = -\mathbf{F}\cdot\mathbf{v} = -(1/2)mv^2 n_{Coul}[\phi(y) - y\phi'(y)]. \tag{6.33}$$

If this cooling rate is expressed in terms of a collision frequency, provided that the energetic electrons are much faster than the lower temperature electrons, we would have as an estimate for the rate at which energetic electrons loose energy to the cold electrons,

$$dW/dt = -\nu_{Coul} W. \tag{6.34}$$

The general problem addressed by the Fokker–Planck equation is to determine the distribution functions, $f_s(\mathbf{r}, \mathbf{v}, t)$, for each species, s, resulting from the operative dynamical processes and all sources and sinks of charged particles and energy. In seeking to apply this approach to plasmas heated by electron cyclotron power and confined in various magnetic configurations, many workers have simplified the complete description of the plasma by employing a succession of reductions. In almost all instances, the plasma ions have been treated as an immobile charge-neutralizing background, leaving only the electron dynamics to be considered. Then either steady-state or quasi steady-state solutions have been sought: $f_o(\mathbf{r}, \mathbf{v}, t) \Rightarrow f_o(\mathbf{r}, \mathbf{v}) + f_1$ where f_1 describes, for example, the high-frequency currents resulting from ECH. Since the unperturbed electron motions are generally assumed to be adiabatic, it is possible to reduce the six-dimensional representation in **r** and **v** to bounce-averaged dynamics of the electron guiding center on specified magnetic lines of force labeled by their Clebsch representation, (α,β). Then $f_o(\mathbf{r}, \mathbf{v}) \Rightarrow f_o(\alpha, \beta, \varepsilon, \mu)$. Bernstein and Baxter [16] have formulated a relativistic treatment of such a reduced Fokker–Planck equation that has been incorporated into sophisticated and powerful geometrical optics computer codes that are currently in widespread use in conjunction with large magnetic confinement facilities. For our present pedagogical purposes, we shall consider one further reduction of the theory using models that mimic in one dimension the important features of electron dynamics in the two-dimensional (ε,μ) space [17].

We use a one-dimensional, steady-state Fokker–Planck equation to represent in an approximate way the equilibrium resulting from the competing dynamical processes that govern the formation of relatively low-density, hot-electron plasma components (Groups 2 and 3) that can result when plasmas are suitably illuminated by ECH power:

$$\partial f(u,t)/\partial t = -\partial[\langle \Delta u/\Delta t\rangle f(u,t)]/\partial u + (1/2)\partial^2[\langle \Delta u \Delta u/\Delta t\rangle f(u,t)]/\partial u^2 = 0. \tag{6.35}$$

Because most of the laboratory experience with the hot-electron plasmas we seek to understand has been in magnetic-mirror configurations, we include a scattering process that mimics nonadiabatic changes in the electron pitch angle leading to loss of the high-energy electrons by scattering into the loss cone. This process is generally negligible in large toroidal magnetic configurations. The resulting Fokker–Planck equation can be solved by quadrature and the solution and its moments can be numerically evaluated to exhibit the outcomes of the competition between the four dynamical processes acting on the higher-energy electrons that we have labeled Groups 2 and 3. These consist of the following four processes:

The Coulomb scattering of relativistic electrons is represented by an effective scattering rate, ν_{cc}, given by

$$\nu_{cc} = 4[(1/4\pi)(e^2/\varepsilon_o)^2 n_c \ln\Lambda/(m^2u^3)]\gamma^2[\phi(s)-s\phi'(s)], \tag{6.36}$$

where n_c and T_c are the density and temperature of the lower-temperature electrons, respectively, the parameter $s = u/u_c$, and $u_c = (2T_c/m)^{1/2}$, where u is the magnitude of the momentum per unit rest mass, and $\phi(s)$ is the error function.

Following Bernstein and Baxter [16], the synchrotron radiation is represented in the one-dimensional model by a dynamical friction term with an effective collision rate, ν_{sr}, given by

$$\nu_{sr} = (\gamma/6\pi)(e^2/\varepsilon_0)\Omega^2/mc^3. \tag{6.37}$$

The nonadiabatic scattering is also treated as a dynamical friction and its rate is given as before by

$$\nu_{na} = \nu_b(1/2)[(M+1)/(M-1)]A^2\exp(-2\,KL_{\|}/\rho). \tag{6.38}$$

The RF heating rate will depend on the amplitude and polarization of the electromagnetic fields in ways that we will consider later in connection with particular experiments. For the present discussion, we simply represent it as a diffusion term with an effective rate, ν_{rf}. The one-dimensional steady-state Fokker–Planck equation then takes the form

$$(\nu_{cc}+\nu_{na}+\nu_{sr})uf + u_c^2(\gamma\nu_{cc}/2+\nu_{rf})\partial f/\partial u = 0, \tag{6.39}$$

and the solution is given by integrating from zero to s to obtain

$$f(s) = C\exp\left\{-\int 2s'ds'[(\nu_{cc}+\nu_{na}+\nu_{sr})/(\gamma\nu_{cc}+2\nu_{rf})]\right\}. \tag{6.40}$$

The normalization constant, C, is evaluated in terms of the specified background electron density, n_c:

$$n_c = 4\pi C u_c^3 \int s^2 f(s)ds.$$

The integral is taken from zero to an upper limit that is large enough to ensure that C is approximately independent of this limit. We shall return to this model in late chapters when it will be applied to interpret several different experiments.

6.6
Ad Hoc Velocity–Space Models of Anisotropic Hot-Electron Equilibria

An important aspect of the hot-electron population that is expected generally in ECH experiments has to do with anisotropy in the hot-electron velocities perpendicular and parallel to the static magnetic field. As we discussed earlier, because the resonant ECH interaction changes primarily the perpendicular velocity of the heated electron, the magnetic intensity at the electron turning point, B_t, changes at resonance according to

$$\Delta B_t = (1 - B_t/B_{res})\Delta W_\perp/\mu.$$

The turning point of an electron that gained energy as it passed through resonance therefore moves closer to the resonance surface on its next encounter with the resonance surface. As we have seen earlier, the electron turning points will undergo a diffusive process except at the resonance surface where $\Delta B_t = 0$. For a population of electrons that have undergone many such encounters with the resonance surfaces, we expect the average parallel kinetic energy to be less than the average perpendicular kinetic energy by an amount that can be estimated as follows:

$$\begin{aligned}\langle W_\parallel\rangle/\langle W_\perp\rangle &\sim \langle(\varepsilon - \mu B_o)/\mu B_o\rangle \le \langle B_t/B_o - 1\rangle \\ &\le B_{res}/B_o - 1.\end{aligned} \tag{6.41}$$

In simple magnetic-mirror configurations like those discussed earlier, the resonance surfaces can be chosen to be open hyperboloids for which the ratio B_{res}/B_o increases with radial distance from the axis of symmetry. Temperature anisotropy can generally be controlled in these configurations by choosing the location of the resonance surfaces so that on the axis of symmetry B_{res}/B_o exceeds any recognized critical values. Critical values of anisotropy for equilibrium, for example, have been derived under very general assumptions using guiding-center fluid models of the plasma [17]. In addition, the microscopic character of the hot-electron equilibrium, specifically the anisotropy in velocity space, is critical for determining the stability with respect to high-frequency modes to be discussed in Chapter 7.

A useful class of distribution functions frequently employed to describe energetic plasmas with local properties similar to those of plasmas confined in magnetic mirrors has the following form [19]:

$$\begin{aligned}f_0(v_\perp v_\parallel) &= \sum C_m \left(v_\perp^2/\alpha_\perp^2\right)^m \exp[-(v_\perp/\alpha_\perp)^2 - (v_\parallel/\alpha_\parallel)^2], \quad \text{outside the loss cone} \\ &= 0, \text{inside the loss cone.}\end{aligned} \tag{6.42}$$

Recall that the location of the loss cone in velocity space was given earlier in terms of the mirror ratio, M: $v_\parallel^2/v_\perp^2]_{z=0} = M - 1$ on the loss cone. The $m = 0$ term of Eq. (6.42) describes a bi-Gaussian distribution with an anisotropic temperature specified by the parameter $\eta = (\alpha_\perp/\alpha_\parallel)^2$. The $m = 1$ term mimics at least qualitatively the effects of the loss cone in velocity space inherent in magnetic-mirror confinement. In Chapter 7, we will use these distribution functions to investigate instabilities associated with temperature anisotropies and the loss-cone nature of plasma confinement in magnetic mirrors. At this point, we examine some of the equilibrium plasma properties that are described by these distribution functions. We start by showing how these local distribution functions can be generalized to describe the [illegible]iation of plasma properties along a magnetic line of force by expressing the [illegible]bution function in terms of constants of the motion, namely, the particle energy and magnetic moment, ε and μ:

$$\mu = mv_\perp^2(0)/2B(0) = mv_\perp^2(z)/2B(z),$$

so that

$$v_\perp^2(0) = v_\perp^2(z)B(0)/B(z) = v_\perp^2(z)/b(z), \tag{6.43}$$

where $b(z) \equiv B(z)/B(0)$. Similarly, in the absence of any electrostatic potentials,

$$\varepsilon = mv_\perp^2(0)/2 + mv_\parallel^2(0)/2 = mv_\perp^2(z)/2 + mv_\parallel^2(z)/2,$$

whence

$$v_\parallel^2(0) = v_\perp^2(z) + v_\parallel^2(z) - v_\perp^2(0) = v_\perp^2(z) + v_\parallel^2(z) - v_\perp^2(z)/b,$$

and so

$$v_\parallel^2(0) = v_\parallel^2(z) + v_\perp^2(z)(1-1/b). \tag{6.44}$$

In this way, we can relate the velocity-space variables on the midplane, $z = 0$, to the corresponding values at arbitrary points along the same magnetic line of force. Our distribution function becomes

$$\begin{aligned} f_0(v_\perp, v_\parallel, z) &= \sum C_m \left(v_\perp^2/b\alpha_\perp^2\right)^m \\ &\quad \times \exp\left[-(v_\parallel/\alpha_\parallel)^2 - (b-1)\left(v_\perp^2/b\alpha_\parallel^2\right) - v_\perp^2/b\alpha_\perp^2\right], \\ &\quad \text{outside the loss cone} \\ &= 0, \text{ inside the loss cone.} \end{aligned} \tag{6.45}$$

Note that the loss cone itself is dependent on the position along the magnetic line of force. Since the loss cone was defined by the condition that the parallel velocity vanish at the mirror throat, $z = L$, we have

$$\begin{aligned} v_\parallel^2(L) = 0 &= v^2 - 2\mu B_{max}/[mB(z)] \\ &= v_\parallel^2(z) + v_\perp^2(z) - v_\perp^2(z)B_{max}/B(z) \end{aligned}$$

so that on the loss cone,

$$v_\parallel^2(z)/v_\perp^2(z) = B_{max}/B(z) - 1 = M/b - 1. \tag{6.46}$$

To obtain a more compact form for the z-dependent distribution function, we define a z-dependent parameter $\alpha_\perp^2(z)$ as follows:

$$\text{If we let } \; 1/\alpha_\perp^2(z) = 1/b\alpha_\perp^2 + (b-1)/b\alpha_\parallel^2, \quad \text{then}$$

$$\begin{aligned} f_0(v_\perp, v_\parallel, z) &= \sum C_m \left(v_\perp^2/b\alpha_\perp^2\right)^m \exp\{-[v_\perp/\alpha_\perp(z)]^2 - (v_\parallel/\alpha_\parallel)^2\}, \\ &\qquad v_\perp^2 > v_\parallel^2/(M/b-1) \\ &= 0, \qquad \text{for} \quad v_\perp^2 < v_\parallel^2/(M/b-1). \end{aligned} \tag{6.47}$$

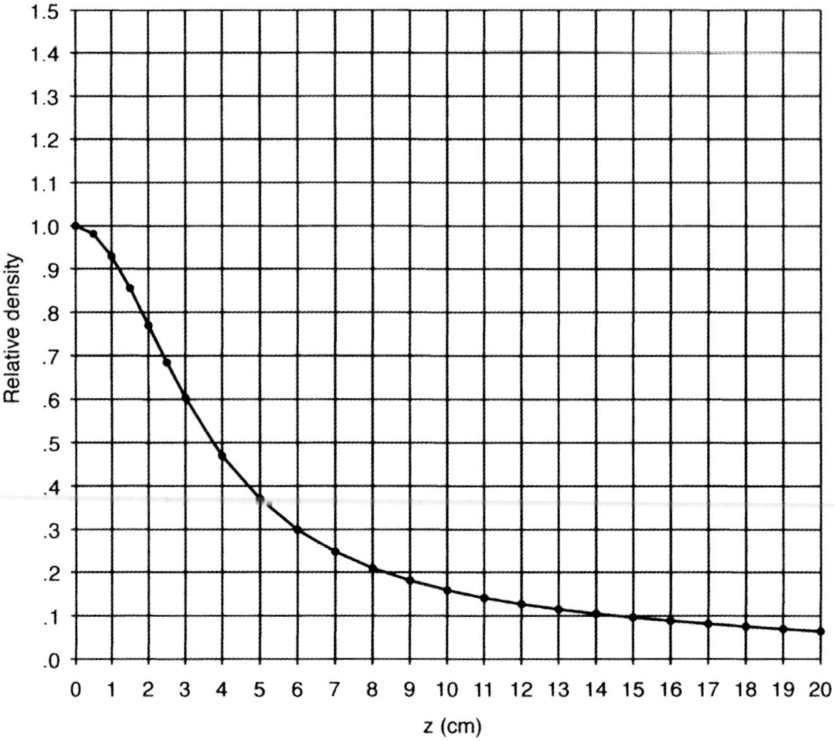

Figure 6.3 An example of the axial dependence of electron density calculated using the bi-Maxwellian *ad hoc* equilibrium distribution function with anisotropy parameter $\eta = 5$.

As an example, we calculate the density as a function of distance along the magnetic line of force for a simple bi-Gaussian distribution function, $m = 0$:

$$n(z) = \pi C_o \int dv_\| \exp[-(v_\|/\alpha_\|)^2] \int dv_\perp^2 \exp\{-[v_\perp/\alpha_\perp(z)]^2\},$$

where the integral over $v_\|$ is from $-\infty$ to $+\infty$, while the integral over $v_\perp^2$ is from $v_\|^2/(M/b-1)$ to $+\infty$. The result is

$$\begin{aligned} n(z) &= \pi^{3/2} C_o \alpha_\| \alpha_\perp(z)^2 \{1 + [\alpha_\|^2/\alpha_\perp(z)^2] b(z)/[M-b(z)]\}^{-1/2} \\ &= \pi^{3/2} C_o \alpha_\|^3 [(M-1+1/\eta)/(M-b)]^{-1/2} \eta b/[1+\eta(b-1)] \end{aligned}$$

Relative to the density on the midplane, the z-dependent density is given by

$$n(z)/n(0) = \{b/[1+\eta(b-1)]\}[(M-b)/(M-1)]^{1/2}. \tag{6.48}$$

Figure 6.3 displays an example in which $\eta = 5$, showing the concentration of the plasma near the midplane.

References

1 L. Spitzer, Jr., *Physics of Fully Ionized Gases* (2nd edition), Interscience Publishers, New York, pp. 21–23 (1967).

2 C.L. Longmire, *Elementary Plasma Physics*, Interscience Publishers, New York, Chapter 10 (1967).

3 F.F. Chen, *Introduction to Plasma Physics*, Plenum Press, New York, pp. 83–84 (1974).

4 C.F. Barnett, *A Physicist's Desk Reference* (2nd edition), H.L. Anderson, Editor in Chief, American Institute of Physics, New York (1989) Section 4.00. See also R.L. Freeman and E.M. Jones, Report No. CLM-R137, Culham Laboratory, 1974.

5 *Handbook of Mathematical Functions*, Milton Abramowitz and Irene A. Stegun, eds., National Bureau of Standards Applied Mathematics Series 55, Washington, D.C. (1964), p. 227 ff especially Eq. (5.1.4) (1964).

6 R.A. Dandl and G.E. Guest, "On the low-pressure mode transition in electron cyclotron heated plasmas," *J. Vac. Sci. Technol.* **A9**(6), pp. 3119–3125 (1991).

7 J.T. Gudmundsson,Report RH-21-2002, Science Institute, University of Iceland, Reykjavik, Iceland.

8 K.S. Gulyaev, A.V. Kasheev, A.S. Kovalev, N.V. Suetin, and A.N. Vasilieva, *J. Phys. D: Appl. Phys.* **27**, 2349 (1994).

9 R.A. Dandl, H.O. Eason, P.H. Edmonds, and A.C. England, *Nucl. Fusion* **11**, 411 (1971).

10 D.L. Book, *Revised and Enlarged Collection of Plasma Physics Formulas and Data*, Naval Research Laboratory Memorandum Report 3332, p. 53 ff (1977).

11 D.J. Rose and M. Clark, Jr., *Plasmas and Controlled Fusion*, MIT Press and John Wiley & Sons, New York, p. 165 ff (1961).

12 G. Bekefi and A.H. Barrett, *Electromagnetic Vibrations, Waves, and Radiation*, MIT Press, Cambridge, MA, p. 298 (1977).

13 R.H. Cohen, G. Rowlands, and J.H. Foote, *Phys. Fluids* **21**, 627 (1978).

14 S. Chandrasekhar, *Rev. Mod. Phys.* **15**, 1 (1943).

15 M.N. Rosenbluth, W.M. MacDonald, and D. Judd, *Phys. Rev.* **107**, 1 (1957).

16 I.B. Bernstein and D.C. Baxter, *Phys. Fluids* **24**, 108 (1981).

17 G.E. Guest, R.L. Miller, and C.S. Chang, *Nucl. Fusion* **27**, 1245 (1987).

18 M.D. Kruskal and C.R. Oberman, *Phys. Fluids* **1**, 275 (1958); J.B. Taylor and R.J. Hastie, *Phys. Fluids* **8**, 323 (1965); H. Grad, *Phys. Fluids* **9**, 225 (1966); for additional references. see G.E. Guest and C.L. Hedrick, "Elements of the Guiding-Center Plasma," ORNL-TM-3943, September 1972.

19 G.E. Guest and R.A. Dory, *Phys. Fluids* **8**, 1853 (1965).

Exercises

6.1 *Calculate the ionization rate constant, $\langle\sigma_{ioniz}v_e\rangle$, for atomic hydrogen using the formulas summarized by Barnett [4].*

6.2 *Using the result from Exercise 1, calculate the electron temperature as a function of the product $p_o L_{esc}$ in a hydrogen plasma.*

6.3 *Show that if the particle balance condition is satisfied, the electron density is related to the microwave power density by*

$$n_e/\mathcal{P}_\mu L_{esc} = (M_i/2T_e)^{1/2}(E_{ion} + 3T_e\tau_e/2\tau_E)^{-1}.$$

If the ionization energy $E_{ion} \approx 40$ eV for $T_e = 10$ eV, estimate the electron density sustained by $\mathcal{P}_\mu = 1\ \mathrm{W/cm^3}$ in a plasma whose characteristic length is 10 cm.

6.4 *For the parameters listed following Eq. (6.16), determine the maximum cooling rate experienced by Group 1 electrons. Let the heating rate exceed this by 20% and starting from $\theta_i = \pi/2$, determine $\theta_f(E) - \theta_i$ for this heating rate for Group 2 electrons.*

6.5 *Consider a 100 keV electron whose initial pitch angle is around 90°. If this electron is confined in a magnetic mirror with a mirror ratio of $M = 2$, calculate the product $n\tau$, where τ is the time required for the electron to be scattered into the mirror loss cone.*

7
Stability

Electron cyclotron heated plasmas have been observed to support various types of plasma instabilities (as will be described in Chapter 8), which can adversely affect the properties of the plasma. Deleterious instabilities occur not only in hot-electron plasmas confined in simple magnetic-mirror configurations but also in overdense plasmas where the bulk electron temperature is only a few electron volts. Nonetheless it is possible to stabilize virtually all modes of instability in many electron cyclotron heated (ECH) plasmas, including, most notably, relativistic-electron plasmas with very high energy densities. In what follows, we discuss the dominant stability properties of ECH plasmas by examining several particularly relevant modes of instability and attempting to identify potentially operative stabilizing mechanism for each of these.

7.1
Interchange Instabilities

ECH plasmas confined in simple magnetic-mirror configurations are generally predicted to be susceptible to a class of macroscopic plasma instabilities linked to the radial gradient in the magnetic intensity. Specifically, instability is expected to result if the radial gradient of plasma pressure is parallel to the radial gradient of magnetic intensity – which is typically the case. The unstable perturbations in the plasma density extend along the magnetic field like flutes on a column (thus the name "flute-like") and cause rapid radial transport of plasma. They are often called interchange instabilities because they effectively interchange regions of higher plasma density with adjacent regions of lower density. Their existence is predicted by a wide range of plasma models and confirmed in numerous experiments [1].

Kadomtsev [2] gave a general stability criterion for interchange modes in an ideal, scalar-pressure plasma confined in a toroidal magnetic trap in which the magnetic lines of force close upon themselves. Although this is not the situation that characterizes open-ended magnetic mirrors, it is exactly the case for plasmas confined in a bumpy torus, which we will discuss in Chapter 10. Underlying Kadomtsev's stability criterion is an energy principle; namely, if there are physically permissible

Electron Cyclotron Heating of Plasmas. Gareth Guest

ISBN: 978-3-527-40916-7

perturbations that can lower the plasma's energy, these perturbations can grow in time without limit. To assess the energetics of interchange perturbations, we imagine a virtual interchange of plasma and the associated magnetic flux between two neighboring infinitesimal flux tubes. If the total energy of the system is decreased by this interchange, the plasma can be unstable to such perturbations. The change in the energy resulting from the interchange can be evaluated as follows.

We label the two neighboring flux tubes as 1 and 2 and stipulate that no heat enters or leaves the flux tubes during the interchange. The plasma then follows a simple equation of state: $pV^{\gamma} = \text{constant}$, where V is the volume of the flux tube and γ is the usual ratio of specific heats. The change in energy resulting from interchanging the two flux tubes is given by

$$\begin{aligned}\delta W = \delta W_p + \delta W_{mag} &= \int \delta^3 r[(p_{1f} - p_{1i})/(\gamma - 1) + (B_{1f}^2 - B_{1i}^2)/(2\mu_o)] \\ &+ \int d^3 r[(p_{2f} - p_{2i})/(\gamma - 1) + (B_{2f}^2 - B_{2i}^2)/(2\mu_o)].\end{aligned} \tag{7.1}$$

The interchange of the two flux tubes means that $p_{2f}V_2^{\gamma} = p_{1i}V_1^{\gamma}$ and $p_{1f}V_1^{\gamma} = p_{2i}V_2^{\gamma}$ We set $p = p_{1i}$ and $V = V_1$, and since the flux tubes are neighboring we can express the parameters in the neighboring flux tubes as $p_{2i} = p + \delta p$ and $V_2 = V + \delta V$. The change in energy associated with the plasma itself is then given approximately for this interchange by

$$(\gamma - 1)\delta W_p = p_{1f}V_1 + p_{2f}V_2 - p_{1i}V_1 - p_{2i}V_2,$$

which becomes, after a modest bit of algebra,

$$\delta W_p = \delta p \delta V + (\gamma p/V)(\delta V)^2.$$

The volume of a flux tube is given in terms of the enclosed magnetic flux, ψ, by

$$V = \int d^2 r ds = \int d^2 r B\, ds/B = \psi \int ds/B = \psi U.$$

The quantity $U = V/\psi$ is often called the specific volume of the flux tube. In a bumpy torus, just as in a simple magnetic mirror, U increases monotonically with distance from the axis: $\partial U/\partial r \geq 0$. Near the surface of any magnetically confined plasma where the plasma pressure is small, it is evidently necessary for the pressure to decrease with increasing radius: $\partial p/\partial r \leq 0$. Then

$$\delta W_p = [(\partial p/\partial r)(\partial U/\partial r) + (\gamma p/U)(\partial U/\partial r)^2](\delta r)^2 \psi \tag{7.2}$$

becomes negative, indicating that the plasma can spontaneously decrease its energy by interchanging the plasma in the two neighboring flux tubes near the surface where the second term in brackets is less than the magnitude of the (negative) first term in the brackets. The magnetic energy density will already be at the lowest possible value unless the kinetic pressure in the plasma is a significant fraction of the energy density in the magnetostatic field. Even when the finite plasma pressure is taken into account, the qualitative conclusion is not changed.

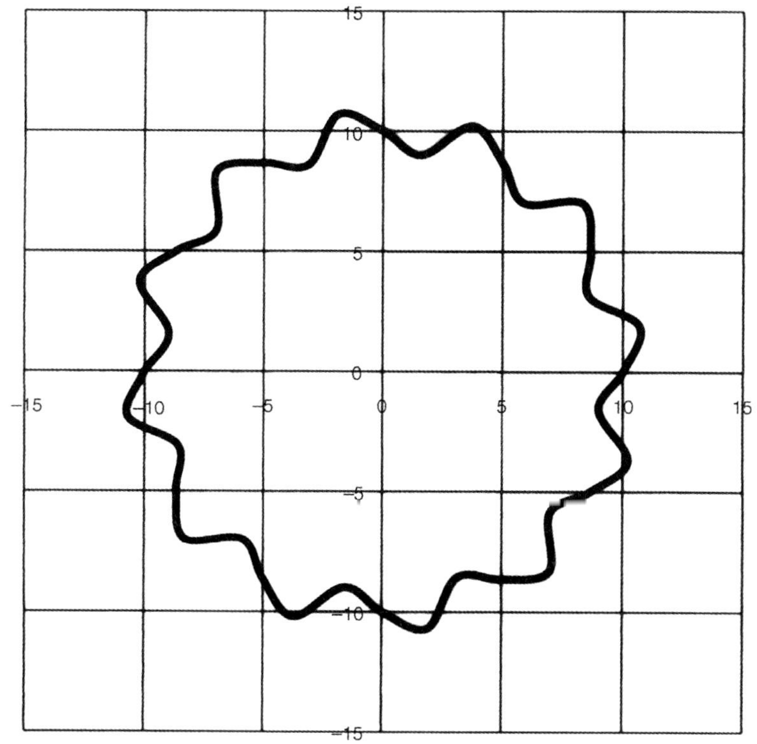

Figure 7.1 An artist's conception of the cross section of a plasma column whose surface is distorted by an $\ell = 12$ flute mode. The magnetic field is directed out of the page and electrons drift in the counterclockwise direction while ions drift in the clockwise direction.

The basic destabilizing mechanism can readily be visualized in terms of drift motions of plasma ions and electrons, as illustrated very schematically in Figure 7.1.

Here the magnetic field is directed out of the plane of the drawing and the plasma initially had a sharp surface boundary at $r = r_S$. The plasma density and the magnetic intensity are both higher for $r < r_S$ and lower for $r > r_S$. Because of the gradient in the magnetic intensity, electrons will drift in the counterclockwise direction while ions will drift in the clockwise direction. If there is an initial localized perturbation in the plasma density, it will spread along the magnetic lines of force on the bounce time scale. On the slower drift time scale, the perturbation will become electrically polarized in the azimuthal direction because of the oppositely directed electron and ion drifts. The resulting azimuthal electric field will induce a radially outward $\mathbf{E} \times \mathbf{B}$ plasma flow that increases the density of the perturbation, as indicated in Figure 7.1. The perturbation will grow exponentially, since the instantaneous rate of growth is proportional to the perturbed density. This growth will continue until the radial gradient in the density vanishes.

For a more quantitative microscopic view of this picture, we recall from Chapter 3 that the bounce-averaged drift velocities of the ions and electrons were given by

$$r\Omega_p = [2(\varepsilon - q\phi) - \mu B]/(qB\langle R_c\rangle),$$

where the bounce-averaged radius of curvature in the simple magnetic mirror is

$$\langle R_c\rangle = R_{co}(r)[1 + 2(\varepsilon - q\phi - \mu B)/\mu B]K(\zeta^2)/[2E(\zeta^2) - K(\zeta^2)],$$

and $\zeta^2 = [2(\varepsilon - q\phi - \mu B)/\mu B]/(M - 1)$. For the moment, we will neglect the dependence of Ω_p on the energy and pitch angle of the particles and focus on the response of the plasma to electrostatic waves propagating in the azimuthal direction and localized in radius at the surface of the plasma but extending indefinitely along the magnetic lines of force. We can evaluate the perturbed velocity of each species using the Lorentz force, $M d\mathbf{v}/dt = q(\mathbf{E} + \mathbf{v} \times \mathbf{B})$. All wave properties will be assumed to vary in azimuth and time as $\exp(i\ell\theta - i\omega t)$ so that the dynamical equation for each species becomes

$$-i\omega'\mathbf{v} = (q_s/M_s)\mathbf{E} + \Omega_s \mathbf{v} \times \mathbf{b},$$

where $\omega' = \omega - \ell\Omega_p$ is the Doppler-shifted frequency and Ω_s is the gyrofrequency of species s. The components of $\mathbf{v}$ perpendicular to the magnetic field are then given by

$$-(i\omega'/\Omega_s)\mathbf{b} \times \mathbf{v} = \mathbf{b} \times \mathbf{E}/B + \mathbf{v} - (\mathbf{b} \cdot \mathbf{v})\mathbf{b}.$$

But since the dynamical equation requires that $-(i\omega'/\Omega_s)\mathbf{v} = \mathbf{E}/B + \mathbf{v} \times \mathbf{b}$, we obtain for the perturbed velocity

$$[1 - (\omega'/\Omega_s)^2]\mathbf{v} - (\mathbf{b} \cdot \mathbf{v})\mathbf{b} = \mathbf{E} \times \mathbf{b}/B - (i\omega'/\Omega_s)\mathbf{E}/B.$$

Because the drift frequencies are several orders of magnitude smaller than the gyrofrequencies, we can neglect $(\omega'/\Omega_s)^2 \ll 1$ and the perpendicular components of the perturbed velocity are then given approximately by

$$\mathbf{v}_\perp = \mathbf{E} \times \mathbf{b}/B - (i\omega'/\Omega_s)\mathbf{E}/B. \tag{7.3}$$

The electric fields of these electrostatic waves are given in terms of their potential by

$$\mathbf{E} = -\nabla\phi = -i(\ell/r)\phi\mathbf{u}_\theta.$$

We now use the first-order continuity equation to determine the perturbed density of each species associated with this wave:

$$\partial n/\partial t + \nabla \cdot (N\mathbf{v} + n\mathbf{v}_o) = 0,$$

where N(r) is the equilibrium plasma density and $\mathbf{v}_o = r\Omega_p\mathbf{u}_\theta$. Thus,

$$-i\omega n + \nabla N \cdot \mathbf{v} + N\nabla \cdot \mathbf{v} + \nabla n \cdot \mathbf{v}_o = 0, \quad \text{or}$$
$$-i\omega n + (dN/dr)v_r + N(i\ell v_\theta/r) + i\ell\Omega_p n = 0.$$

Substituting for the components of the perturbed velocity from Eq. (7.3) we obtain

$$n = -[(1/\omega')(dN/dr)(\ell/r) + (\ell/r)^2(N/\Omega)](\phi/B). \tag{7.4}$$

The potential, ϕ, is the solution to Poisson's equation with the charge densities given by the sum over both species of the perturbed density:

$$\varepsilon_o \nabla \cdot \mathbf{E} = \varepsilon_o \nabla \cdot (-\nabla\phi) = \varepsilon_o (\ell/r)^2 \phi = \rho = e(n_i - n_e)$$
$$\varepsilon_o (l/r)^2 \phi = e(n_i - n_e) = e\{-[(1/\omega'_i)(dN/dr)(\ell/r) + (\ell/r)^2 (N/\Omega_i)] + [(1/\omega'_e)(dN/dr)(\ell/r) + (\ell/r)^2 (N/\Omega_e)]\}(\phi/B).$$

The resulting dispersion relation is

$$-\Omega_i (r/\ell)(N^{-1} dN/dr)[(1/\omega'_i) - (1/\omega'_e)] = 1 + \varepsilon_o B^2/(M_i N) = 1 + (c_A/c)^2, \tag{7.5}$$

where c_A is the Alfven speed and $(c_A/c)^2 \ll 1$ for reasonably dense plasmas.

We can illustrate the implications of this dispersion relation conveniently in an isothermal plasma, $T_i = T_e$ so that $-\Omega_{pi} = \Omega_{pe}$ and the dispersion relation becomes $[(1/\omega'_i) - (1/\omega'_e)] = -2\Omega_{pe}/(\omega^2 - \Omega_{pe}{}^2)$. If we define a scale length characterizing the density gradient, $L_N{}^{-1} \equiv -(N^{-1} dN/dr) > 0$, then

$$\omega^2 = -2\Omega_i \Omega_{pe} r/(\ell L_N) + \Omega_{pe}^2 \approx -2\Omega_i \Omega_{pe} r/(\ell L_N)$$

and for this isothermal plasma the flute instability will have a growth rate given by

$$\gamma = [2\Omega_i \Omega_{pe} r/(\ell L_N)]^{1/2}. \tag{7.6}$$

The tendency of the perturbed electric charge to disperse by flowing along the magnetic field to conducting metallic endplates provides an important stabilizing mechanism. This mechanism is particularly effective if the plasma contains a significant population of cold electrons confined in an electrostatic ambipolar potential. Since hot-electron ECH plasmas confined in simple magnetic-mirror configurations generally contain a substantial cold-electron component, flute-like interchange modes usually occur in these plasmas only when the ambient gas pressure is very near the critical value discussed in Chapter 6. A convenient stability criterion incorporating the plasma connection with conducting end plates is [3]

$$\beta \le (4R_p R_c/L_h L_c)\left[\omega_{pe,\,cold}^2 / \left(k_\perp^2 c^2 + \omega_{pe,\,cold}^2\right)\right]. \tag{7.7}$$

Here β is the ratio of the plasma kinetic pressure, nkT, to the magnetostatic pressure of the static magnetic field:

$$\beta = nkT/(B^2/2\mu_o),$$

where k is the Boltzmann constant. R_p and R_c are lengths that characterize the radial gradients of plasma pressure and the static magnetic field, respectively:

$$R_p^{-1} = -d\ln(nkT)/dr \quad \text{and} \quad R_c^{-1} = -d\ln B/dr.$$

L_h is the axial length of the magnetic-mirror confined hot plasma; and L_c is the length of the cold-plasma region separating the hot plasma from the conducting end walls of the vacuum chamber. $\omega_{pe,\,cold}$ is the cold-electron plasma frequency and measures the cold-plasma density. As we have seen, the perpendicular propagation vector for the interchange mode is given by $k_\perp = \ell/r_s$, where

$\ell = 2, 3, 4, \ldots$, and r_s is the radius of the plasma surface. c is the speed of light in vacuum. At the boundary separating stable from unstable conditions, the cold-electron density is given by

$$\begin{aligned} n_{cold} &= (m_e/\mu_o e^2)(\ell/r_s)^2[(4R_pR_c/\beta L_h L_c)-1]^{-1} \\ &= 2.8 \times 10^{11}\ \mathrm{cm}^{-1}(\ell/r_s)^2[(4R_pR_c/\beta L_h L_c)-1]^{-1}. \end{aligned} \tag{7.8}$$

As β approaches $4R_pR_c/L_hL_c$ the critical cold-plasma density becomes infinite; i.e., no amount of cold plasma can stabilize the interchange modes above this limiting value of beta. But for lower values of beta interchange modes are predicted to be stabilized by comparatively low densities of cold plasma. For example, if $\beta = 2R_pR_c/L_hL_c$, in a plasma that is 10 cm in radius, all modes up to $\ell = 6$ are stabilized by a cold plasma whose density is 10^{11} cm^{-3}, and all modes up to $\ell = 20$ are stabilized if the cold-plasma density is 10^{12} cm^{-3}. We shall examine more of the implications of cold-plasma stabilization of interchange modes in connection with particular experiments.

7.2 Electrostatic Velocity–Space Instabilities Driven by Wave-Particle Interactions

Hot-electron plasmas generated by electron cyclotron heating can support unstable plasma waves at frequencies around the electron gyrofrequency if energy stored in the hot-electron population can be transferred to the waves; i.e., the reverse of the heating process. In addition to the electromagnetic waves discussed in Chapter 4, ECH plasmas also exhibit two distinct longitudinal electrostatic waves that can grow by exchanging energy with the hot electrons. The energy available to drive such unstable growth is the free energy associated with the non-Maxwellian nature of the anisotropic hot-electron equilibria confined in magnetic mirror configurations.

As an introduction to the discussion of electrostatic instabilities, we first consider an idealized one-dimensional example. The unperturbed particle orbits are then simply $z(t) = z(0) + vt$, and in what follows we choose $z(0) = 0$. The force exerted on the particles by the electric field of the electrostatic wave under consideration is

$$F = qE\exp[i(kz-\omega t)] = qE\exp[i(kv-\omega)t].$$

The perturbed distribution function describing these particles at the time $t = 0$ can be obtained from the Vlasov equation by integrating along the unperturbed orbit from remote earlier times to the time of observation:

$$\begin{aligned} f_1(0) &= -(q/m)E\,\partial f_o/\partial v \int dt\,\exp[i(kv-\omega)t] \\ &= i(q/m)E\,\partial f_o/\partial v(1/k)(v-\omega/k)^{-1}. \end{aligned}$$

The current associated with this perturbation is given by

$$j(0) = q\int dv\,v f_1(0) = i(q^2/m)E(\omega/k^2)\int dv\,\partial f_o/\partial v(v-\omega/k)^{-1},$$

where the integral is from $-\infty$ to $+\infty$ and f_o vanishes at both limits. Upon integrating once by parts, we obtain

$$j = i(q^2/m)E(\omega/k^2)\int dv f_o(v)(v-\omega/k)^{-2} = \sigma E$$

so that the dielectric constant, $\kappa = 1 + i\sigma/\omega\varepsilon_o$ is given by

$$\kappa = 1-(q^2/m\varepsilon_o)(1/k^2)\int dv f_o(v)(v-\omega/k)^{-2}.$$

If we now specialize to a mono-energetic stream distribution function, $f_o(v) = N\delta(v-u_o)$, the dielectric constant takes the form

$$\kappa = 1-(\omega_p^2)/(\omega-ku_o)^2,$$

and the dispersion relation, $\kappa = 0$, has two solutions:

$$\omega/k = u_o \pm \omega_p/k. \tag{7.9}$$

The fluctuating densities and velocities associated with these two waves must satisfy the continuity condition, $\partial N/\partial t + \nabla\cdot(N\mathbf{v}) = 0$. For the present case this becomes

$$(\omega-ku_o)n_1 = kN_o v_1 \quad \text{or}$$
$$n_1/N_o = \pm kv_1/\omega_p.$$

For the fast wave (with phase velocity greater than u_o), corresponding to the upper sign in Eq. (7.9), the density and velocity fluctuations are in phase. Thus, where the density is greatest the velocity is also at its maximum. The wave has, therefore, increased the energy density of the plasma; it could only be excited if positive work were done on the plasma, and it is, therefore, denoted a "positive-energy wave." For the slow wave (lower sign), the velocity and density fluctuations are 180° out of phase. Thus, where the density is greatest the velocity is at its minimum. The wave has, therefore, decreased the energy density of the plasma; it could only be excited if negative work were done on the plasma; i.e., work were done on the wave by the plasma. For this reason, waves of this type, "negative-energy waves," are only found in plasmas with excess free energy. Note that if a plasma can support negative energy waves, any positive dissipation can permit them to grow in amplitude. Alternatively, if a negative-energy wave can couple to a positive-energy wave the resulting coupled wave can grow. The general expression for the fluctuating energy density associated with an electrostatic wave is $U_1 = (\varepsilon_o/4)|E|^2\omega(\partial\kappa_\ell/\omega)$, which must be evaluated at the zeros of the longitudinal dielectric constant, κ_ℓ. In the case of the flute modes described earlier, the instability resulted from the coupling of a negative-energy wave and a positive-energy wave. In what follows we will deal mostly with wave particle interactions that transfer energy from non-Maxwellian plasmas to positive-energy waves to cause them to grow.

The general dispersion relation describing the propagation of electrostatic waves through a collisionless homogeneous Maxwellian plasma in a uniform magnetic field was first derived by Bernstein [4]. Harris [5] gave a parallel derivation in which the equilibrium distribution functions were allowed to have arbitrary, non-Maxwellian forms. For most circumstances relating to hot-electron ECH plasmas, the propagation of the electrostatic waves is governed by the cold-electron population. We, therefore, start our discussion of unstable electrostatic waves with a recapitulation of the cold-electron dispersion relation given by zeros of the longitudinal dielectric constant, κ_ℓ [6]

$$\kappa_\ell = 1-(\omega^2_{\mathrm{pe,\,cold}}/\omega^2)(k^2_{||}/k^2)-[\omega^2_{\mathrm{pe,\,cold}}/(\omega^2-\Omega^2)](k^2_{\perp}/k^2) = 0. \qquad (7.10)$$

The condition that the longitudinal dielectric constant must vanish results from the requirement that electrostatic waves satisfy $\nabla\cdot\mathbf{D} = -\nabla\cdot\varepsilon_o\kappa\nabla\phi = 0$, where ϕ is the electrostatic potential of the wave. If we once again employ the collisionless Langevin equation to describe the cold-electron dynamical response to the electric field of the waves represented by $\mathbf{E}\exp(i\mathbf{k}\cdot\mathbf{r} - i\omega t)$, we have $m d\mathbf{v}/dt = -i\omega m\mathbf{v} = -e(\mathbf{E} + \mathbf{v}\times\mathbf{B})$. With $\mathbf{B}$ in the z-direction and $\Omega = eB/m$ this becomes

$$-i\omega(v_x\mathbf{u}_x + v_y\mathbf{u}_y + v_z\mathbf{u}_z) = -(e/m)\mathbf{E}-\Omega(v_y\mathbf{u}_x - v_x\mathbf{u}_y).$$

The solutions can be written in the matrix form as follows:

$$\begin{pmatrix} v_x \\ v_y \\ v_z \end{pmatrix} = -(e/m)\begin{pmatrix} i\omega/(\omega^2-\Omega^2) & \Omega/(\omega^2-\Omega^2) & 0 \\ -\Omega/(\omega^2-\Omega^2) & i\omega/(\omega^2-\Omega^2) & 0 \\ 0 & 0 & i/\omega \end{pmatrix}\begin{pmatrix} E_x \\ E_y \\ E_z \end{pmatrix}$$

and setting $\mathbf{j} = -en\mathbf{v} = \sigma\cdot\,\mathbf{E}$ gives

$$\begin{pmatrix} j_x \\ j_y \\ j_z \end{pmatrix} = \varepsilon_0\,\omega^2_{\mathrm{pe}}\begin{pmatrix} i\omega/(\omega^2-\Omega^2) & \Omega/(\omega^2-\Omega^2) & 0 \\ -\Omega/(\omega^2-\Omega^2) & i\omega(\omega^2-\Omega^2) & 0 \\ 0 & 0 & i/\omega \end{pmatrix}\begin{pmatrix} E_x \\ E_y \\ E_z \end{pmatrix}$$

Since $\kappa = \mathbf{1} + i\sigma/\varepsilon_0\omega$, we have for the dielectric tensor (in this Cartesian representation)

$$\kappa = \begin{pmatrix} 1-\omega^2_{\mathrm{pe}}/(\omega^2-\Omega^2) & (i\Omega/\omega)\omega^2_{\mathrm{pe}}/(\omega^2-\Omega^2) & 0 \\ (-i\Omega/\omega)\omega^2_{\mathrm{pe}}/(\omega^2-\Omega^2) & 1-\omega^2_{\mathrm{pe}}/(\omega^2-\Omega^2) & 0 \\ 0 & 0 & 1-\omega^2_{\mathrm{pe}}/\omega^2 \end{pmatrix}$$

As mentioned earlier, the dispersion relation for electrostatic waves results from the condition that $-\nabla\cdot\varepsilon_o\kappa\nabla\phi = \varepsilon_0\mathbf{k}\cdot\kappa\cdot\mathbf{k}\phi = 0$ has nontrivial solutions if and only if

$$\kappa_\ell = \mathbf{k}\cdot\kappa\cdot\mathbf{k}/\mathbf{k}^2 = 0.$$

With our usual choice for $\mathbf{k} = k_\perp\mathbf{u}_x + k_{||}\mathbf{u}_z$ we obtain the cold-plasma electrostatic dispersion relation cited above in Eq. (7.10):

$$1-(k^2_{||}/k^2)\omega^2_{\mathrm{pe}}/\omega^2-(k^2_\perp/k^2)[\omega^2_{\mathrm{pe}}/(\omega^2-\Omega^2)] = 0.$$

For perpendicular propagation, the two roots are $\omega^2 = 0$ and $\omega^2 = \omega^2_{\mathrm{pe}} + \Omega^2$, the upper hybrid frequency, while for parallel propagation the roots are $\omega^2 = \omega^2_{\mathrm{pe}}$

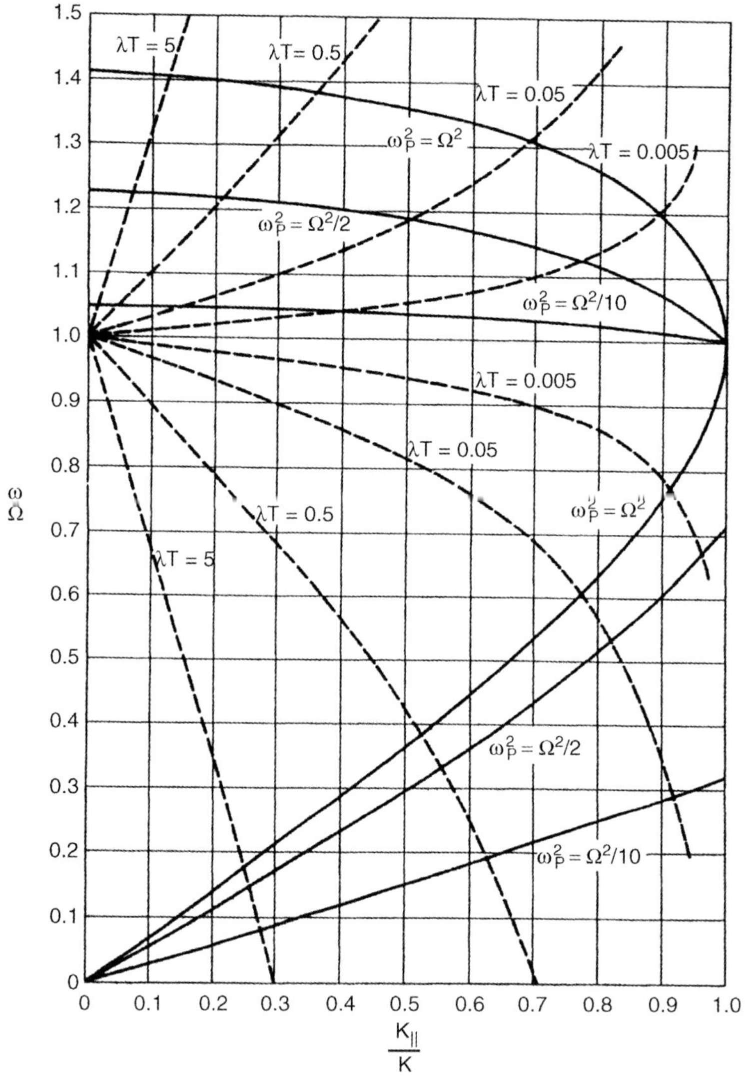

Figure 7.2 Cold-electron electrostatic plasma waves (solid lines) for three densities: $(\omega_{pe}/\Omega)^2 = 0.1$, 0.5, and 1.0. The dashed lines show the conditions for Doppler-shifted resonance with $\lambda_T = k_\perp^2 \alpha_{||}^2 / 2\Omega^2$ as a parameter.

and $\omega^2 = \Omega^2$. For oblique propagation, the two solutions of this bi-quadratic dispersion relation join these limiting cases continuously as illustrated in Figure 7.2.

Both of the waves are positive-energy waves, as one would expect for a cold-electron plasma. If these positive-energy waves propagate in a dissipative medium, the wave amplitude will decrease in time as the electric field of the wave does (positive) work on the medium at a rate given by $\langle \mathbf{j} \cdot \mathbf{E} \rangle$, where $\langle \mathbf{j} \cdot \mathbf{E} \rangle = \langle \sigma_\ell E_\ell^2 \rangle$. Here $\mathbf{j}$ is the current density that flows in response to the electrostatic ("longitudinal") field of the wave, $\mathbf{E}_\ell$, and σ_ℓ is the longitudinal conductivity, $\mathbf{k}\cdot\sigma\cdot\mathbf{k}/k^2$. Because of the non-Maxwellian character of the hot-electron population, the longitudinal conductivity can be negative, and the hot-electron plasma component can do net work on the wave, in which case the (positive-energy) wave will grow in amplitude. The real part of the longitudinal conductivity associated with the hot-electron component can be determined using the Vlasov equation and the *ad hoc* equilibrium distribution functions that describe at least qualitatively the anisotropic, mirror-confined electrons. For the class of *ad hoc* equilibrium distribution functions discussed in Chapter 6, Re σ_ℓ is given by

$$\begin{aligned} \mathrm{Re}\,\sigma_\ell &= 2\sqrt{\pi}\omega\varepsilon_0\left(\varepsilon^2_{\mathrm{pe,\,hot}}/k^2\alpha^2_{||}\right)(\Omega/k_{||}\alpha_{||}) \\ &\times\sum\exp\{-[(\omega-n\Omega)/k_{||}\alpha_{||}]^2\}\left[(\omega-n\Omega)C_n(\lambda)+nD_n(\lambda)\left(\alpha^2_{||}/\alpha^2_\perp\right)\right]. \end{aligned} \tag{7.11}$$

Here $\alpha_{||}$ and $\alpha_\perp$ are the average speeds of hot electrons parallel and perpendicular, respectively, to the magnetic field; the sum is over all values of the index n ranging from $-\infty$ to $+\infty$. The functions $C_n(\lambda)$ and $D_n(\lambda)$ are the following moments of the hot-electron distribution function, $f_o(v_\perp, v_{||}) = g_o(v_\perp)\exp(-v^2_{||}/\alpha^2_{||})/\alpha_{||}\sqrt{\pi}$:

$$\begin{aligned} C_n(\lambda) &= 2\pi\int v_\perp dv_\perp J^2_n(k_\perp v_\perp/\Omega)g_o(v_\perp) \\ D_n(\lambda) &= -(\alpha^2_\perp/2)2\pi\int v_\perp dv_\perp J^2_n(k_\perp v_\perp/\Omega)v_\perp^{-1}dg_o(v_\perp)/dv_\perp \end{aligned} \tag{7.12}$$

The parameter $\lambda \equiv k^2_\perp\alpha^2_\perp/2\Omega^2$ and J_n is the Bessel function. The properties of these moments of the distribution function have been discussed at length by Guest and Dory [7]. The functions $C_n(\lambda)$ are manifestly positive, but the functions $D_n(\lambda)$ can be negative if the distribution function describes a hot-electron population confined in a magnetic mirror. Such a population may contain more electrons with high perpendicular speeds than with lower perpendicular speeds so that $dg_o(v_\perp)/dv_\perp > 0$ over much of the hot-electron population. A model commonly used to simulate this aspect of magnetic-mirror confinement is the $m = 1$ term from Eq. (6.42):

$$g_o(v_\perp) = \left(\pi\alpha^2_\perp\right)^{-1}(v_\perp/\alpha_\perp)^2\exp\left(-v^2_\perp/\alpha^2_\perp\right). \tag{7.13}$$

The functions $C_n(\lambda)$ and $D_n(\lambda)$ for this distribution function are displayed in Figure 7.3(a) and (b) for $n = 0$–3.

We consider each of the two electrostatic waves in turn, starting with the forward wave. Although the frequency of the forward wave ($d\omega/dk_{||} > 0$) is less than the hot-electron gyrofrequency, the wave frequency can be up-shifted to resonate with the electron gyration by virtue of the (average electron) Doppler shift if $\omega + k_{||}\alpha_{||} \approx \Omega$. Energy can then be transferred from the hot-electron population to the wave if the real part of the longitudinal conductivity is negative. We are thus led to search for negative extrema in Re σ_ℓ for frequencies below the electron gyrofrequency. The dominant contributions to Re σ_ℓ then arise from the terms with $n = 0$ and 1; and provided $\alpha^2_{||}/\alpha^2_\perp \ll 1$ these negative extrema can be located approximately by first evaluating the following quantity:

$$Q = \exp[-(\omega/k_{||}\alpha_{||})^2]\omega C_0(\lambda) + \exp\{-[(\omega-\Omega)/k_{||}\alpha_{||}]^2\}[(\omega-\Omega)C_1(\lambda)]. \tag{7.14}$$

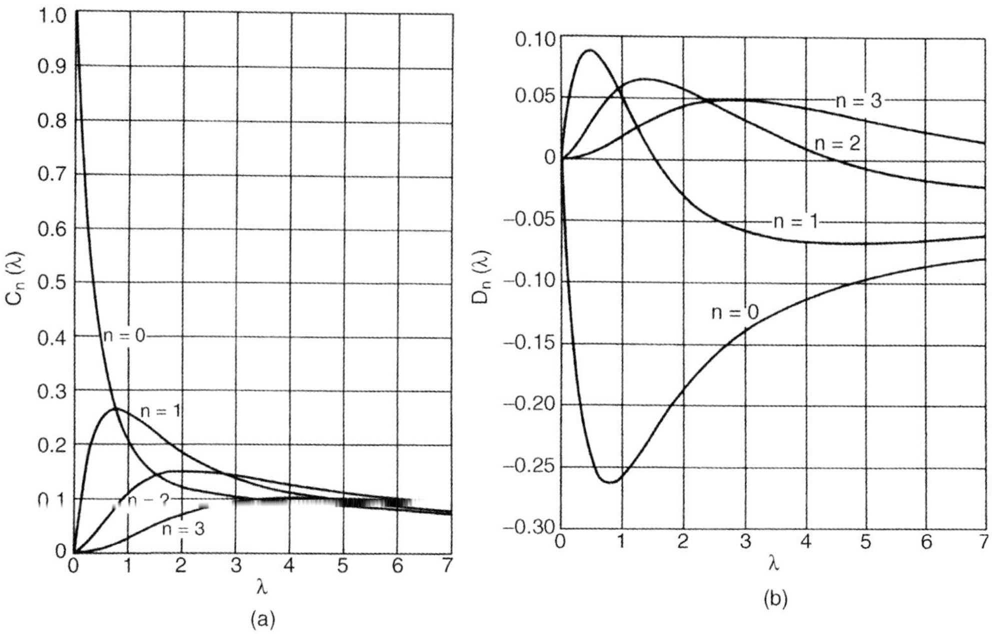

Figure 7.3 (a) The functions $C_n(\lambda)$ with $n = 0$–3 for the distribution function given by Eq. (7.13). (b) The functions $D_n(\lambda)$ with $n = 0$–3 for the distribution function given by Eq. (7.13).

The first term is always positive or stabilizing, whereas the second term is negative for all frequencies below the gyrofrequency. For our model distribution function, the function $C_1(\lambda)$ has its maximum value at $\lambda = 0.79$; and $C_0(\lambda) = C_1(\lambda)$ for that value of lambda. We can readily locate the zeros of Q with the results shown in Figure 7.4.

Here we have defined $\eta \equiv \omega/\Omega - 1/2$ and $H \equiv k_{||}\alpha_{||}/\Omega$. Then provided we can neglect the contribution from D_1, the zeros of Re σ_ℓ are given by

$$H^2 = 2\eta\{\ln[(2\eta + 1)C_0/(1-2\eta)C_1]\}^{-1}.$$

For the model hot-electron distribution function used here, the real part of the longitudinal conductivity takes on its most negative value for waves with frequencies $\omega = 0.755\Omega$ and parallel propagation vector $k_{||} = 0.4\Omega/\alpha_{||}$. Nonetheless, the real part of the conductivity remains negative, corresponding to the possibility of growing electrostatic waves, over a substantial range of values of the wave parameters:

$$\Omega/2 < \omega < \Omega \text{ and } 0 < k_{||} < 0.7\Omega/\alpha_{||}. \tag{7.15}$$

Potentially unstable waves in this frequency-wave number range will propagate only in plasmas whose properties, such as density and temperature anisotropy,

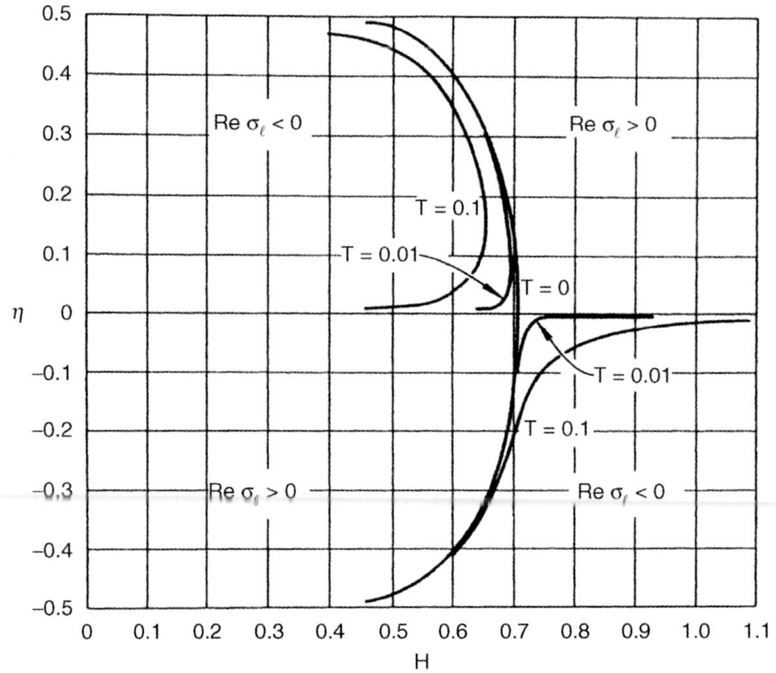

Figure 7.4 Locus of zeros of the real part of the longitudinal conductivity, where $\eta \equiv \omega/\Omega - 1/2$ and $H \equiv k_{||}\alpha_{||}/\Omega$.

satisfy the cold-plasma dispersion relation. These plasma properties can be conveniently displayed by plotting $\omega^2_{pe,\,cold}/\omega^2$ versus $\alpha^2_{||}/\alpha^2_{\perp}$, as in the upper curve of Figure 7.5.

This curve is generated for a specific pair of wave parameters, ω and $k_{||}$. Note that instability is only possible for these waves if the temperature anisotropy is high and provided the cold-electron density exceeds a threshold value given approximately by $\omega^2_{pe,\,cold}/\omega^2 > 1/2$; for more moderate values of anisotropy, the threshold density exceeds the cutoff value.

The frequency of the backward electrostatic wave ($d\omega/dk_{||} < 0$) exceeds the electron gyrofrequency but can resonate with electron gyration if $\omega - k_{||}\alpha_{||} \approx \Omega$.

The real part of the longitudinal conductivity can be negative in this frequency range if $D_1(\lambda) < 0$. In fact, for unstable growth the condition is $(\omega - \Omega)C_1(\lambda) < -D_1(\lambda)\alpha^2_{||}/\alpha^2_{\perp}$.

If we choose $\lambda = 4.6$ to minimize $D_1(\lambda)$, then for our model distribution function we have $D_1(\lambda) = -0.07$ and $C_1(\lambda) = 0.1$. Under these conditions, growth is possible only if $\omega/\Omega - 1 < 0.7\alpha^2_{||}/\alpha^2_{\perp}$. As in the case of the forward wave, these waves must satisfy the cold-plasma dispersion relation. The predicted maximum density for growth of the backward electrostatic wave is proportional to $\alpha^2_{||}/\alpha^2_{\perp}$, as illustrated by the lower curve in Figure 7.5. Note that there is an intermediate range of stable densities separating the two electrostatic modes where neither mode is expected to be unstable.

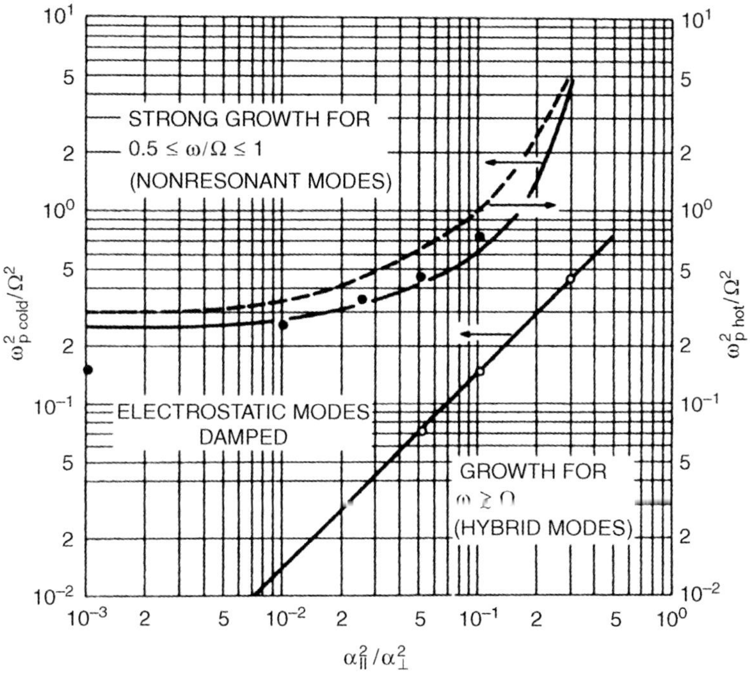

Figure 7.5 The two critical cold-plasma densities (solid lines) versus the anisotropy parameter $T = \alpha_{\parallel}^2/\alpha_{\perp}^2$. The dashed line shows the corresponding hot-electron limit of the temperature-anisotropy mode.

7.3
Electromagnetic Velocity Space Instabilities

Of the two cold-fluid electromagnetic waves propagating parallel to the magnetic field discussed in Chapter 4, only the right-hand circularly polarized component can resonate with the electron gyration. This wave, the now familiar whistler wave [8], is a positive-energy wave. In Chapter 4, we examined the damping of this wave in a Maxwellian plasma. Here we shall examine its growth in non-Maxwellian plasmas, and we can, therefore, restrict our stability analysis to situations in which the plasma electrons can do net work on the wave in order to amplify it.

The time-averaged work done on the plasma electrons by the RF electric field of the wave is $1/2\mathrm{Re}(\mathbf{E}^* \cdot \mathbf{j})$. Here the RF current is given by $\mathbf{j} = -e\int d^3v\mathbf{v}f_1$ and the perturbed distribution of electrons, f_1, can be obtained by integrating the linearized Vlasov equation in time along the unperturbed electron orbits from $t = -\infty$ to $t = 0$:

$$f_1 = (e/m)\int dt\,(\mathbf{E} + \mathbf{v} \times \mathbf{B}) \cdot \partial f_0/\partial \mathbf{v}.$$

For the case of parallel propagation, the electric and magnetic fields of the wave are assumed to depend on space and time through transverse eigenmode expressions of the form E(x,y)exp[i(kz − ωt)]. From Faraday's law, the wave electric and magnetic fields are related by $\nabla \times \mathbf{E} = -\partial \mathbf{B}/\partial t = i\omega \mathbf{B}$ whence

$$\begin{aligned} i\omega\mathbf{B} &= ik\mathbf{u}_z \times \mathbf{E} + [\mathbf{u}_x \partial E_z/\partial y - \mathbf{u}_y \partial E_z/\partial x + \mathbf{u}_z(\partial E_y/\partial x - \partial E_x/\partial y)] \\ &\approx ik\mathbf{u}_z \times \mathbf{E}. \end{aligned}$$

With this approximation for **B**, we have $\mathbf{E} + \mathbf{v} \times \mathbf{B} = (1 - kv_z/\omega)\mathbf{E} + (k/\omega)\mathbf{v}\cdot\mathbf{E}\mathbf{u}_z$.

Since the equilibrium distribution function can depend on any constants of the motion, we choose the electron energy, ε, and magnetic moment, μ, just as before so that

$$\partial f_0/\partial \mathbf{v} = m[(\partial f_0/\partial \varepsilon)\mathbf{v} + B^{-1}(\partial f_0/\partial \mu)\mathbf{v}_\perp].$$

With these substitutions, our expression for the perturbed distribution function becomes

$$f_1 = e\int dt\,[(\partial f_0/\partial \varepsilon)\mathbf{E}\cdot\mathbf{v} + B^{-1}(\partial f_0/\partial \mu)(1 - kv_z/\omega)\mathbf{E}\cdot\mathbf{v}_\perp].$$

The integration in time is over the unperturbed orbits of the electrons, given in a uniform magnetic field by $\mathbf{v} = v_\perp[\mathbf{u}_x \cos\phi(t) + \mathbf{u}_y \sin\phi(t)] + v_z\mathbf{u}_z$. In the uniform magnetic field approximation the gyrophase, ϕ(t), increases linearly with time: $\phi(t) = \phi(0) + \Omega t$ so that

$$\begin{aligned} \mathbf{E}\cdot\mathbf{v} &= E_x v_\perp \cos\phi(t) + E_y v_\perp \sin\phi(t) + E_z v_z \\ &= (v_\perp/2)(E_x - iE_y)\exp[i\phi(0) + i(kv_z - \omega + \Omega)t] \\ &\quad + (v_\perp/2)(E_x + iE_y)\exp[-i\phi(0) + i(kv_z - \omega - \Omega)t] + E_z v_z \exp[i(kv_z - \omega)t]. \end{aligned}$$

After integrating over time, we obtain for the perturbed distribution function

$$\begin{aligned} f_1 &= e[(\partial f_0/\partial \varepsilon) + B^{-1}(\partial f_0/\partial \mu)(1 - kv_z/\omega)] \\ &\quad \times (v_\perp/2)[(E_x - iE_y)\exp i\phi(0)/i(kv_z - \omega + \Omega) \\ &\quad + (v_\perp/2)(E_x + iE_y)\exp i\phi(0)/i(kv_z - \omega - \Omega)] + e(\partial f_0/\partial \varepsilon)E_z v_z/i(kv_z - \omega), \end{aligned} \tag{7.16}$$

and the RF current is

$$\begin{aligned} \mathbf{j} &= -e\int d^3v\,\{v_\perp[\mathbf{u}_x \cos\phi(0) + \mathbf{u}_y \sin\phi(0)] + v_z\mathbf{u}_z\} f_1 \\ &= -\pi e^2/2 \int v_\perp dv_\perp \int dv_z\,\{e[(\partial f_0/\partial \varepsilon) + B^{-1}(\partial f_0/\partial \mu)(1 - kv_z/\omega)] \\ &\quad \times [(E_x - iE_y)v_\perp^2(\mathbf{u}_x + i\mathbf{u}_y)/i(kv_z - \omega + \Omega) \\ &\quad + (E_x + iE_y)v_\perp^2(\mathbf{u}_x - i\mathbf{u}_y)/i(kv_z - \omega - \Omega)] + 2(\partial f_0/\partial \varepsilon)E_z v_z \mathbf{u}_z/i(kv_z - \omega)\}. \end{aligned} \tag{7.17}$$

We can now form

$$\begin{aligned} \mathbf{E}^*\cdot\mathbf{j} &= (E_x^*\mathbf{u}_x + E_y^*\mathbf{u}_y + E_z^*\mathbf{u}_z)\cdot\mathbf{j} \\ &= i\pi e^2 \int v_\perp dv_\perp \int dv_z\,\{[(\partial f_0/\partial \varepsilon) + B^{-1}(\partial f_0/\partial \mu)(1 - kv_z/\omega)]v_\perp^2 \\ &\quad \times [|E_-|^2/(kv_z - \omega + \Omega) + |E_+|^2/(kv_z - \omega - \Omega)] \\ &\quad + 2(\partial f_0/\partial \varepsilon)|E_z|^2 v_z^2/(kv_z - \omega)\}. \end{aligned} \tag{7.18}$$

Here as before $E_\pm = (E_x \pm iE_y)/\sqrt{2}$, and the time-averaged rate at which the wave does work on the plasma is given by

$$
\begin{aligned}
1/2\,\mathrm{Re}\,\mathbf{E}^* \cdot \mathbf{j} = -\pi e^2/2\,\mathrm{Im} \int v_\perp\, dv_\perp \int dv_z \\
\times \{[(\partial f_0/\partial\varepsilon) + B^{-1}(\partial f_0/\partial\mu)(1 - kv_z/\omega)]v_\perp^2 \\
\times [|E_-|^2/(kv_z - \omega + \Omega) + |E_+|^2/(kv_z - \omega - \Omega)] \\
+ 2(\partial f_0/\partial\varepsilon)|E_z|^2 v_z^2/(kv_z - \omega)\}
\end{aligned}
$$

so that

$$
\begin{aligned}
1/2\,\mathrm{Re}\,\mathbf{E}^* \cdot \mathbf{j} = -\pi e^2/2k \int v_\perp dv_\perp \\
\times \{v_\perp^2 |E_-|^2[(\partial f_0/\partial\varepsilon) + B^{-1}(\partial f_0/\partial\mu)\Omega/\omega]|_{v_z=(\omega-\Omega)/k} \\
+ v_\perp^2 |E_+|^2[(\partial f_0/\partial\varepsilon) + B^{-1}(\partial f_0/\partial\mu)\Omega/\omega]|_{v_z=(\omega+\Omega)/k} \\
+ 2(\partial f_0/\partial\varepsilon)|E_z|^2 v_z^2|_{v_z=\omega/k}\},
\end{aligned}
\qquad (7.19)
$$

where we have employed the familiar Plemelj formula in evaluating the imaginary part of the integral over v_z. The electron gyroresonance term is

$$1/2\,\mathrm{Re}\,\mathbf{E}^* \cdot \mathbf{j} = -\pi e^2/2k \int v_\perp^3 dv_\perp \times |E_-|^2[(\partial f_0/\partial\varepsilon) + B^{-1}(\partial f_0/\partial\mu)\Omega/\omega]|_{v_z=(\omega-\Omega)/k.}$$

If $\partial f_0/\varepsilon$ and $\partial f_0/\mu$ were both negative for resonant electrons with $v_z = (\omega - \Omega)/k$, the time-averaged work done on the plasma electrons by the wave would be positive and the wave would be damped in time. But if the quantity in brackets can change sign and become positive, then the plasma electrons can do net work on the wave and amplify it in time. We can illustrate the conditions for such amplification by assuming the hot-electron distribution to be an aniostropic bi-Maxwellian, the $m = 0$ term of Eq. (6.42):

$$f_0 = n_h(\pi^{3/2}\alpha_\perp^2 \alpha_{||})^{-1} \exp(-v_\perp^2/\alpha_\perp^2 - v_{||}^2/\alpha_{||}^2).$$

The average rate of energy transfer from the wave to the plasma electrons is then given by

$$1/2\,\mathrm{Re}\,\mathbf{E}^* \cdot \mathbf{j} = -\sqrt{\pi}(\omega_{pe,\,hot}^2/k\alpha_{||})(\varepsilon_0|E_-|^2/2)(Ax - 1)\exp(-x^2), \qquad (7.20)$$

where

$$A \equiv (T_\perp/T_{||} - 1)k\alpha_{||}/\omega, \qquad (7.21)$$

and

$$x \equiv (\Omega - \omega)/k\alpha_{||}. \qquad (7.22)$$

We choose this sign for x so that x is positive for wave frequencies less than the local gyrofrequency. The function $(Ax - 1)\exp(-x^2)$ has a zero at $x_o = 1/A$, and extrema at

$$x_{ext} = [1 \pm (1 + 2A^2)^{1/2}]/2A. \qquad (7.23)$$

The upper sign corresponds to the conditions for maximum growth rate; i.e., the maximum rate at which electrons can transfer energy to the wave. The lower sign gives the conditions for maximum damping of the wave; i.e., the maximum rate at which the wave can transfer energy to the electrons. The zero, $x_o = 1/A$, gives the conditions for marginal stability; i.e., the conditions for which no net work is done on the wave or the plasma. The marginally stable frequency, ω_o, is given by the condition that

$$x_o = (\Omega - \omega_o)/k\alpha_{||} = 1/A = 1/[(T_\perp/T_{||} - 1)k\alpha_{||}/\omega_o] \tag{7.24}$$

whence,

$$\omega_o/\Omega = 1 - T_{||}/T_\perp.$$

Note that $\Omega = \Omega_o/\gamma$ is the appropriate relativistic gyrofrequency of the resonant energetic electrons, so the frequency of the marginally stable wave is $\omega_o = \Omega_o(1 - T_{||}/T_\perp)/\gamma$. Waves with frequencies less than ω_o will grow, while those with frequencies greater than ω_o will be damped.

The situation for the extrema is more complicated. For example, if the plasma parameters are such that $A = 2$, the maximum growth rate occurs for $x_{max} = 1$ and the maximum damping rate for $x_{min} = -1/2$. This is the case shown in Figure 7.6.

The corresponding frequencies are then, for growth, $\omega_{max} = \Omega - k\alpha_{||}$ and for damping $\omega_{min} = \Omega + k\alpha_{||}/2$. But note that A and x depend on the wave frequency ω and wavenumber k, and these must satisfy an appropriate dispersion relation, in this instance, the whistler dispersion relation

$$n^2 = 1 + \omega_{pe}^2/[\omega(\Omega - \omega)].$$

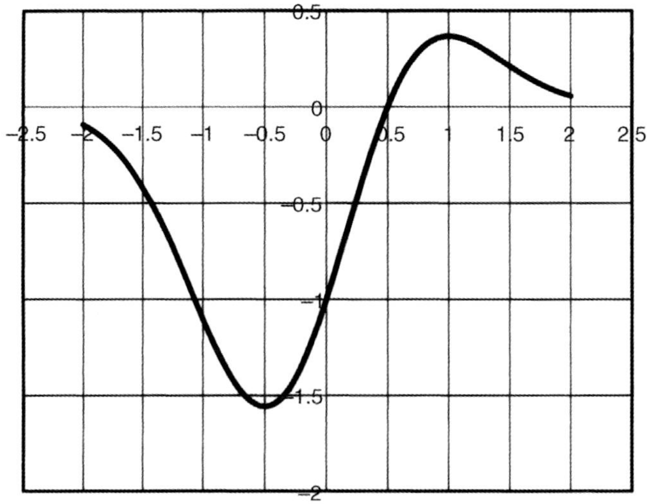

Figure 7.6 The function $(Ax - 1)\exp(-x^2)$ for $A = 2$.

Neglecting for the moment the relativistic increase in mass, we can replace $(\Omega - \omega)$ by $xk\alpha_{||}$ so that the whistler dispersion relation becomes

$$n^2 = 1 + \omega_{pe}^2/(\omega xk\,\alpha_{||}) = 1 + \left(\omega_{pe}^2/\omega^2\right)/(nx\alpha_{||}/c),$$

giving the following cubic equation for the index of refraction:

$$n^3 - n - \left(\omega_{pe}^2/\omega^2\right)/(x\alpha_{||}/c) = 0.$$

By way of illustration, to find the physical conditions for maximum growth rate, we set

$$x = x_{max} = \left[1 + (1 + 2A^2)^{1/2}\right]/2A,$$

with

$$A = (T_\perp/T_{||} - 1)n(\alpha_{||}/c).$$

This set of three coupled equations for n, x_{max}, and A can be solved by employing a simple iterative scheme if ω_{pe}/ω, $\alpha_{||}/c$, and $T_\perp/T_{||}$ are specified. Then with an arbitrary initial guess for n, which we designate as n_1, we can evaluate A_1 and $x_{max,1}$ and solve the cubic dispersion relation for $n = n_2$. We repeat this cycle until $n_{i+1} - n_i = 0$, which then gives the solution to the set of three equations. This procedure yields self-consistent values for the index of refraction, n, the anisotropy parameter, A, and the value of x for maximum growth rate, x_{max}, and the various related functions of interest, such as the extremal value of $(Ax - 1)\exp(-x^2)$. In Figure 7.7, we display curves of ω_{max}/Ω_o versus ω_{pe}/ω, for fixed values of $T_{||}$ and $T_\perp/T_{||}$.

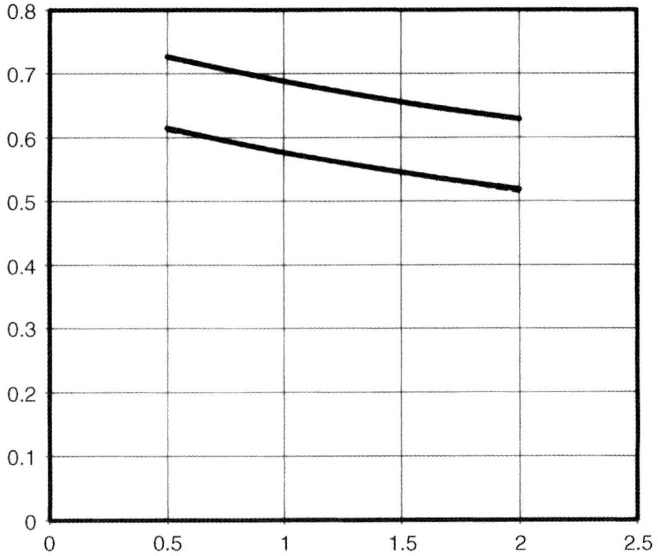

Figure 7.7 The frequency for maximum growth rate, ω_{max}/Ω_o, versus ω_{pe}/Ω_o for $T_\perp/T_{||} = 15$ and $T_{||} = 5$ keV (upper curve) and 10 keV (lower curve).

In anticipation of some of the experimental results to be discussed in Chapter 8, we observe an interesting possibility that arises in ECH plasmas with separate warm- and hot-electron groups. The frequency for maximum growth of whistlers for the warm-electron population may coincide with a frequency for damping by the more relativistic hot-electron population. For example, in the extreme case of maximum damping by the relativistic electrons,

$$\omega_{max,2} = \Omega_o/\gamma_2 - k\alpha_{\|2} = \omega_{min,3} = \Omega_o/\gamma_3 + k\alpha_{\|3}/2.$$

Here Ω_o is the nonrelativistic gyrofrequency and γ is the usual relativistic factor. The necessary value of γ_3 for maximum damping is then given by

$$\gamma_3 = \gamma_2[1-(\gamma_2 k/\Omega_o)(\alpha_{\|2} + \alpha_{\|3}/2)]^{-1}.$$

In this case, the relativistic-electron population might act to stabilize the warm-electron group against these electromagnetic instabilities. Several authors [9] have discussed the possibility of stabilizing tendencies due to relativistic hot-electron populations.

References

1 H.P. Furth, *Phys. Fluids* **6**, 48 (1963).

2 B.B. Kadomtsev, in *Plasma Physics and the Problem of Controlled Thermonuclear Reactions*, (M.A. Leontovich and J. Turkevich, eds.), Vol. 4, Pergamon Press, Oxford, (1960), p. 450.

3 G.E. Guest, R.L. Miller, and M.Z. Caponi, Inductive inhibition of cold-plasma stabilization of curvature-driven modes in finite-length plasmas, *Phys. Fluids* **29**, 2556 (1986) and references cited there.

4 I.B. Bernstein, *Phys. Rev.* **109**, 10 (1958).

5 E.G. Harris, *J. Nucl. Energy* **C2**, 138 (1961).

6 G.E. Guest and D.J. Sigmar, *Nucl. Fusion* **11**, 151 (1971).

7 G.E. Guest and R.A. Dory, *Phys. Fluids* **8**, 1853 (1965).

8 R.N. Sudan, *Phys. Fluids* **6**, 57 (1963).

9 N.T. Gladd, *Phys. Fluids* **26**, 974 (1983); andG.E. Guest and R.L. Miller, *Nucl. Fusion* **28**, 419 (1989).

Exercises

7.1 *In relating the flute instabilities to idealized two-stream electrostatic instabilites, we ignored the energy and pitch-angle dependence of the poloidal drift speed. With this in mind, consider the following equilibrium distribution function:*

$$f_o(v) = (u_o/2\pi)\{[(v - v_o)^2 + u_o^2]^{-1} + [(v + v_o)^2 + u_o^2]^{-1}\}.$$

(a) *Show that* $f_o(v)$ *has two distinct peaks only if* $v_o^2 > u_o^2/3$.

(b) *Since* $f_o(v)$ *is symmetric about* $v = 0$*, it is reasonable to assume that when instability occurs* $\mathrm{Re}\,\omega = 0$*, as was the case with the flute modes described earlier. Show that the conditions for the onset of instability are then given by*

$$k^2/\phi_{po}^2 = (v_o^2 - u_o^2)/(v_o^2 + u_o^2)$$

and

$$v_o^2 > u_o^2 > u_o^2/3.$$

7.2 *Verify Eq. (7.11) and find the values of* η *and H for which the real part of the longitudinal conductivity takes on its most negative value when* $\lambda = 0.79$ *and* $(\alpha_{||}/\alpha_{\perp})^2 \ll 1$.

7.3 *For the ad hoc* equilibrium distribution function of Eq. (7.13), derive an expression for $D_1(\lambda)$ and evaluate it for $\lambda = 4.6$.

7.4 *Estimate the frequency for maximum growth rate,* ω_{max}/Ω_o *of whistler modes in a plasma whose hot-electron component has a density* $\omega_p^2/\omega^2 = 0.2$*, a parallel temperature* $T_{||} = 25$ *keV and temperature anisotropy* $T_{\perp}/T_{||} = 6$.

8
Experimental Results in Magnetic Mirrors

In this chapter we summarize the results of a number of experiments on electron cyclotron heating (ECH) of plasmas confined in magnetic mirror configurations. These specific experiments have permitted interpretations that either validated important theoretical models or helped guide the further development of more comprehensive analytical models and through this, they have made significant contributions to the development of the science of ECH.

8.1
Hot-Electron Experiments in "Physics Test Facility" and EPA [1–3]

The magnetic field in Physics Test Facility (PTF) was a simple magnetic mirror configuration with a 2:1 mirror ratio. The field was generated by two DC coils whose centers were separated axially by 55 cm. Up to 5 kW (cw) of microwave power at 10.6 GHz was launched into the plasma chamber by one or sometimes two waveguide inputs at the midplane. The waveguides were typically, but not necessarily, oriented so that the microwave electric field was perpendicular to the static magnetic field. The basic features of PTF are indicated schematically in Figure 8.1.

Not shown in the figure are additional diagnostic devices including a diamagnetic loop in the form of a coil 12 cm in radius encircling the microwave cavity and displaced axially 12 cm off the midplane. The diamagnetic loop was used to determine the total plasma kinetic energy in motion perpendicular to the magnetic field by measuring the change in the magnetic flux through the loop after turning off the microwave power. Also not shown in the figure is the bremsstrahlung diagnostic apparatus used to determine the spectrum of x-rays in the MeV energy range arising from free–free bremsstrahlung. Details of these and other diagnostic devices are given in the references. Of special interest is the detailed analysis of neutron production in the PTF plasmas, assumed to result from the electron-dissociation of deuterium nuclei. This diagnostic provided useful information on the highest energy electrons present in the plasma; namely, those with energies above the 2.2 MeV binding energy of the deuterium nucleus.

Electron Cyclotron Heating of Plasmas. Gareth Guest

ISBN: 978-3-527-40916-7

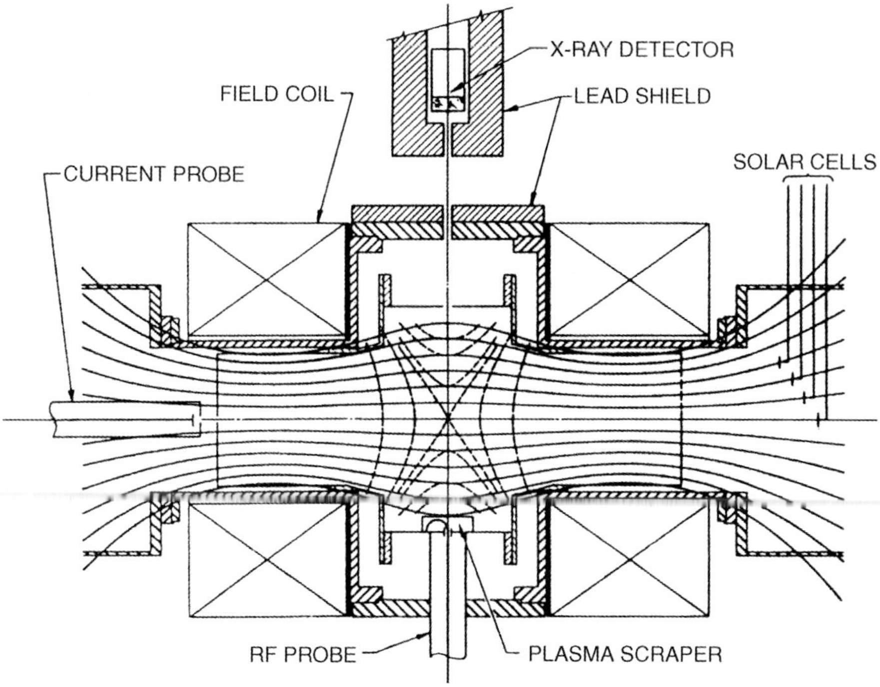

Figure 8.1 Schematic representation of the Physics Test Facility (PTF).

EPA was a larger, 3 : 1 magnetic mirror (although other magnetic configurations were also explored in EPA) in which the DC magnetic coils were separated by 36 in. (91 cm). Up to 50 kW (cw) of 10.6 GHz power was fed into the EPA cavity through a somewhat more elaborate array of waveguide couplers. Typical experimental plasma parameters obtained in PTF were as follows:

(1) A hot-electron component with temperatures of 50–100 keV and densities around $1.5\text{–}3 \times 10^{11}\,\text{cm}^{-3}$; and a group of relativistic electrons with energies above 2 MeV and densities of roughly $10^{9}\,\text{cm}^{-3}$.
(2) Diamagnetic stored energies ranging from 1.2 J at 1 kW microwave power to 9 J at 3 kW.
(3) Axial current densities due to a cold-electron component of $15\,\text{mA}\,\text{cm}^{-2}$ with the microwave power on and $5\,\text{mA}\,\text{cm}^{-2}$ after turning off the microwave power.
(4) An initial rapid decay of the plasma density with a characteristic time of 100–200 μs following the turnoff of microwave power, leading a much longer decay time of some tenths of seconds.
(5) An effective plasma radius around 10 cm.
(6) A plasma build-up time of 100 ms.

In these experiments, the pressure of the deuterium fill gas was around $2\text{–}5 \times 10^{-5}$ Torr, corresponding to molecular concentrations around $0.7\text{–}1.8 \times 10^{12}\,\text{cm}^{-3}$. Note that the experiments in PTF and EPA were steady-state rather than pulsed, and the final plasma state was achieved by careful adjustment of magnetic field strength, gas feed rate, and microwave power during operation. In addition to establishing these plasma parameters, Dandl and his co-workers [2] investigated two distinct types of plasma instabilities that could be excited in these plasmas.

We now consider the theoretical interpretation of the experimental observations in the context of the basic theoretical topics discussed in the earlier chapters, starting with the equilibrium. As described in Chapter 6, we separate the full distribution of electrons in energy into distinct groups that reflect in an approximate way the different production and loss processes that characterize each group:

(1) a cold-electron population with energies of 10s of eV confined by the electrostatic ambipolar potential;
(2) a warm-electron population with energies greater than a few keV;
(3) a hot-electron population with average energies of 100s of keV; and
(4) highly relativistic electrons with energies in the MeV range.

The spatially averaged conditions for steady-state particle balance for the first three of these groups were given in Chapter 6:

$$\begin{aligned}
dn_{e1}/dt &= n_{e1}n_o\langle\sigma_{ion}v_e\rangle_1 + n_{e2}n_o\langle\sigma_{ion}v_e\rangle_2 + n_{e3}n_o\langle\sigma_{ion}v_e\rangle_3 - n_{e1}/\tau_1 \\
&\quad - n_{e1}(dW/dt)_1/\Delta W_{2,1} = 0 \\
dn_{e2}/dt &= n_{e1}(dW/dt)_1/\Delta W_{2,1} - n_{e2}/\tau_2 - n_{e2}(dW/dt)_2/\Delta W_{3,2} = 0 \\
dn_{e3}/dt &= n_{e2}(dW/dt)_2/\Delta W_{3,2} - n_{e3}/\tau_3 - n_{e3}(dW/dt)_3/\Delta W_{escape} = 0
\end{aligned}$$

Recall that n_{e1}, n_{e2}, n_{e3}, and n_o are, respectively, the densities of the cold-electron group, the warm-electron group, the hot-electron group, and the (neutral) deuterium molecules. The ionization rates for electrons in the three groups are $\langle\sigma_{ion}v_e\rangle_1$, $\langle\sigma_{ion}v_e\rangle_2$, and $\langle\sigma_{ion}v_e\rangle_3$. In what follows we shall adopt the values for these ionization rates given by Freeman and Jones [4], as discussed in Chapter 6. The average electron confinement times for each of the three groups are τ_1, τ_2, and τ_3; and the average heating rates for each group are $(dW/dt)_1$, $(dW/dt)_2$, and $(dW/dt)_3$. The energy separating groups 1 and 2, denoted by $\Delta W_{2,1}$, is typically of the order of 1–10 keV. The energy gap separating the warm- and hot-electron groups, $\Delta W_{3,2}$, is typically several orders of magnitude larger, 100–200 keV. ΔW_{escape} is the energy increase required for a hot electron to become nonadiabatic and is several MeV in these experiments.

The bremsstrahlung spectra in the hard x-ray energy range were the primary diagnostic for determination of the density and temperature of the hot-electron group, and these are the values cited earlier. The density of the hot-electron group was also deduced from the density of the current flowing along the axis of the chamber after turning off the microwave power and the subsequent rapid decay of the warm-electron population within 0.1–0.2 ms. Since the measured current density, **j**, at this time is maintained solely by ionization due to the long-lived hot-electron group, we have

$$\int \mathbf{j}\cdot d\mathbf{A} = e\int dV n_{e3} n_o \langle\sigma_{ion}v_e\rangle_3.$$

Assuming the measured current density to be representative of the entire cross section of the plasma column intercepted by the current-collecting probe, the spatially averaged values of the hot-electron density are given by

$$n_{e3} = 2j_{||}(eL\, n_o \langle\sigma_{ion}v_e\rangle_3)^{-1}$$

with $j_{||} = 5\,\text{mA cm}^{-2}$, $n_o = 10^{12}\,\text{cm}^{-3}$, $\langle\sigma_{ion}v_e\rangle_3 = 10^8\,\text{cm}^3\,\text{s}^{-1}$ and $L = 20\,\text{cm}$, we obtain $n_{e3} = 3 \times 10^{11}\,\text{cm}^{-3}$, in reasonable agreement with results from the free–free bremsstrahlung measurements. Here L, the effective length of the hot-electron plasma column in PTF, is assumed to be approximately equal to the distance between the fundamental resonance surfaces.

The radial distribution of plasma kinetic energy in motion perpendicular to the magnetic field, $W_\perp$, stored in the hot-electron plasma, was deduced by dropping a small stainless steel ball, 1/8 in. in diameter, through the plasma and measuring the emf induced in the diamagnetic loop as the falling ball collected energetic electrons. The ball fell about $0.3\,\text{cm ms}^{-1}$, slowly enough to ensure that all energetic electrons in a radial layer roughly equal to the ball's diameter would be collected by the ball. The resulting time-dependent emf showed a single extremum when the radial position of the ball was 6 cm above the axis; otherwise the emf signal was devoid of significant structures. These features can be reproduced if the transverse stored energy is modeled as a parabolic function of radius with an effective radius, $a = 10.5\,\text{cm}$:

$$W_\perp(r) = W_\perp(0)[1-(r/a)^2].$$

Using this radial profile, we can relate the central values of the hot-electron density and (perpendicular) temperature to the total diamagnetic stored energy:

$$W_\perp(\text{total}) = 1/2\,\pi a^2\, L n_{e3}(0) T_{\perp e3}(0).$$

For example, if the hot-electron density on axis is $n_{e3}(0) = 2 \times 10^{11}\,\text{cm}^{-3}$ and the diamagnetic stored energy is 6 J, the present model would lead to an estimate for the (perpendicular) temperature of the hot-electrons of 54 keV.

After the microwave power is switched off, the lifetimes of the warm and hot electrons, τ_2 and τ_3, are governed by scattering into the mirror loss cone, but at rates that are significantly affected by slowing down through inelastic collisions with deuterium molecules. This process was simulated numerically by Ard *et al.* [5] for the hot electrons in PTF. They noted that the specific inelastic collision rate constant decreases with electron energy approximately as $E^{-1/3}$; and since this is the dominant process by which energetic electrons slow down in the PTF plasma, the time dependence of the electron energy is given by

$$dE/dt = -\Delta E n_o \sigma_{ion} v_e = -CE^{-1/3}$$

where ΔE is the energy lost per collision. Thus, the energetic electron energy will vary in time after turnoff of the microwave power as

$$E(t) = [E(0)^{4/3} - 4/3\,Ct]^{3/4}.$$

In Ref. [5] Ard *et al.* demonstrate a reasonable agreement between experimentally measured time-dependent hot-electron distributions in energy and results from their simulation model provided $C = 4\,\mathrm{keV}^{4/3}$ and $\ln\Lambda = 10$. Here Λ is the ratio of the maximum and minimum impact parameters in the Rutherford scattering formula. In a fully ionized plasma, the maximum impact parameter is set by the Debye length, and one would expect to have $\ln\Lambda = 17$ for plasma parameters similar to those in PTF. However, in the PTF plasmas, the density of gas molecules is approximately equal to the charged-particle density and the maximum impact parameter is the much smaller atomic radius. These considerations are essential to reconcile the experimentally observed confinement times with the observed electron energies.

One of the most striking observations reported by Dandl's group in Ref. [3] was the emission of a substantial flux of neutrons with the application of 2 kW of 10.6 GHz power and suitable adjustment of the DC magnetic field strength for quiescent steady-state operation. Four different diagnostic methods were employed to determine the energy spectrum of the neutrons, most of which were found to have energies in the keV range. The most plausible explanation for the production of these neutrons appeared to be electron-dissociation of deuterium by electrons with energies greater than 2.2 MeV, the binding energy of the deuterium nucleus. Indeed, x-ray spectra indicated the presence of roughly 10^9 electrons per cm^3 with energies above 2.2 MeV. We note in passing that if the lifetime of these hot electrons is some tenths of seconds, the implied heating rate would be around $10\,\mathrm{MeV\,s^{-1}}$, a typical theoretical estimate for the conditions in PTF.

These observations raise the issue of the maximum energy for which electrons can be adiabatically confined in PTF. From the work of Cohen *et al.* [6], we estimate the limiting electron energies for radii up to 10 cm and with two different values of the DC magnetic field strength, as displayed in Figure 8.2.

Inside a radius of 3–4 cm, electrons with energies greater than 2.2 MeV are expected to exhibit adiabatic confinement for DC magnetic field strengths on the midplane at the axis, $B(0,0) > 0.2\,\mathrm{T}$. This may indicate that the volume within which neutrons are produced is substantially less than the 6 l as assumed in Ref. [3]. The rate of the electron-dissociation process in the hot-electron plasma may therefore be greater than that predicted by binary-collision theories, which do not take into account the possibility that collective, many-body effects may occur in the plasma medium.

Two distinct modes of instability were studied in PTF and reported in Ref. [2]. The lower frequency mode was identified as a flute-like instability associated with the unfavorable guiding-center drift of energetic electrons in the magnetic-mirror field. This mode was effectively stabilized by increasing the gas pressure and thereby increasing the density of the cold-electron group, given by our condition for steady-state particle balance as

$$n_{e1} = n_o(n_{e2}\langle\sigma_{ion}v_e\rangle_2 + n_{e3}\langle\sigma_{ion}v_e\rangle_3)\tau_1[1 + \tau_1(dW/dt)_1/\Delta W_{2,1}]^{-1}.$$

When this type of instability occurred, plasma was rapidly transported across the confining magnetic field and detected by observing x-rays produced when electrons with energies around 100 keV impacted a radial limiter. The instabilities were accompanied by oscillations in the 3–30 MHz frequency range. We first note that

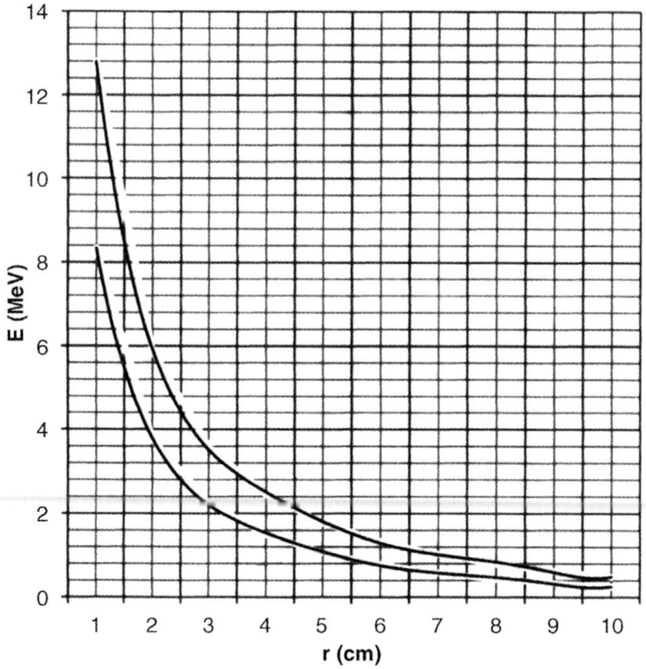

Figure 8.2 Maximum energy for adiabatic confinement in PTF versus the midplane radius for B(0,0) = 0.2 T (lower curve) and 0.3 T (upper curve).

the predicted frequencies for this mode of instability are given approximately by

$$f_\ell = (\ell/2\pi)T_{e3}/(eBrR_c),$$

where $\ell = 2, 3, 4, \ldots$ is the azimuthal mode number, T_{e3} is the average energy of the hot-electron group, B is the magnetic intensity at radius r, the plasma surface, and R_c is the radius of curvature of the magnetic field lines. As we saw earlier, on the midplane the product rR_c is roughly constant in simple magnetic mirror fields. For the fields in PTF we estimate $rR_c = 0.03\,m^2$. If the average energy of the hot-electron group is around 100 keV and the magnetic intensity at the plasma surface is 0.15 T, we would anticipate the frequencies to be roughly $f_\ell = 3.5\ell$ MHz, about as good an agreement as can be expected without properly averaging the magnetic field parameters along the lines of force.

As discussed in Chapter 7, the density of cold plasma required to stabilize this flute-like mode is predicted to be [7]

$$\begin{aligned} n_{e1} &= 2.8 \times 10^{11}\,cm^{-1}(\ell/r_s)^2[(4rR_c/\beta LL_c)-1]^{-1} \\ &\approx 2.8 \times 10^{11}\,cm^{-1}(\ell/r_s)^2\beta(a)L_hL_c/4rR_c \end{aligned}$$

In Ref. [2], the average value of beta was estimated to be 5–6%, assuming a plasma volume of 5 l, but the local value of beta near the plasma surface may be significantly higher. For example, if the local density is around $10^{11}\,cm^{-3}$ with a temperature of 10^5 eV, in a region where the local magnetic intensity is around 0.15 T, the local value of beta, $\beta(a)$, can approach 20%. Although it was not measured directly, we can estimate the cold-electron density crudely from our equilibrium model to be around $5 \times 10^{10}\,\text{cm}^{-3}$, which is predicted to be adequate to stabilize low-order modes with $\ell \leq 6$-9 for beta values less than 20%. Observation of rapidly growing modes in this frequency range that are stabilized by small increases in gas pressure (and hence cold-electron densities) is generally consistent with theoretical expectations; but the nonlinear development of the instability is not understood. Indeed, it may be likely that the onset of a high-order mode could lead to a reduction in the density of cold plasma and permit an avalanche of lower order modes to be destabilized. This remains largely speculative, since the actual experimental evolution of unstable modes was too rapid to follow in this level of detail.

The second mode of instability observed in PTF occurred if the magnetic intensity on axis at the midplane exceeded a threshold value given by $B(0,0) = 2985$ G. Since the resonant magnetic intensity for the 10.6 GHz power is 3786 G, this threshold corresponds to a mirror ratio on axis at resonance, $M_{res} = 1.27$. The threshold magnetic field was observed to depend only weakly on input power and gas pressure. In keeping with the discussion of Chapter 5 [8], we assume that the group of moderately energetic electrons with density n_{e2} are turning inside the resonance surfaces and will, therefore, have a temperature anisotropy at the midplane such that $T_{\parallel}/T_{\perp} \leq 0.27$. If our estimate of the cold-electron density of roughly $5 \times 10^{10}\,\text{cm}^{-3}$ is correct, then $\omega_{pecold}^2/\Omega^2 \approx 0.1$ and we can anticipate that modes at the upper hybrid frequency can be driven unstable by the loss-cone character of the warm-electron distribution. It may also be possible that whistler waves could be weakly unstable, depending on the actual value of beta, although the presence of the substantial number of relativistic electrons could damp whistler modes. Higher cold-electron densities are predicted to stabilize the electrostatic loss-cone modes but not the electromagnetic whistlers. For either the electrostatic or electromagnetic mode, growth of the instability would extract perpendicular kinetic energy from the anisotropic electrons and permit them to escape through the loss cone, as observed. Neither mode, however, would be expected to have a frequency of 5.3 GHz, that is, exactly half of the frequency of the 10.6 GHz microwave power, which was observed at all values of the magnetic intensity above the threshold value. We note in passing that the cavity eigenmodes in PTF had frequencies of some hundreds of MHz. Only the repetition rate of the bursts of this instability increased as the resonance surfaces were moved closer to the midplane and/or the heating power was increased. The authors of Ref. [2] did not preclude the possibility that the observed 5.3 GHz signal may have been triggered by the input microwave power. The actual instability mechanism corresponding to this frequency remains to be identified.

In summary, the experiments in PTF and EPA established the existence of stable, steady-state, high energy density, hot-electron plasmas in simple magnetic mirrors provided the gas pressure was high enough to stabilize flute-like instabilities and the mirror ratio at resonance was high enough to keep the hot-electron temperature anisotropy below the threshold for velocity–space microinstabilities. The fact that PTF contained a substantial population of electrons with energies greater than

2.2 MeV demonstrated that highly relativistic electrons could be heated even if there were no fundamental resonance surfaces inside the cavity. Diamagnetic stored energies approached 10 J in these PTF experiments.

8.2
High-Beta Experiments in ELMO [9]

Heating at frequencies well above the fundamental electron gyrofrequency was demonstrated experimentally in PTF by the presence in the plasma of a substantial group of highly relativistic multi-MeV electrons. A theoretical analysis by Grawe [10] also lent support to the potential importance of nonresonant heating of energetic electrons. In April of 1965, Dandl and his co-workers began experiments in the ELMO device employing independent sources of resonant and nonresonant microwave power to clarify these different modes of ECH. The results obtained with less than 1 kW (cw) of resonant power at 35.7 GHz (8.4 mm wavelength) together with 1 kW of nonresonant power at 55 GHz (5.5 mm) were remarkable: hot-electron temperatures exceeded 1 MeV and stored energies as high as 400 J were observed. Most significantly, the nonresonant heating, referred to as "upper off resonant heating" (UORH), apparently eliminated the velocity–space instabilities thought to be driven by temperature anisotropy even at relatively moderate levels of the UORH power. The hot-electron component of the plasma was subsequently shown to be contained in a short, cylindrical shell in which the plasma pressure was comparable to the magnetostatic pressure of the magnetic field. The estimated value of beta – the ratio of the diamagnetic pressure to the magnetostatic pressure – was $\beta \approx 0.75$. The diamagnetic currents associated with this high-beta plasma profoundly modified the magnetostatic field to create a local, axisymmetric magnetic well in the interior of the hot-electron shell. The main features of the ELMO device are indicated schematically in Figure 8.3.

Three independent sources of microwave power were available in ELMO: up to 3 kW (cw) at 10.6 GHz, 1.8 kW (cw) at 35.7 GHz, and roughly 5 kW (cw) at 55 GHz. The magnetic configuration, generated by two pairs of coaxial DC magnetic coils, could be varied continuously from a magnetic-mirror shape with a range of possible mirror ratios to a folded cusp configuration. The outer pair of coils (the "flat-field coils") were in the Helmholtz geometry, whereas the inner pair of coils operated alone would produce a magnetic mirror field with a mirror ratio of 3.3 : 1. The axial separation of both pairs of coils was 16 in. or 40.64 cm. The high energy density plasmas were obtained in a magnetic mirror configuration with mirror ratios around 2.1–2.2. Great care was taken to ensure that the plasma chamber had the maximum possible microwave integrity. Typical results of two-frequency heating are shown in Figure 8.4.

In these experiments, 880 W of 8 mm power provided the resonant heating and up to 1 kW of 5.5 mm power provided the UORH. Note that the hot-electron *temperature* changes only slowly with increasing UORH power, whereas the hot-electron *density, the stored energy, and the neutron flux* increase rapidly. In this particular case, the

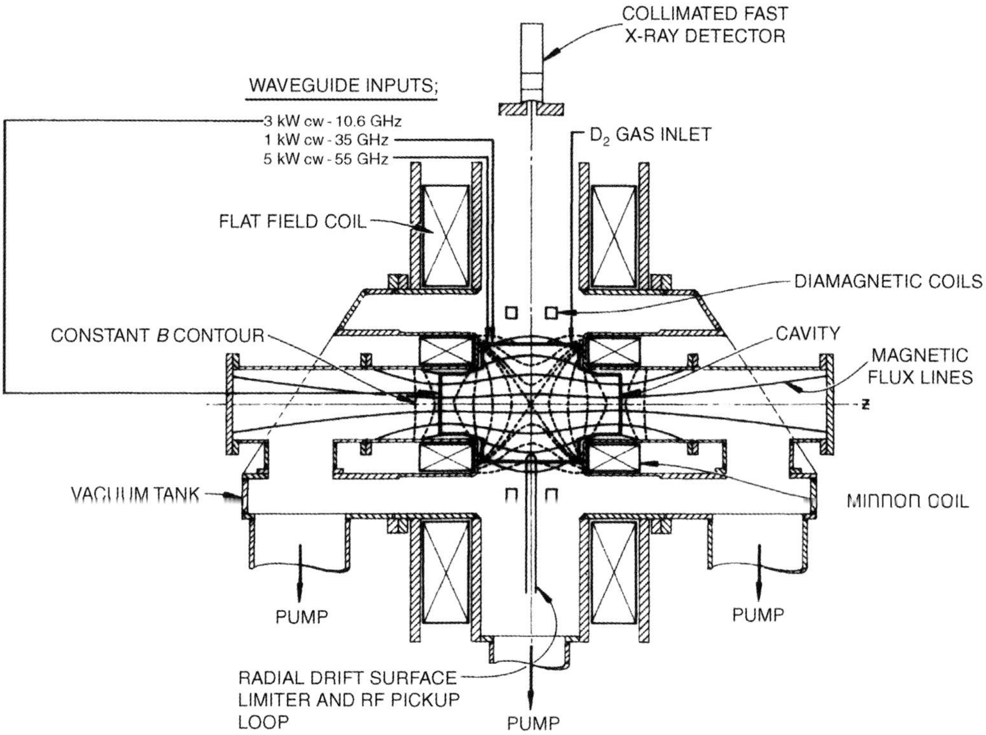

Figure 8.3 Schematic representation of the ELMO Facility.

diamagnetic stored energy reached 150 J and the rate of neutron production was as high as $5 \times 10^9\ s^{-1}$. As in the earlier PTF experiments, the neutron flux was attributed to Coulomb dissociation of deuterons by relativistic electrons.

Dandl's explanation for the increase in hot-electron density with the UORH was based on the assumption that velocity–space microinstabilities were limiting the confinement time of the warm-electron population and the experimental observation that these instabilities were suppressed by the UORH. From the conditions for steady-state particle balance in the three groups of electrons we obtain, approximately:

$$n_{e1} = n_o[n_{e2}\langle\sigma_{ion}v_e\rangle_2 + n_{e3}\langle\sigma_{ion}v_e\rangle_3]\tau_1[1 + \tau_1(dW/dt)_1/\Delta W_{2,1}]^{-1}$$

$$n_{e2} = n_{e1}\tau_2[(dW/dt)_1/\Delta W_{2,1}][1 + \tau_2(dW/dt)_2/\Delta W_{3,2}]^{-1}$$

$$n_{e3} = n_{e2}\tau_3[(dW/dt)_2/\Delta W_{3,2}][1 + \tau_3(dW/dt)_3/\Delta W_{escape}]^{-1}$$

If the warm-electron loss rate, $1/\tau_2$, is reduced when the instabilities are suppressed, both n_{e2} and n_{e3} will increase, as will the density of the cold-electron group, n_{e1}, assuming that their (resonant) heating rate, $(dW/dt)_2$, is not significantly increased by the UORH. Dandl carried out experiments to quantify the conditions for suppression of these instabilities with the results shown in Figures 8.5 and 8.6.

Just as in the PTF instability observations, the appearance of bursts of radiation at half the resonant-heating frequency was taken as one signature of the instability; and

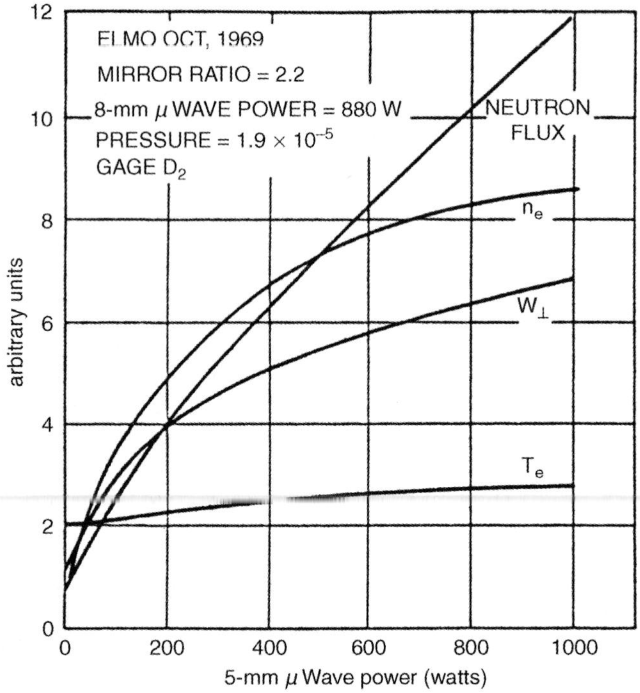

Figure 8.4 Response of the plasma parameters to two-frequency heating in ELMO.

the UORH power needed to suppress this mode was determined as a function of resonant heating power for two different gas pressures, as displayed in Figure 8.5. The resulting changes in stored energy are then shown in Figure 8.6.

In order to determine the radial distribution of the hot-electron density, a water-cooled skimmer probe was inserted radially from outside the plasma. As it was moved progressively deeper into the plasma, the experimental indicators of hot-electron density such as the magnetic flux change, the x-ray intensity, the microwave noise, and the neutron flux were monitored. In this way, it was found that the hot-electron plasma was localized largely within a hollow annulus and had the shape of a short cylindrical shell. More refined measurements of the magnetic flux change were then undertaken using the original skimmer probe plus a second, L-shaped skimmer probe mounted eccentric to the axis of the chamber. The tip of this second "dog-leg" probe could be moved outward in radius from 6.3 cm to the cavity wall at a radius of 15.2 cm. Results of measurements with both skimmer probes are shown in Figure 8.7.

Here the stored energy obtained from diamagnetic measurements is displayed as (1) the outer probe is moved inward with the inner probe removed; (2) the inner probe is moved outward with the outer probe removed; and (3) the outer probe is moved inward with the inner probe tip at a radius of 7 cm. The difference in the radial boundaries of the annulus as deduced from the first two scans – roughly 2 cm – was attributed to the gyrodiameter of the energetic electrons making up the shell. Thus, if

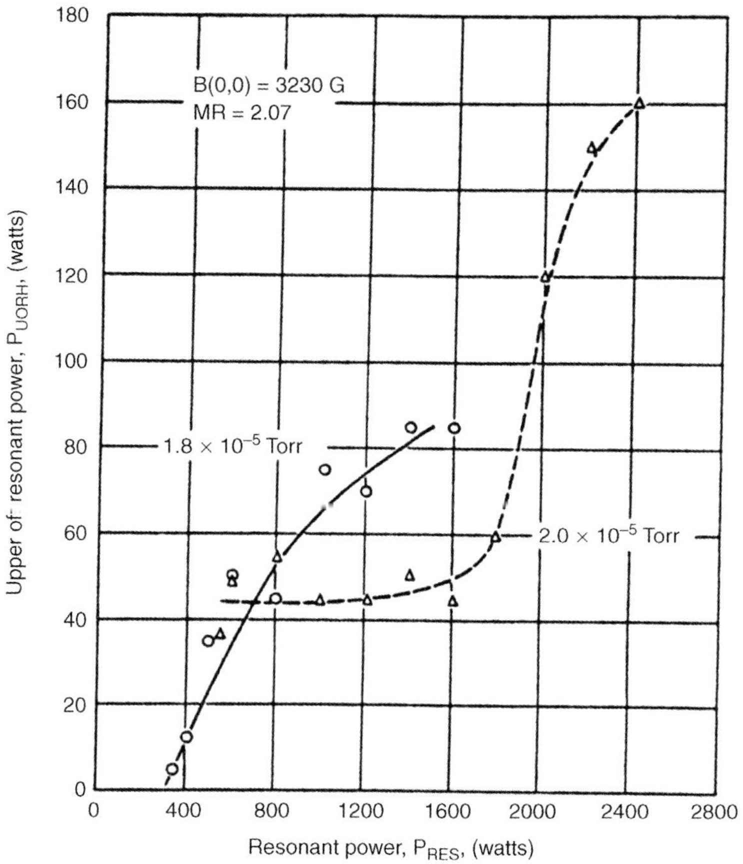

Figure 8.5 UORH power required to suppress instabilities in ELMO.

the average gyrodiameter of the energetic electrons were 2 cm, both probes would indicate the annulus to extend radially from an inner surface at a radius of 6 cm to an outer surface at a radius of 12 cm. Since the strength of the vacuum magnetic field in the region occupied by this annulus is roughly 0.7 T in the case shown here, electrons with 1 cm gyroradius would have energies around 1.7 MeV.

Because the skimmer probes caused major perturbations to the hot-electron plasma, these inferred values of the mean radius and thickness of the hot-electron annulus were only regarded as semiquantitative. Subsequently, the diamagnetic field produced by the plasma was measured at many closely spaced points along the axis as well as outside the cavity using Hall probes. Although there is in principle no unique set of currents corresponding to the measured magnetic fields, it was possible to construct plausible spatial distributions of circular current elements that were consistent with all of the diamagnetic measurements. This approach to determining the spatial distribution of the hot-electron pressure was refined in later investigations using guiding-center fluid models of the plasma equilibrium rather than discrete current loops as the basis for computing the magnetic fields.

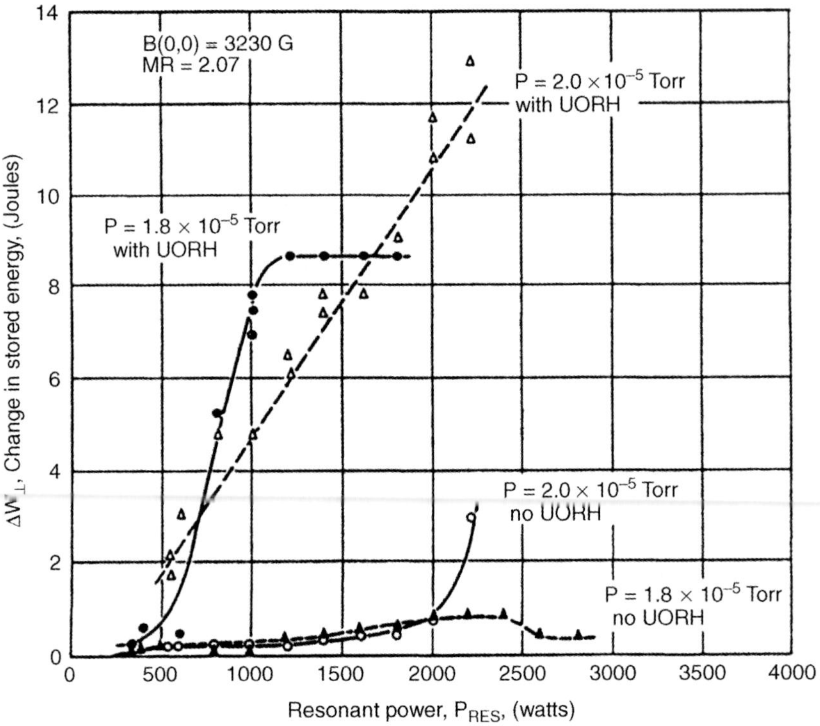

Figure 8.6 Increase in the stored energy resulting from UORH in ELMO.

In all cases, the annular structure of the hot-electron plasma was unambiguously confirmed.

The reason for the annular hot-electron plasma in ELMO was not entirely clear. In later experiments, notably Test Plasma by Microwaves (TPM) and SM-1, where similar annular hot-electron plasmas were produced, it appeared that the annulus formed at radii associated with second-harmonic heating. However, in ELMO the radial profile of magnetic intensity was much flatter than that in the simple magnetic-mirror fields, and the nonrelativistic second-harmonic resonance surface was just barely inside the cavity and, therefore, at a significantly larger radius than the annulus. Dandl speculated that the radial position of the annulus was probably due to the conditions for cold-plasma stabilization of the flute-like modes of instability, together with the local diamagnetic reduction of the magnetic intensity that would permit second-harmonic heating of the relativistic electrons. The plasma current distribution consistent with all magnetic measurements in ELMO yielded the striking self-consistent magnetic field displayed in the upper half of the drawing in Figure 8.8.

The contours of constant magnetic intensity form nested closed surfaces characteristic of a magnetic well. In a sense, the annular hot-electron plasmas, often called "ELMO rings," acted as transparent magnetic coils. The possibility of using the diamagnetic properties of these ELMO rings to stabilize a bumpy torus plasma

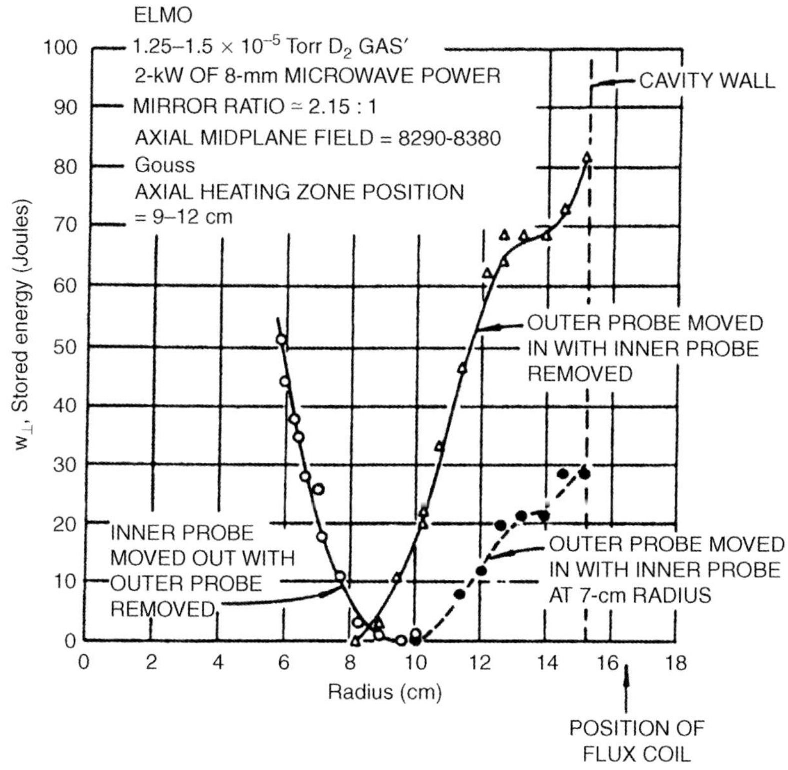

Figure 8.7 Skimmer probe measurements of the radial thickness of the hot-electron annulus in ELMO.

confinement device gave rise to the ELMO Bumpy Torus concept to be discussed later [11].

8.3 Unstable Electromagnetic Waves in the TPM [12]

In the preceding sections, we have seen how experiments in PTF exhibited a type of high-frequency microinstability when a threshold was exceeded that was plausibly related to the temperature anisotropy of the warm-electron group. Later experiments in ELMO showed that virtually all high-frequency instabilities could be suppressed by UORH, resulting in large increases in the hot-electron density and the diamagnetic stored energy. Experiments conducted in the TPM facility at the Institute of Plasma Physics in Nagoya, Japan, under the direction of H. Ikegami, provided a relatively detailed description of one type of high-frequency instability that could be triggered in the afterglow of a pulsed ECH discharge, heated at 6.4 GHz, under circumstances that were once again related to temperature anisotropy of the hot-electron component of the plasma. This instability is of particular interest in that it led to large amplitude

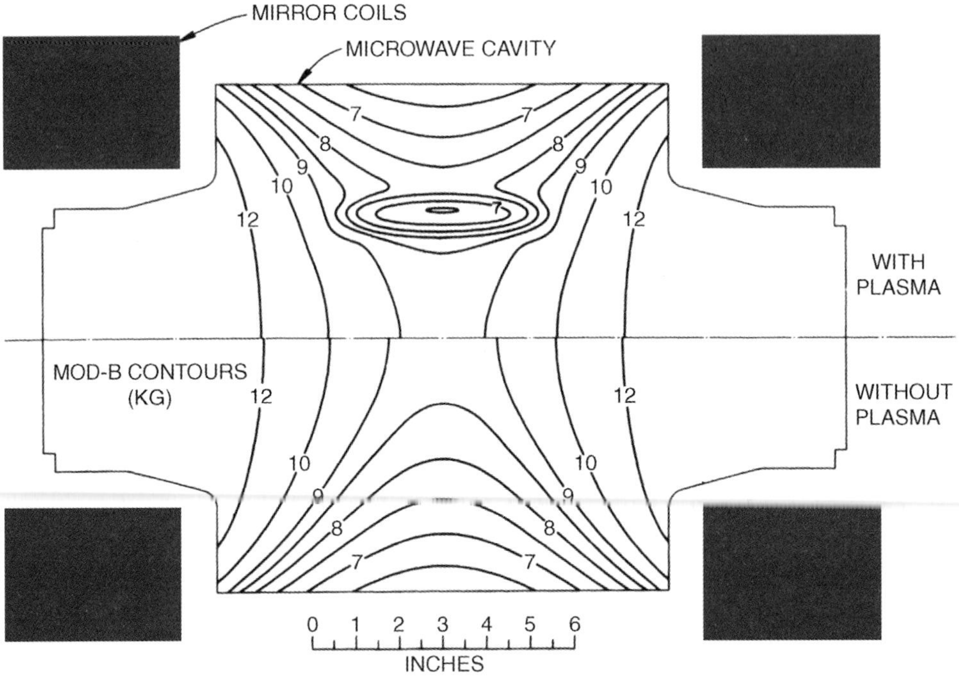

Mod-B Contours for ELMO System with Plasma Profile Derived from Diamagnetic Data.

Figure 8.8 Diamagnetic changes in the ELMO magnetic fields.

standing electromagnetic waves whose axial wavelength and frequency were measured directly in a series of experiments. The results have proven to be of interest not only for their implications with regard to laboratory hot-electron ECH plasmas but also for space plasmas.

We begin with some considerations that follow directly from the magnetic configuration in TPM, a simple magnetic mirror with a mirror ratio of 3.4:1. The plasma was contained in a cylindrical vacuum vessel whose center section was 40 cm long and 30 cm in inside diameter. A schematic diagram of TPM is shown in Figure 8.9 and details of the magnetic field are shown in Figure 8.10.

The fundamental resonance surface, where B = 2280 G, lies just inside the end wall of the cavity; and a second-harmonic resonance surface, where B = 1140 G, extends inward to within 5–6 cm of the axis on the midplane. Of particular significance is the flux surface shown in Figure 8.10 that extends from the second-harmonic resonance zone at the midplane to the "corner" of the vacuum chamber, where the 30-cm ID section ends and the 10-cm ID sections begin. Plasma formed outside this flux surface can escape much more rapidly than the plasma inside it, and in this way the corner, together with the fundamental resonance surface, acts much as a radial limiter. It also provides an effective means for stabilizing flute-like instabilities by ensuring a low-impedance path for

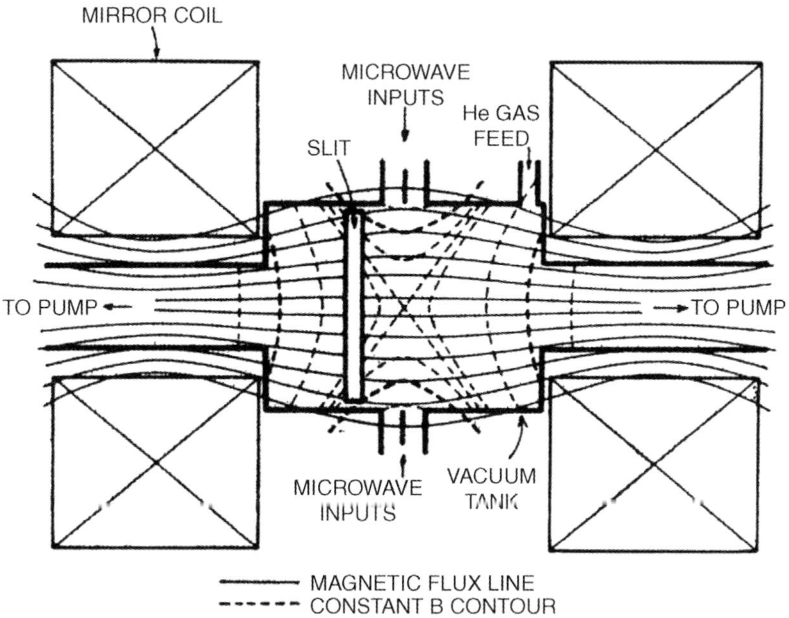

Figure 8.9 Schematic representation of the Test Plasma by Microwaves (TPM) Facility.

cold-plasma currents flowing along magnetic lines of force into the conducting end walls just at the radial position where the plasma pressure gradient is at its most negative value.

The magnetic lines of force are tangential to the second-harmonic resonance surface at a radius on the midplane of approximately 6 cm. Consequently, second-harmonic heating of warm electrons can be very rapid in this region where $\partial B/\partial s = 0$. This rapid heating will yield a high-temperature anisotropy, since the second-harmonic resonance surface is at or near the midplane throughout the volume inside the limiting flux surface. In effect, the second-harmonic heating in TPM increases only the perpendicular kinetic energy of the warm electrons with negligible change in their parallel kinetic energy. The axial extent of the second-harmonic heating zone is only 5 cm at the limiting flux surface; the experimentally observed length of the hot-electron annulus, 10 cm, is about twice the separation of the resonance zones. Under typical heating conditions, the hot-electron component occupied an annular shell with an inner diameter of roughly 10 cm. The location of this annulus was found to correspond with the innermost position of the second-harmonic resonance surface; namely, the surface on which the magnetic intensity was 1140 G. The ratio of the field at the second-harmonic resonance surface to the field at the midplane in this heating zone does not exceed 1.06 so that $T_\perp/T_\| > (M-1)^{-1}$ could exceed 15. The z-dependent hot-electron density described in Chapter 6 is plotted in Figure 8.11 for $T_\perp/T_\| = 15$ and with the mirror ratio set at 3.634, the value on the limiting flux surface of the second harmonic resonance zone. The strong localization of the hot-electron population for $z < 10$ cm is evident.

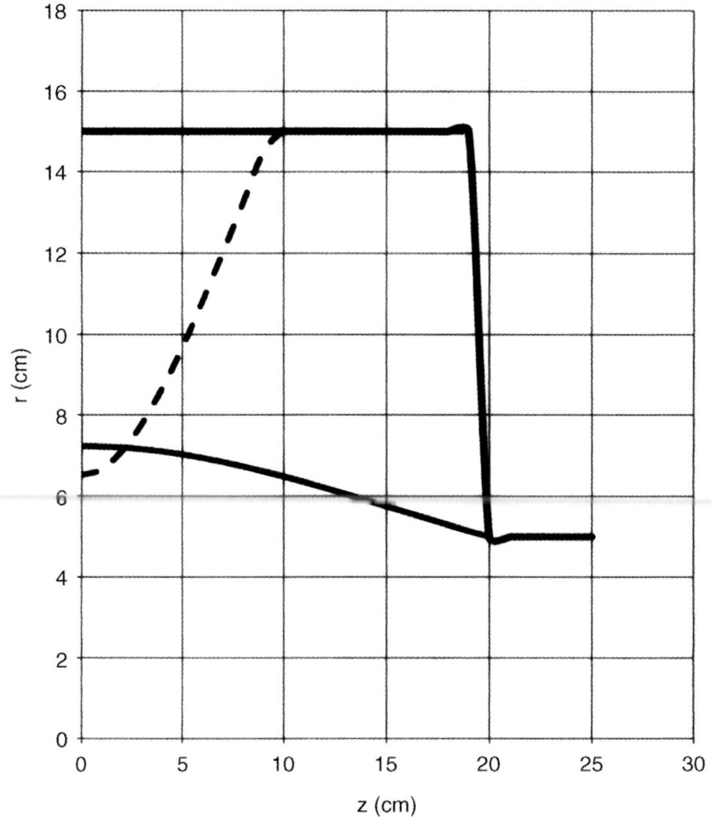

Figure 8.10 Detail of the TPM magnetic field showing the limiting flux surface and the second-harmonic heating zone relative to the vacuum chamber walls.

The characteristic gradient scale length of the magnetic intensity in the second-harmonic resonance zone is

$$L = \left|\partial \ln B/\partial r\right|^{-1} = 68\ \text{cm}.$$

If the limit for adiabatic confinement is $\rho_{max} = (0.05–0.06)L = 3.4–4.1$ cm, then the maximum electron energy for adiabatic confinement is $E_{max} = 0.767–0.985$ MeV, in keeping with the experimental fact that no electrons with energies above 1 MeV were observed. Note also that since the average hot-electron gyrodiameter was 2–3 cm, the radial positions of the electrons cannot be localized more precisely than this. The hot-electron annulus is, therefore, only a few gyrodiameters thick in the radial direction.

The composition of the afterglow plasma differs significantly from that of the initial TPM plasma, which was generated by launching a 20-ms pulse of 6.4 GHz microwave power at levels up to 5 kW from two waveguide ports, spaced 180° apart in the radial wall of the stainless steel vacuum chamber, as indicated in Figure 8.9. This initial plasma consists of the three groups of electrons discussed earlier: a cold-

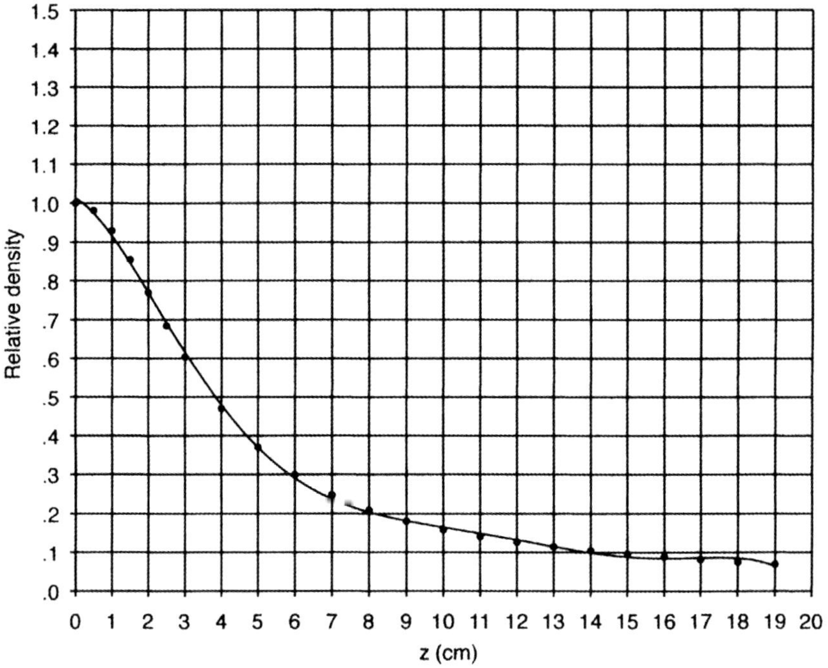

Figure 8.11 Axial distribution of hot-electron density in TPM.

electron component with a temperature of 20 eV and a hot-electron population with a temperature of 100–150 keV and a density around $10^{10}\,cm^{-3}$. We can infer the existence of a warm-electron component with a temperature around 10 keV from the 5 ms decay time observed following the end of the heating pulse, when roughly 80% of the initial plasma was lost. Since the experiments were typically conducted at a helium pressure of $2.5–3 \times 10^{-4}$ Torr, corresponding to a neutral density of roughly $9 \times 10^{12}\,cm^{-3}$, 5 ms would be the Coulomb scattering time for 12 keV electrons, keeping in mind that $\ln\Lambda \approx 10$ for these neutral dominated plasmas. The corresponding decay time for the 100 keV hot electrons is in excess of 100 ms. Recall that it is the population of warm electrons, sometimes called "feed electrons," that supplies the hot-electron group.

After the decay of the warm-electron component, the afterglow plasma contains only two groups of electrons: the 100–150 keV hot electrons with a density around $n_{e3} \approx 10^{10}\,cm^{-3}$, and the cold electrons with density n_{e1}, resulting from the ionizing collisions of hot electrons with helium atoms. Since there is no microwave heating of the afterglow plasma, the cold electrons remain at their initial temperature and exert negligible pressure on the cold ions. We can extrapolate data from McDaniels *et al.* [13] on scattering of He ions in helium gas to lower energies and estimate the corresponding cross section to be less than $10^{-14}\,cm^2$; this will result in the mean free path for scattering of the He ions to be roughly 10 cm. We, therefore, discount the possibility of diffusion and assume that cold ions drift out of the vacuum chamber with their thermal speed, $1.1 \times 10^5\,cm\,s^{-1}$, corresponding to their temperature, 1/40 eV. The loss time is thus given by

$$\tau \approx 20\,\text{cm}/(1.1 \times 10^5\,\text{cm/s}^{-1}) \approx 200\,\mu\text{s}.$$

Following the rapid loss of hot electrons resulting from a burst of instability, the cold plasma was indeed observed to decay in roughly 200 μs, supporting our assumptions regarding cold-plasma transport in the afterglow. With this model, we can readily estimate the relative cold-plasma density as

$$n_{e1}/n_{e3} = n_o \langle \sigma_{ion} v_e \rangle_3 \tau_1$$

We display a plot of n_{e1}/n_{e3} versus the helium pressure, p_0, in Figure 8.12.

Note that for p_0 around 2.5–3 × 10^{-4} Torr (0.25–0.3 μm), the typical operating range, the relative cold-plasma density is predicted to be $n_{e1}/n_{e3} \approx 10$, as observed experimentally.

The high-frequency instability studied in TPM occurred spontaneously under some conditions at pressures below 10^{-4} Torr, and could be triggered artificially for pressures ranging from (1–8) × 10^{-4} Torr by injecting a short pulse of 6.4 GHz power from 1–50 ms after the end of the main heating pulse. At pressures above 8 × 10^{-4} Torr, it was not possible to trigger the instability. Typically the triggering pulse would have a duration of a few tens of microseconds with a power level less than 100 W. The microwave radiation associated with the occurrence of the instability took the form of standing electromagnetic waves with an axial wavelength, $\lambda = 19$ cm and a fundamental frequency $f = 2.1$ GHz, corresponding to an index of refraction $n = 0.75$.

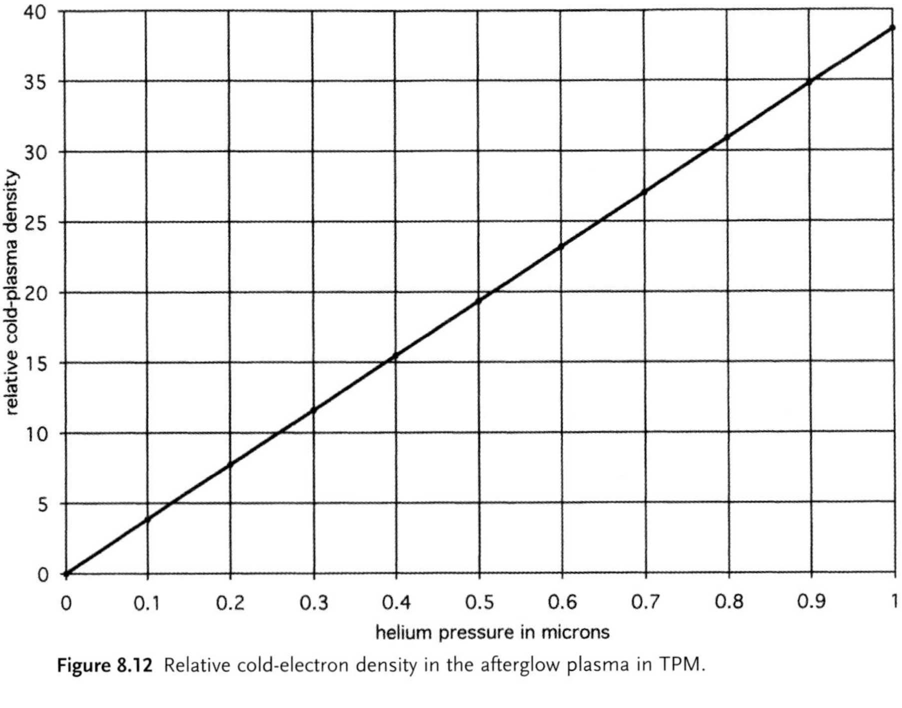

Figure 8.12 Relative cold-electron density in the afterglow plasma in TPM.

To interpret this instability theoretically, the TPM scientists identified the growing wave as a whistler while recognizing the apparent discrepancy between the observed wavelength and that predicted by the customary whistler dispersion relation. At one point, they suggested that because of the presence of nearby conducting vacuum chamber walls, the observed standing wave was possibly a cylindrical waveguide eigenmode. The following alternative explanation may be more plausible.

If the waves are constrained by the conducting boundaries to propagate along the axis of the chamber, their propagation cannot be exactly parallel to the magnetic lines of force – this is because the field lines passing through the hot-electron annulus region are inclined at an average angle of 10° to the axis. Waves propagating at even so small an angle from parallel are no longer purely left- or right-circularly polarized, but instead contain a mixture of all three components. Indeed, if the plasma density varies along the magnetic lines of force, the polarization can change substantially and even reverse at the O-mode cutoff. We illustrate this "cross-over" phenomenon [14] by plotting the fast- and slow-wave indices of refraction against the density-related parameter, ω_{pe}/ω over the experimental range of interest in TPM for $\theta = 10°$.

Everywhere except in the neighborhood of the O-mode cut off, $\omega_{pe}/\omega = 1$, the two waves have indices of refraction that are close to the corresponding values for parallel propagation. But at $\omega_{pe}/\omega = 1$, the fast wave is cut off while the slow-wave index drops to unity, the free space value, and then merges with the $\theta = 0$ fast-wave index of refraction. Just above a narrow interval of evanescence, the former fast wave resumes propagation with approximately the $\theta = 0$ whistler index of refraction.

The polarizations of the two waves undergo a related change near cutoff: if $\omega_{pe}/\omega < 1$ or >1, the slow wave is predominantly right-hand and the fast wave is predominantly left-hand circularly polarized; but in the neighborhood of $\omega_{pe}/\omega = 1$, the relative amplitudes of the right- and left-hand components change to match the reconnection of the slow wave to the fast wave. The slow wave emerges from the crossover as a mainly left-hand fast wave and the fast wave reappears after the evanescent interval as a mainly right-hand slow wave. We conclude that in a plasma where the density varies along the magnetic field, the two waves are likely to be coupled unless they can propagate exactly parallel to the magnetic field.

As mentioned earlier, the frequency and wavelength of the electromagnetic standing wave observed in the TPM afterglow experiments were 2.1 GHz and 19 cm, corresponding to an index of refraction of 0.75. From Figure 8.13, we see that this is the index of refraction for a fast wave propagating at an angle of 10° in a plasma whose density is such that $\omega_{pe}/\omega = 1.1$, which in this case is a density of 6.5×10^{10} electrons cm^{-3}. Under the same conditions, the slow-wave index of refraction is 2.25. From the discussion in Chapter 7, we can evaluate the anisotropy parameter, A, and the frequency for maximum growth rate, x_{max}:

$$A = (T_{\perp}/T_{||} - 1)n(\alpha_{||}/c) = 4.725$$

$$x_{max} = [1 + (1 + 2A^2)^{1/2}]/(2A) = 0.82$$

leading to

$$\omega_{max}/\Omega = 0.78;$$

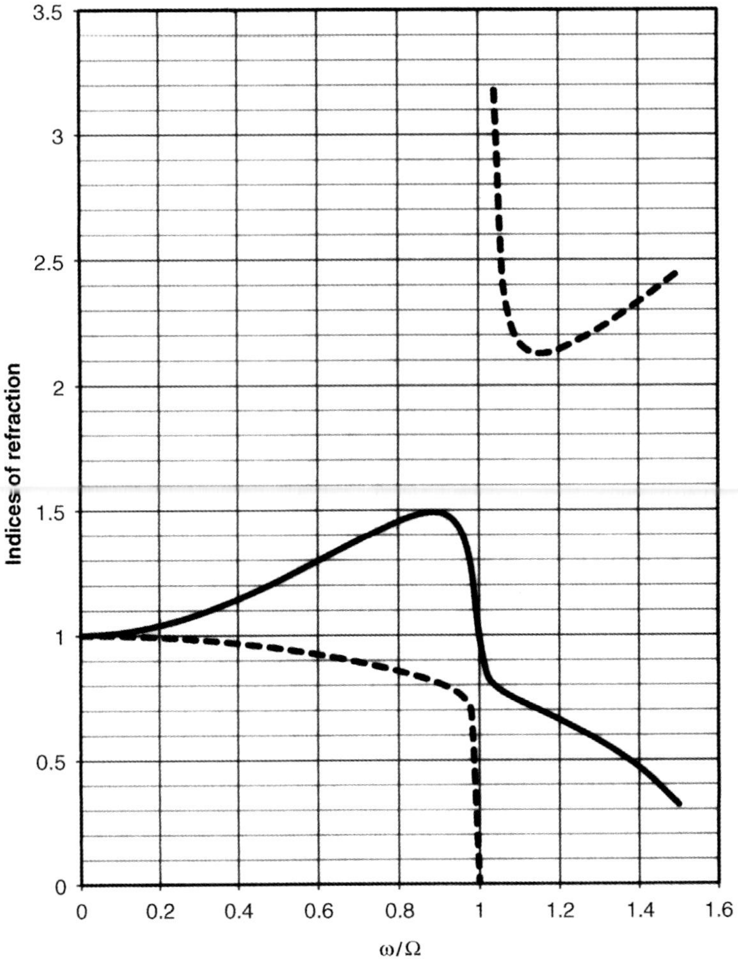

Figure 8.13 Indices of refraction for the cold-plasma electromagnetic waves in TPM versus the ratio of plasma frequency to wave frequency for $\omega/\Omega = 2/3$.

and since $\omega_{max}/\Omega_o = (1/\gamma)\omega_{max}/\Omega$, the observed and predicted frequencies are equal for $\gamma = 1.17$, or electron energies of 90 keV, in reasonable agreement with the 100 keV observed in the experiments.

These experiments illustrate an important point regarding the assumption that the waves of interest propagate in finite-size systems as though they were propagating in an infinite homogeneous plasma immersed in a uniform magnetic field, sometimes referred to as the quasioptical assumption. For the waves in TPM, as in many small, low magnetic-field experiments, wavelengths are comparable to the dimensions of the system and cavity eigenmodes are a more realistic model than plane waves. Nonetheless, the conditions for optimum energy transfer between the wave and the plasma remain valid.

8.4 Heating Experiments in AMPHED [15]

In 1988–1989, Quon and Dandl [16] carried out a series of experiments designed to elucidate different heating dynamics underlying three distinct ECH methodologies, namely, high-field launch whistler-wave heating, low-field launch O-mode heating, and UORH. These experiments demonstrated how ECH could be used for preferential heating of particular classes of electrons in plasmas containing two or more groups of electrons with different average energies.

The experiments were performed in the AMPHED facility using a 2 : 1 magnetic-mirror configuration with a 60-cm long center section in which the magnetic field was essentially uniform. The DC current energizing the magnetic coils could be adjusted precisely to place the resonance surface at selected positions in this magnetic field. The ratio of the magnetic intensity on axis, B(z), to the current in the coils, I_c, is displayed in Figure 8.14.

Whistler waves were launched by a 800 W, 2.45 GHz commercial microwave oven coupled into the high-field region at the mirror throat using a dielectric-loaded C-band waveguide terminated in a tapered Teflon slab. The local wavelengths of the resulting waves were evaluated by measuring the spatial correlation function E(z)cos [k(z)z] at closely spaced points along the axis of the magnetic field. In the presence of

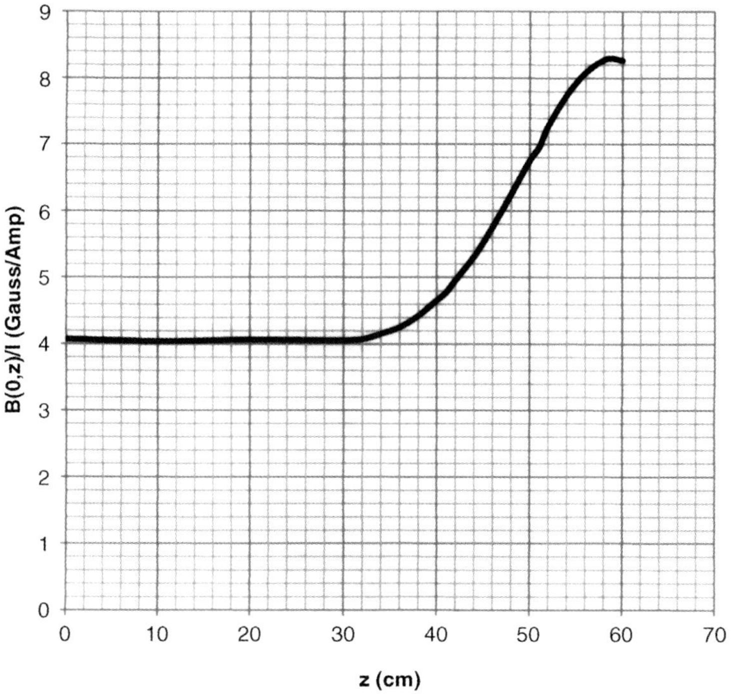

Figure 8.14 The ratio of the magnetic field on axis to the coil current in the "flat-field" configuration of AMPHED.

over-dense plasmas with $n_e \approx 1.5 \times 10^{12}\,cm^{-3}$ the index of refraction inferred from the measured wavelengths ranged from 3.7 at a point where $\Omega/\omega = 1.6$ to 10.2 where $\Omega/\omega = 1.275$. These values are consistent with those predicted by the cold-fluid limit of the whistler-wave dispersion relation for $\omega_{pe}^2/\omega^2 \approx 10\text{–}20$. The wave amplitude was observed to decrease as the wave approached the resonance surface and was unobservably small beyond resonance.

We can account reasonably well for the experimentally observed values of electron density and temperature using the particle and power balance model described earlier in Chapter 6. For argon gas, the model predicts a minimum pressure above which equilibria can exist given by the condition $p_o L \geq 3.3 \times 10^{-4}$ Torr cm, where L, the half length of the plasma column, is approximately 41 cm in the AMPHED experiments cited here. Thus, for pressures greater than 8×10^{-6} Torr the electron temperature is predicted to decrease from roughly 25 eV at 0.01 mTorr to roughly 4 eV at 0.1 mTorr. The predicted electron temperatures agree with the experimental results obtained by Quon and Dandl [16] in the pressure range between 0.01 and 0.1 mTorr. However, the theoretical model predicts values of the electron density that are somewhat higher than the experimental values if it is assumed that all of the incident 800 W of microwave power is absorbed by the electrons we have labeled as "Group 1." The discrepancy suggests that some fraction of the incident power, perhaps as much as 20%, is lost through processes not included in the Group 1 particle and power balance model.

Quon and Dandl [16] used a multigrid energy analyzer to measure the distribution of electrons in parallel energy at a point where the magnetic intensity was reduced locally to 600 G. If the magnetic moment and total energy are constants of the electron motion, the distribution in parallel energy can be related to a distribution in perpendicular energy. The experiments revealed two distinct groups of electrons: a low-energy group with pressure-dependent temperatures similar to the temperatures determined by the Langmuir probe measurements; and a higher temperature group whose average parallel energy was between 50 and 70 eV and independent of pressure. At the resonance surface, the perpendicular energy of these electrons was estimated from the adiabatic invariance argument to be between 160 and 220 eV. Since none of these electrons would have been reflected from the grounded energy analyzer structure, Quon and Dandl [16] concluded that these electrons had gained roughly 200 eV of perpendicular energy in a single pass through the resonance layer. Moreover, these energetic electrons were shown to play a critical role in producing the over-dense plasma observed when the ambient gas pressure exceeded a value somewhere between 0.2 and 0.3 mTorr.

The condition under which the energetic electrons have an ionizing collision with an argon atom before they return to the resonance surface that is illuminated by microwave power is $n_o \sigma_i v \tau \geq 1$. Here n_o is the density of argon atoms, σ_i is the cross section for electron impact ionization of argon, v is the speed of the energetic electron, and τ is the time for the electron to make one complete bounce through the magnetic mirror. For these experiments, the magnetic intensity in the 60-cm long flat field region was 732 G and the (parallel) velocity in this region is given by

$$v_{||} = [(2\varepsilon/m)(1-732/875)]^{1/2} = v(1-732/875)^{1/2}.$$

We approximate the average parallel velocity in the 11 cm long regions between the resonance surfaces and the flat-field region as half of the value in the flat field region. In this way, we estimate $\tau = 208\,\text{cm}/v_{||}$ and $v\tau = 208\,\text{cm}\; v/v_{||}$. The ionization cross section in this energy range is approximately $3 \times 10^{-16}\,\text{cm}^2$ giving for our critical condition

$$n_o \geq 6.5 \times 10^{12}\,\text{cm}^{-3} \quad \text{or} \quad p_o \geq 0.2\,\text{mTorr},$$

in good agreement with the experiment.

We can estimate the fractional density of the fast-electron group if we assume that they provide most of the ionization to sustain the high-density (thermal) plasma in steady state:

$$n_{e,fast} n_o \sigma_i v_{fast} \approx n_{e,thermal}/\tau_{thermal}$$

As before, we assume that the thermal electron lifetime is given by the half length of the plasma column and the ion acoustic speed:

$$\tau_{thermal} = L/c_s$$

With these assumptions, we find that the required density of fast electrons is about 1% of the total density:

$$n_{e,fast} \approx 1.6 \times 10^{10}\,\text{cm}^{-3}$$

Our earlier discussion in Chapters 4 and 5 regarding the absorption of whistler waves and the heating of individual electrons suggests a plausible explanation for the source of the fast electrons. Recall that the onset of damping of the whistler wave propagating along the magnetic field toward the resonance surface was determined by electrons in the tail of the energy distribution of the thermal electrons with velocities antiparallel to the wave vector: $\omega - kv_{||} = \Omega$. These electrons resonate with the wave before it has been heavily damped by the bulk of the thermal electron population and consequently may be given disproportionately larger energy increments. We can estimate the maximum energy these electrons can gain in a single transit through the resonance layer using results from the analysis presented in Chapter 5.

Since $k_\perp = 0$ for the whistler waves, the time-averaged change in the electron's perpendicular energy after a single transit of the resonance layer is predicted to be given by [17]

$$\Delta W_\perp = G^2/mc^2 - G \cos\phi_{res} v_\perp(0)/c,$$

where

$$G \equiv eE_\perp t_{eff} c/2.$$

We can estimate $E_\perp$ from the Poynting vector as in Chapter 4, using the experimental value of the index of refraction, n:

$$E_\perp^2 = (P/A)/(\varepsilon_o cn),$$

where $P = 800$ W is the incident microwave power and A is the area of the resonance surface. In the flat-field region, $B = 732$ G and $r = 15$ cm in these experiments. Therefore at the resonance surface where $B = 875$ G, we estimate that the area of the resonance surface is approximately $A = \pi(15\,\text{cm})^2(732/875) = 591\,\text{cm}^2$. If $n = 3.8$, we find $E_\perp = 11.6\,\text{V cm}^{-1}$.

To estimate the duration of resonance, t_{eff}, we assume that the electrons gaining the greatest energy are turning just beyond the resonance surface so that

$$t_{eff} \approx 2[(3\pi/2)/|v''(t_{res})|]^{1/3}$$

where

$$v''(t_{res}) \approx [(dv_{||}/dt)(d\Omega/dz)]_{res}.$$

Assuming $m(dv_{||}/dt)_{res} = \mu(\partial B/\partial z)_{res} = \varepsilon(\partial \ln B/\partial z)_{res}$ with $\Omega_{res} = 2\pi f$, we find

$$v''(t_{res}) \approx 2\pi f(\varepsilon/m)/L_B^2.$$

We estimate that the magnetic scale length at the resonance surface is $L_B = 34$ cm in these experiments and the microwave frequency is 2.45 GHz. The thermal electron temperature at the higher pressures where over-dense plasmas are formed is around 4 eV, but we are concerned here with electrons in the tail of the thermal energy distribution. In fact, if the relative density of fast electrons is about 1% of the thermal population, we are lead to estimate that $\varepsilon \approx (1.82)^2 \times 4\,\text{eV} \approx 13\,\text{eV}$ giving an estimated value for the duration of resonance $t_{eff} \approx 5 \times 10^{-8}$ s. Finally we obtain $G \approx 8700$ eV so that $\Delta W_\perp = 148\,\text{eV} - 62\,\text{eV}\cos\phi_{res}$ or

$$86\,\text{eV} < \Delta W_\perp < 210\,\text{eV}$$

in reasonable agreement with the experimental result that the maximum value of $\Delta W_\perp$ was between 100 and 200 eV.

Although coherent bursts of energetic electrons resulted from whistler-wave heating when the resonance zone was in the uniform magnetic field region, no relativistic electron plasmas could be formed with high-field launch, whistler-wave heating. On the other hand, launching microwave power with the O-mode polarization in the low-field region at the midplane of the magnetic mirror produced plasmas with substantial populations of electrons having energies in the range of 30–150 keV. The generation of these relativistic-electron plasmas was found to be optimal under two different conditions of magnetic field strength; namely, with DC currents in the field coils between 110 and 130 A and also between 180 and 200 A. For coil currents of 120 A, the fundamental resonance surface is roughly 52 cm from the midplane and near the position of the mirror throat at 59 cm. Relativistic electrons with $\gamma = 1.12456$, corresponding to a kinetic energy of 64 keV, would then experience second harmonic resonance in the uniform field region. For coil currents of 190 A, the (nonrelativistic) fundamental resonance is around 39 cm from the midplane and less than 10 cm from the uniform field region.

In the course of these experiments, the heating rate of electrons with energies below 1.56 keV was measured by exploiting the large difference between the electron bounce frequency and the much slower azimuthal drift, as discussed in Chapter 3. A probe inserted radially inward from the surface of the plasma will "skim" off the outer layer of plasma, since the rapid bounce motion ensures that all electrons will be intercepted by this "skimmer probe" before their precession carries them azimuthally past the probe. Quon and Dandl used a fixed skimmer probe to collect all electrons with radial positions greater than 12 cm and then used an azimuthally movable ionization chamber probe, whose sensitive element was in the radial interval intercepted by the skimmer probe, to record the azimuthal flux of electrons with energies greater than 1.56 keV – the threshold energy of the detector. The signal from the detector showed an onset at an angle of 230° with respect to the skimmer probe, permitting the following plausible albeit approximate unfolding of the average electron heating rate (dW/dt).

The azimuthal guiding center drift speed was given in Chapter 3 by Eq. (3.32):

$$v_{\perp gc} = (2W_{\|} + W_{\perp})\nabla B \times \mathbf{B}/(eB^3),$$

which we approximate as

$$r d\phi/dt \approx W/(eBR_c) \approx [W_o + (dW/dt)t]/(eBR_c).$$

Thus, the time-dependent azimuthal displacement will have the form

$$\Delta\phi = \phi(t) - \phi_o \approx [W_o t + (dW/dt)t^2/2]/(eBrR_c).$$

The electron energy at the conclusion of this azimuthal displacement is

$$W = W_o + (dW/dt)t.$$

In this way, Quon and Dandl estimate the heating rate in the low-field launch experiments in AMPHED to be 60 MeV s^{-1}. Since the density of the relativistic electrons in these experiments is around 10^{10} cm^{-3}, one can estimate that the power (density) absorbed by these electrons is roughly 0.1 W cm^{-3}. The incident 2.45 GHz power is only 800 W and the total volume of the AMPHED cavity is 57 l, suggesting that the power is not distributed uniformly throughout the cavity, but rather is concentrated in the local region where the experimental measurements were made. Accordingly, Quon and Dandl hypothesized that the plasma filled cavity acts as an open resonator in a whispering gallery mode, an effect that they then demonstrated in an atmospheric pressure mock up. The observed heating rate is consistent with an RF electric field strength $E_{\perp} \approx 35$ V cm^{-1}.

References

1 R.A. Dandl *et al.*, *Nucl. Fusion* **4**, 344 (1964).

2 W.B. Ard, R.A. Dandl, and R.F. Stetson, *Phys. Fluids* **9**, 1498 (1966).

3 W.B. Ard, M.C. Becker, R.A. Dandl, H.O. Eason, A.C. England, and R.J. Kerr, *Phys. Rev. Lett.* **10**, 87 (1963).

4 R.L. Freeman, and E.M. Jones, "Atomic Collision Processes in Plasma Physics Experiments", Culham Laboratory Report CLM-R-137 (1974).

5 W.B. Ard, W.A. Hogan, and R.F. Stetson, "Electron Distribution Measurements in a

Hot Electron, Mirror Contained Plasma", Oak Ridge National Laboratory Technical Report ORNL-TM-3302, February (1971).

6 R.H. Cohen, G. Rowlands, and J.H. Foote, *Phys. Fluids* **2**, 627 (1978).

7 G.E. Guest, R.L. Miller, and M.Z. Caponi, *Phys. Fluids* **29**, 2556 (1986).

8 G.E. Guest, and D.J. Sigmar, *Nucl. Fusion* **11**, 151 (1971).

9 R.A. Dandl, H.O. Eason, P.H. Edmonds, and A.C. England, *Nucl. Fusion* **11**, 411 (1971).

10 H. Grawe, *Plasma Phys.* **11**, 151 (1969).

11 R.A. Dandl *et al.*, in *Plasma Physics and Controlled Nuclear Fusion Research (Proc. Conf. Madison, 1970)* **2**, IAEA, Vienna (1971) 607.

12 H. Ikegami, H. Ikezi, M. Hosokawa, S. Tanaka, and K. Takayama, *Phys. Rev. Lett.* **19**, 778 (1967); H. Ikegami, H. Ikezi, M. Hosokawa, K. Takayama, and S. Tanaka, *Phys. Fluids* **11**, 1061 (1968).

13 E.W. McDaniel, J.B.A. Mitchell, and M.E. Rudd, *Atomic Collisions: Heavy Particle Projectiles*, Wiley, New York (1993).

14 For an analogous crossover phenomenon, see D.A. Gurnett, and A. Bhattacharjee, *Introduction to Plasma Physics*, Cambridge University Press, New York (2005) pp. 104–107.

15 For a more complete description of the AMPHED facility, see C. Bobeldijk, Special Supplement, *Nucl. Fusion* **26**, 184 (1986).

16 B.H. Quon, and R.A. Dandl, *Phys. Fluids* **B1**, 2010 (1989).

17 G.E. Guest, M.E. Fetzer, and R.A. Dandl, *Phys. Fluids B* **2**, 1210 (1990).

Exercises

8.1. *Consider the instability observed to occur in PTF with a frequency of 5.3 GHz when the mirror ratio to the resonance surface was less than 1.27. Assume that the hot-electron temperature anisotropy satisfies our estimate that* $T_{\perp}/T_{||} \geq (M_{res} - 1)^{-1} \approx 4$.

(a) *If the free energy of the anisotropic hot electrons is transferred to electromagnetic waves propagating in the whistler mode, what is the marginally stable frequency relative to the electron gyrofrequency on the midplane and to the heating frequency?*

(b) *If* $T_{||} = 20$ *keV,* $T_{\perp} = 80$ *keV, and* $\omega_{pe}/\omega = 0.9$*, what is the frequency for which energy transferred to the waves is at its maximum rate?*

(c) *Is it possible for energy to be transferred to electromagnetic waves propagating in the O-mode?*

8.2. *Calculate the polarizations of the two waves whose indices of refraction are shown in Figure 8.13 for* $\omega_{pe}/\omega = 0.6, 0.8, 0.9, 1.1$*, and 1.2.*

8.3. *Consider the two-frequency heating experiments in ELMO using 35.7 GHz and 55 GHz power with* $B(0,0) = 8300$ *G and the mirror ratio equal to 2. The minimum and maximum magnetic intensities for good confinement are then given by* $B_{min} \approx 0.664$ *T and* $B_{max} = 1.66$ *T. Calculate the range of relativistic-electron energies (in terms of* γ*) for which the 55 GHz power can resonate at the* $n = 2, 3, 4, 5, 6, 7, \ldots$ *harmonics of the electron gyrofrequency. Show that for* $\gamma > 1.4$*, there are three or more pairs of resonance surfaces for this UORH power present simultaneously in the confined region.*

9
Electron Cyclotron Heating in Tokamaks

In this chapter, much as was discussed in Chapter 8 for magnetic-mirror confined plasmas, we summarize a number of electron cyclotron heating (ECH) experiments in tokamaks that can be interpreted using theories that were presented in earlier chapters or that suggest directions for further development of the theory. As was discussed in Chapter 4, the condition under which microwave power coupled into O-modes propagating perpendicular to the static magnetic field can illuminate the resonance surface in tokamaks is simply that the electron density should be less than the O-mode cutoff value as specified by the condition $\omega_{pe}^2 < \omega^2$. If this condition is satisfied, O-modes can reach the fundamental resonance surface if launched from either the low-field or the high-field side. Since low-field launched O-modes entail fewer technological difficulties, it has generally been considered the preferred approach for many applications of ECH, especially in large tokamaks. Accordingly, we begin this chapter with an experimental study of O-mode absorption in the Princeton Large Torus (PLT).

9.1
Ordinary-Mode Fundamental ECH Absorption in PLT

The PLT [1] was a conventional circular cross-sectional tokamak with major and minor radii of 132 and 40 cm, respectively. The experiments described in Ref. [1] utilized a low-power 71 GHz klystron microwave transmitter and heterodyne receiver. The corresponding critical electron density for O-mode cutoff is $n_e = 6.25 \times 10^{13}\,\mathrm{cm}^{-3}$ and the magnetic intensity for fundamental resonance is 25.36 kG. The low-power microwave beam was launched from the high-field side on the equatorial plane of the torus and the transmitted power was received at the opposite point on the low-field side. The location of the resonance surface, r_{res}, was varied from shot to shot by varying the magnetic intensity, B_o, on the axis of the torus, where $R = R_o : r_{res} = R_o[(B_o/B_{res}) - 1]$.

The fraction of the incident microwave power transmitted through the plasma is expressed in terms of the optical depth, τ, defined in Chapter 4: $P_t/P_i = \exp(-\tau)$. In the PLT experiments, much of the incident power is refracted away from the

Electron Cyclotron Heating of Plasmas. Gareth Guest

ISBN: 978-3-527-40916-7

receiving microwave horn by the plasma rather than being absorbed at the resonance surface. The magnitude of these refraction losses was determined experimentally by lowering the central magnetic intensity to 16.8 kG, thereby moving the resonance surface outside the plasma column. Typically, at least half of the transmission losses were found to be due to refraction with higher refractive losses at the lower central magnetic intensities.

To interpret the experimental results in terms of the theory of O-mode absorption, such as that derived in Chapter 4, it is necessary to know the electron density and temperature at the resonance surface. In the PLT experiments, these were obtained from laser Thomson scattering measurements of the radial profiles of density and temperature. Following the authors of Ref. [1], we compare the measured values of the optical depth with the values obtained from Eq. (4.25):

$$\tau = \int 2k_i dx = (\pi/2)(\omega_{pe}^2/\omega^2)(1-\omega_{pe}^2/\omega^2)^{1/2}(\alpha^2/2c^2)(\omega/c)L.$$

For an isotropic Maxwell–Boltzmann electron distribution, the average kinetic energy of the electrons is $(3/2)m\alpha^2/2 = 3/2kT_e$ so that $\alpha^2/2c^2 = kT_e/mc^2$, where k is Boltzmann's constant. We will generally express T_e in energy units and omit the factor k. For tokamaks, the scale length, L, characterizing the rate of spatial variation of the magnetic field is the major radius, R_o. We display the experimental results and the corresponding calculated values of the optical depth in Table 9.1.

The data were collected during two separate experimental runs and are presented separately in the table. It is clear that the experimental and calculated values of the optical depth are in reasonable agreement. Although the two different experimental runs give slightly different radial profiles for the optical depth, both runs indicate that $\tau > 1$ only in the core of the plasma column; i.e., for $r_{res} < 10$ cm. One might anticipate that because only the core of the plasma is optically thick, it would be only the core of the plasma that would radiate as an ideal black body. In fact, the

Table 9.1 Experimental and Calculated Values of the Optical Depth.

B_o (kG)	r_{res} (cm)	$T_{e,res}$ (eV)	$n_{e,res}$ ($\times 10^{13}$ cm^{-3})	τ_{exp}	τ_{01}
Experimental results					Calculated values
19.1	−2.58	160	0.384	0	0.057
20.2	−26.86	293	0.716	0.09	0.191
21.5	−20.09	240	0.7	0.11	0.153
23.8	−8.12	647	1.09	0.38	0.619
24.9	−2.39	900	1.32	0.876	1.019
21.5	−20.09	270	1.30	0.3	0.302
22.6	−14.37	416	0.96	0.33	0.3547
24.2	−6.04	760	2.59	1.49	1.454
25.0	−1.87	850	2.7	1.89	1.67

experiments in PLT demonstrated that when the microwave power was switched off, the entire plasma column radiated as a black body at the fundamental electron gyrofrequency. Consequently, by sweeping the receiver frequency repeatedly in time, the electron cyclotron emission could be used to measure the radial profile of the electron temperature as it evolves in time during a discharge. The rationale for this conclusion, which is supported by independent measurements of the electron temperature profile, is as follows.

The variation in the intensity of radiation, S, along a ray passing through a slab of plasma and entering the detector is given by

$$dS/ds = -\alpha S + \xi,$$

where the emissivity, ξ, and the absorptivity, α, are related by Kirchoff's law to the black body intensity, S_{bb}:

$$\xi = \alpha S_{bb}.$$

Using Kirchoff's law, we then have

$$dS/ds = -\alpha(S - S_{bb}), \quad \text{or} \quad d\ln(S - S_{bb}) = -\alpha ds.$$

Integrating from an initial position x_i to a final position x_f gives

$$S(x_f) - S_{bb} = [S(x_i) - S_{bb}]\exp(-\tau).$$

In the absence of reflected radiation, we would set $S(x_i) = 0$ to obtain

$$S(x_f) = S_{bb}[1 - \exp(-\tau)].$$

Clearly, if $\tau \gg 1$, the radiation intensity at the final position will be given by the black body value; but if $\tau \ll 1$, as is the case in PLT except in the core of the plasma, the radiation intensity will be much less than the black body value. If, however, some fraction, b, of the radiation is reflected from the stainless steel vacuum vessel, then the intensity at the initial position will be $S(x_i) = bS(x_f)$ and the intensity at the detector will be $S(x_f) = S_{bb}\,[1 - \exp(-\tau)]/[1 - b\exp(-\tau)]$, and if $b \approx 1$, the intensity of radiation reaching the detector can closely approximate the black body value. In the frequency range around the fundamental electron gyrofrequency, S_{bb} follows the Rayleigh–Jeans law to a good approximation:

$$S_{bb} = \omega^2 kT_e/(8\pi^3 c^2).$$

Efthimion *et al.* [1] concluded that $b = 0.95$ in PLT and thus the power reaching their detector is proportional to $T_e\Delta f$, where Δf is the bandwidth of the receiver. By suitably calibrating the receiver and sweeping the receiver frequency in time, they were able to obtain useful space- and time-resolved measurements of the electron temperature. This diagnostic use of electron cyclotron emission is now widely employed in tokamak experiments, together with other corroborating measurements of the electron temperature profile.

9.2
ECH-Assisted Start-up in Tokamaks

In conventional tokamaks, each discharge is initiated by inducing an electric field in the toroidal direction that first ionizes the gas in the vacuum chamber and then drives the (toroidal) current that generates and sustains the poloidal magnetic field. The electric field strength required to break down the gas is typically around 0.5 V/m for initial gas pressures around 3×10^{-5} Torr. In JFT-2 [2], for example, the loop voltage required for reliable breakdown is 20 V. The major radius of JFT-2 is 90 cm, and if the loop voltage for breakdown is proportional to the major radius as expected, then it could approach or exceed 150 V in future large tokamaks such as International Thermonuclear Experimental Reactor (ITER) [3], whose major radius has a design value of 6.2 m. The requirement of a large voltage spike to initiate the plasma in such large tokamaks leads to problematic design constraints. The walls of the vacuum vessel must be made strong enough to withstand the forces caused by disruptions of the plasma current with the resultant collapse of the poloidal magnetic field. Additionally, if there are no insulating gaps in the vacuum vessel, as is the case for ITER, the electrical conductivity of the wall is high enough to slow the rate at which the voltage diffuses through the walls, limiting the induced electric fields to values as low as 0.3 V m^{-1}. The resulting start-up may be less reliable than is required.

It has been widely recognized that ECH could provide an effective means for significantly improving tokamak start-up, and as suitable microwave sources became available, experiments were undertaken to explore the possible advantages of ECH-assisted start-up. In DIII-D [4], for example, the standard start-up (without ECH) required a loop voltage around 10 V; the corresponding toroidal electric field in the center of the plasma is nearly 1 V m^{-1}. With 650–850 kW of ECH power injected into the chamber, reliable breakdown could be achieved with induced electric fields as low as 0.15 V m^{-1}. Moreover, reliable breakdown was achieved over a wider range of initial gas pressures and with less sensitivity to magnetic error fields. The DIII-D experiments employed X-modes launched from the high-field side and propagating at oblique angles with respect to the static magnetic field. Other experiments using various polarizations and launch geometries achieved similar results, namely, significant reductions in the loop voltage for breakdown and an increased range of initial gas pressures over which reliable breakdown resulted. Experiments were undertaken in the TCA tokamak [5] to determine the differences between X- and O-mode breakdowns in tokamak plasmas and how any such differences might affect the subsequent start-up. These TCA experiments provided several useful insights into the phenomenology of ECH-assisted breakdown and start-up, particularly as regards the role of the upper hybrid resonance.

The salient features of the TCA tokamak are major and minor radii, $R_o = 61.5$ cm and $a = 18$ cm with a maximum toroidal magnetic field $B_o = 16$ kG. The induced toroidal electric field is driven by an air-core transformer and the vacuum vessel has two insulating gaps. Typically, up to 125 kW of 39 GHz power could be delivered to the vacuum chamber by a beam with less than 1.5% cross-polarization and reflected into the plasma region by means of an ellipsoidal stainless steel mirror. The resulting microwave beam had a diameter of 7 cm on the axis of the tokamak. Although the mirror could be tilted to launch the microwave beam at oblique angles with respect to the toroidal magnetic field, no significant effects on breakdown were observed, and consequently, the published results utilized injection perpendicular to the magnetic field. The diagnostics for determining the spatial and temporal properties of the breakdown consisted of a 100-channel H_α camera, an optical spectrometer, and a five-channel microwave interferometer. The data acquisition rate for the three systems ranged from 1.5 to 10 kHz, providing better than millisecond temporal resolution. In fact, the H_α camera measured the horizontal H_α emission profile every 650 μs with a spatial resolution of 4 mm in the equatorial plane of the tokamak. In order to make the external conditions similar for both launch modes, the polarization was alternated from X- to O-mode between successive shots.

A comparison of breakdown initiated by X- and O-mode polarizations shows striking differences in the first-pass absorption and the breakdown location. After roughly 800 μs, the X-mode polarization yields localized visible emission profiles indicating a relatively high first pass absorption before the microwave beam is scattered by the vacuum chamber walls. This localized absorption appears to take place at the upper hybrid resonance, not at the cyclotron resonance surface. In contrast, the O-mode polarization yields much broader visible emission profiles indicating very weak first-pass absorption and multiple, polarization-changing reflections from the vacuum chamber walls. Absorption does occur at the cyclotron resonance surface, but to a much lesser extent than at the upper hybrid resonance. The electron density produced by the O-mode polarization is roughly 60–80% of that produced by an equal amount of microwave power in the X-mode polarization. There is also evidence from impurity radiation that the O-mode polarization produces a higher influx of impurities, assumed to be due to the greater microwave power dissipated in the vacuum chamber walls compared with the X-mode polarization, for which the plasma itself absorbs most if not all of the incident microwave power. Apart from these differences, both launch polarizations show that a dominant role is played by the upper hybrid resonance in transforming the incident microwave power into energy density in the plasma. We shall consider this process in greater detail shortly, but first we examine the equilibrium properties of the breakdown plasma in light of the fundamental processes discussed in earlier chapters.

It has long been recognized that the $\nabla B \times \mathbf{B}$ drift of electrons and ions in the inhomogeneous magnetic field of the torus would polarize the breakdown plasma and produce an electrostatic field which would, in turn, give rise to an $\mathbf{E} \times \mathbf{B}$ flow of plasma ions and electrons along the equipotentials of the polarization field into the low-field side of the torus. In TCA, the polarization of the breakdown plasma, initially formed at the cyclotron resonance surface, and the resulting flow of the plasma into the low-field side of the vacuum chamber appear to take place in less than 1 ms. After this initial expansion of the breakdown plasma, the H_α emission contours remain relatively constant throughout the remainder of the microwave pulse with no evidence of macroscopic instability.

The average electron temperature and density during the microwave pulse can be deduced from the ambient gas pressure and the microwave power using the Point

Model from Chapter 6. In steady state, the balance between the ionization rate and the loss rate of charged particles requires that $n_o\langle\sigma_{ion}v_e\rangle = \tau_p^{-1}$. Here, n_o is the density of hydrogen molecules, $\langle\sigma_{ion}v_e\rangle$ is the ionization rate constant, and τ_p^{-1} is the (ambipolar) rate at which ions and electrons are lost from the plasma region. The ionization rate constant can be evaluated using the following expression [6]:

$$\langle\sigma_{ion}v_e\rangle = 371 \times 10^{-9}\,\mathrm{cm^3\,s^{-1}}\exp(-15.6/T_e)T_e^{1/2}(15.6+T_e)^{-1}$$

$$\times\{T_e/(15.6+20T_e)+\ln[(19.5+1.25T_e)/15.6]\}, \qquad (9.1)$$

where T_e is in eV. We shall assume that the dominant loss process is ambipolar flow at the ion sound speed along magnetic lines of force that intersect the limiter because of magnetic error fields. These stray magnetic fields are especially likely in tokamaks like TCA that employ air-core ohmic heating transformers. The loss rate will thus have the form $\tau_p^{-1} = c_s/L$, where $c_s = (2T_e/M)^{1/2}$. The distance along the perturbed magnetic field line to the limiter is not given in Ref. [5] and so we assume that L is equal to the outer circumference of the torus on its equatorial plane: $L = 2\pi(61.5 + 18)$ cm $= 500$ cm. As shown in Chapter 6, the electron temperature is determined by the ambient neutral gas pressure through the condition that $n_oL = (2T_e/M)^{1/2}/\langle\sigma_{ion}v_e\rangle$. With $L = 500$ cm and $n_o = 1.06 \times 10^{12}\,\mathrm{cm^{-3}}$ corresponding to the initial fill pressure 3×10^{-5} Torr of hydrogen gas, we employ Eq. (9.1) and find the predicted electron temperature to be $T_e = 10.7$ eV, the ionization rate constant to be $8.538 \times 10^{-9}\,\mathrm{cm^3\,s^{-1}}$, and the ambipolar lifetime to be $\tau_p = 1.1 \times 10^{-4}$ s, dependent only on the initial fill pressure.

In the point model, the electron density is determined by the power balance condition which we can express as $\langle P_\mu\rangle = n_eW_{ion}/\tau_p$, where $\langle P_\mu\rangle$ is the average microwave power per cubic centimeter deposited in the plasma and W_{ion} is the energy required to form an ion–electron pair. Following the discussion of Chapter 6, we estimate that $W_i = 30$ eV for electrons with $T_e = 10.7$ eV in hydrogen gas. Whaley *et al.* [5] give $\langle P_\mu\rangle = 0.32\,\mathrm{W\,cm^{-3}}$ for an injected microwave power of 125 kW. For the striking experimental results shown in Figure 9 of Ref. [5], the total microwave power was 80 kW, so that to interpret those results, we set $\langle P_\mu\rangle = (80/125) \times 0.32\,\mathrm{W\,cm^{-3}} = 0.2\,\mathrm{W\,cm^{-3}}$ for the X-mode polarization. The resulting density predicted for this case is

$$n_e = 0.2\,\mathrm{W\,cm^{-3}} \times 1.1 \times 10^{-4}\,\mathrm{s}/30\,\mathrm{eV} = 4.6 \times 10^{12}\,\mathrm{cm^{-3}}.$$

The square of the electron plasma frequency for this density is $\omega_{pe}^2 = 1.464 \times 10^{22}\,\mathrm{s^{-2}}$. The frequency of the microwave power is 39 GHz, so that $\omega_o^2 = 6.005 \times 10^{22}\,\mathrm{s^{-2}}$. At the upper hybrid resonance, $\omega_o^2 = \omega_{pe}^2 + \Omega^2$; therefore, the electron gyrofrequency at the upper hybrid resonance must be $\Omega = (\omega_o^2 - \omega_{pe}^2)^{1/2} = 2.13 \times 10^{11}\,\mathrm{s^{-1}}$ corresponding to a magnetic intensity at the upper hybrid resonance of 1.212 T. Since $B_o = 1.38$ T in these experiments, the radial position of the upper hybrid resonance is predicted to be at $x = r/a = 0.475$, in good agreement with the experimental value shown in Figure 9 of Ref. [5].

O-modes do not experience upper hybrid resonance, and power launched in the O-mode polarization must undergo repeated polarization-changing reflections from the vacuum chamber walls if it is to be absorbed at the upper hybrid resonance. The apparent location of this resonance for power injected with the O-mode polarization is at $x = r/a = 0.3$, where the magnetic intensity is 1.269 T, giving a local electron gyrofrequency of $\Omega = 2.23 \times 10^{11}\,s^{-1}$. The corresponding density for upper hybrid resonance is $n_e = 3.23 \times 10^{12}\,cm^{-3}$, and based on the point model, the average microwave power density absorbed in this case is $0.14\,W\,cm^{-3}$ or roughly 70% of the power density absorbed in the X-mode polarization. The remaining power is assumed to be dissipated in the walls of the vacuum chamber. This estimate of the power absorbed when the power is launched with O-mode polarization agrees well with the statement from Ref. [5] that "X-mode launch is seen to produce 20–40% higher average densities than O-mode launch."

We next consider possible mechanisms by which the incident microwave power is absorbed by the plasma. In the experiments reported in Ref. [5], this absorption appears to take place very close to the upper hybrid resonance surface. Recall that in Chapter 4, it was shown that the polarization of X-modes propagating toward the upper hybrid resonance becomes longitudinal at the resonance. That is, the electric field of the wave is parallel to the direction of propagation, **k**, and is thus given by $\mathbf{E} = -\nabla\Phi = -i\mathbf{k}\Phi$ and governed by Poisson's equation: $\nabla\cdot\mathbf{E} = i\mathbf{k}\cdot\mathbf{E} = \mathbf{k}^2\Phi = \rho/\varepsilon_o$. At the upper hybrid resonance surface, the electrons oscillate along **k** at the wave frequency. This oscillating layer of electrical charge can excite any electrostatic waves that can propagate at the driving frequency. If the amplitude of oscillation is great enough, it can excite pairs of electrostatic waves through nonlinear parametric decay processes. In this case, the sum of the frequencies of the driven waves must equal the driving frequency. We first consider the linear regime.

Any electrostatic waves excited by the oscillating charge layer at the upper hybrid resonance surface can be described by the Harris dispersion relation discussed earlier. The general form of the Harris dispersion relation is [7]

$$1 = \omega_p^2/k^2 \sum \int d^3v J_n^2\,(k_\perp v_\perp/\Omega)(k_{||}\partial f_o/\partial v_{||} + n\Omega v_\perp^{-1}\partial f_o/\partial v_\perp)/[k_{||}v_{||} - (\omega - n\Omega)]. \tag{9.2}$$

The sum over n is from $n = -\infty$ to ∞. In general, the right-hand side would be summed over all species of charged particles comprising the plasma; but for frequencies around the electron gyrofrequency, we can consider the ions to be infinitely massive and their contribution then vanishes. If $k_{||} = 0$ and the equilibrium has an isotropic Maxwell–Boltzmann distribution function, then Eq. (9.2) reduces to the Bernstein dispersion relation [8]

$$D_B = 1 - 2\left[\omega_p^2/(k^2\alpha^2)\right] \sum 2e^{-\lambda} I_n(\lambda)[(\omega/n\Omega)^2 - 1]^{-1} = 0. \tag{9.3}$$

Here the sum over n is from $n = 1$ to ∞. For the same isotropic Maxwell–Boltzmann distribution, f_o, and again omitting the sum over all plasma species, the Harris dispersion relation, which is valid for $k_{||} \neq 0$, becomes

$$DIS = 1 + 2(\omega_p^2/k^2\alpha^2)\{1 + (\omega/k_{||}\alpha)\Sigma e^{-\lambda} I_n(\lambda) Z[(\omega - n\Omega)/k_{||}\alpha]\} = 0, \tag{9.4}$$

where $\lambda = k_\perp^2 \alpha^2 / 2\Omega^2$, $I_n(\lambda)$ is the modified Bessel function, α is the electron thermal speed, and Z is the plasma dispersion function. As we saw earlier, in the cold-fluid limit, the real part of Eq. (9.4) reduces to

$$\mathrm{Re\,DIS} = 1 - (\omega_p^2/\omega^2)(k_\parallel^2/k^2) - [\omega_p^2/(\omega^2 - \Omega^2)](k_\perp^2/k^2) = 0. \tag{9.5}$$

The imaginary part of the dispersion relation is entirely due to the finite electron temperature and for frequencies near the electron gyrofrequency is given by

$$\mathrm{Im\,DIS} = 2(\omega_p^2/k^2\alpha^2)e^{-\lambda} I_1(\lambda)(\omega/k_\parallel\alpha)\sqrt{\pi}\exp\{-[(\omega - \Omega)/k_\parallel\alpha]^2\}. \tag{9.6}$$

We assume that any electrostatic waves excited at the upper hybrid resonance surface initially propagate perpendicular to the magnetic field. These are the Bernstein waves. They have vanishingly small group velocities and, since $k_\parallel = 0$, they are not damped. As the waves propagate into the region of increasing magnetic intensity, $k_\parallel/k$ increases as dictated by the real part of the dispersion relation, Eq. (9.5). Their angle of propagation rotates away from perpendicular and as the waves approach the cyclotron resonance surface, their propagation becomes entirely parallel to the magnetic field. In the region between the two resonance surfaces, the propagation is oblique so that neither $k_\parallel$ nor $k_\perp$ vanishes. Here the waves are damped at a rate, γ, that can be estimated by expanding the complete dispersion relation around its real solutions [9]. We write $\mathrm{DIS} = 1 + F(\omega + i\gamma, k_\parallel, k_\perp) = 0$ and expand F about solutions to the real part of the dispersion relation:

$$F \cong F(\omega, k_\parallel, k_\perp) + i\gamma\partial F(\omega, k_\parallel, k_\perp)/\partial\omega = -1. \tag{9.7}$$

Taking the imaginary part of Eq. (9.7) yields the desired estimate of the damping rate:

$$\gamma = -F_i/(\partial F_r/\partial\omega), \tag{9.8}$$

where F_i and $\partial F_r/\partial\omega$ are to be evaluated for values of ω, $k_\parallel$, and $k_\perp$ that satisfy Eq. (9.5). Since $k_\parallel$ vanishes at the upper hybrid resonance surface, F_i and thus γ vanish there. And since $k_\perp$ vanishes at the cyclotron resonance surface, the factor $e^{-\lambda}I_1(\lambda)$ and thus γ vanish there. The damping therefore is confined to the region between the two resonance surfaces. The detailed variation of γ over this region can be determined using Eq. (9.8). Even without carrying out this calculation, we can conclude that the predicted deposition of the power in the region between the upper hybrid and cyclotron resonance surfaces differs from the experiments which clearly show that most of the incident power is absorbed at the upper hybrid resonance surface. This suggests that a stronger absorption mechanism may be acting through collisions between the more energetic electrons oscillating at the upper hybrid resonance and the background electrons whose temperature we have estimated to be around 10 eV. At these low temperatures, Coulomb scattering [10] can be very rapid. Electrons with energy W_h will loose energy to the background electrons, whose temperature is T_c, at a rate given by

$$dW_h/dt = -\nu_\varepsilon W_h,$$

where $\nu_\varepsilon = 7.7 \times 10^{-6}(1 - 0.5T_c/W_h)n_e \ln \Lambda/W_h^{3/2}$ and $\ln \Lambda = 23 - \ln(n_e^{1/2}T_c^{-3/2})$. Using the experimental values of n_e and T_c and considering energetic electrons with $W_h = 25$ eV, we find that $dW_h/dt = -2 \times 10^7$ eV s^{-1}. We choose 25 eV since electrons of energy higher than 30 eV would have been detected by the Thompson scattering diagnostic, and no electrons of this energy were seen. If all of the electrons in the upper hybrid resonance layer were given 25 eV of energy by the electric field of the wave, the power density absorbed there could be as high as 16 W cm^{-3}. This power deposition would need to be very localized if the average power absorbed by the entire plasma is to be 0.2 W cm^{-3}, suggesting that the upper hybrid resonance layer is less than 1 cm in thickness.

When the injected microwave power was increased to 120 kW, launched in the X-mode polarization, there was experimental evidence that some of the plasma ions were heated to energies in excess of 1 keV. These energetic ions were detected as fast neutral atoms produced by charge transfer between the energetic ions and the background hydrogen atoms. This ion heating was assumed by Whaley *et al.* [5] to result from the nonlinear excitation of lower hybrid waves which were subsequently damped by ions in the tail of the distribution. The lower hybrid resonance is described by the Harris dispersion relation when the ion contribution is included. For perpendicular propagation, the cold-plasma limit of the Harris dispersion relation, now including finite-mass ions, is

$$1 = \omega_{pi}^2/(\omega^2 - \Omega_i^2) + \omega_{pe}^2/(\omega^2 - \Omega_e^2). \tag{9.9}$$

The two solutions to Eq. (9.9), $\omega_\pm^2$, are given to a very good approximation by the upper (+) and lower (−) hybrid frequencies, respectively:

$$\omega_+^2 = \omega_{uh}^2 = \omega_{pe}^2 + \Omega_e^2 \quad \text{and} \quad \omega_-^{-2} = \omega_{lh}^{-2} = \left(\omega_{pi}^2 + \Omega_i^2\right)^{-1} + (\Omega_i\Omega_e)^{-1}. \tag{9.10}$$

As we mentioned earlier, the ions do not participate in the upper hybrid resonance since the frequency exceeds the electron gyrofrequency. The lower hybrid frequency is much greater than the ion gyrofrequency and much smaller than the electron gyrofrequency. The electrons will execute an $\mathbf{E} \times \mathbf{B}$ drift in the oscillating electric field at the lower hybrid resonance, whereas the ion orbits will deviate only slightly from straight lines during a period of the oscillation. This nonlinear coupling to the lower hybrid waves can be an additional mechanism for transforming wave energy into kinetic energy of the plasma if the incident microwave power is large enough.

9.3
ECH Suppression of Tearing Modes in Tokamaks

A comprehensive discussion of the stability of tokamak plasmas is entirely outside the scope of this chapter, and the interested reader is urged to consult Wesson [10] and the excellent bibliography included in Chapters 6 and 7 of his compendium. In this section, we will address just one aspect of the macroscopic stability phenomena in tokamak plasmas, namely, the tearing modes that are manifested through the growth of magnetic islands on flux surfaces where the safety factor, q, is given by the ratio of low-order integers, particularly $q = 2/1$ and $3/2$. If the growth of these islands is not suppressed, they can lead to the abrupt disruption of the discharge with potentially damaging effects on the experimental apparatus.

Early theoretical studies [11] using *ad hoc* radial profiles of the plasma current density showed that local modifications of the radial gradient of the plasma current density could stabilize the low-order tearing modes on the flux surfaces where $q = 2$ and $3/2$. Since electron cyclotron heating could plausibly deposit power in the plasma electrons confined in narrow radial intervals, thereby increasing the conductivity, it seemed possible that the kind of local modification of the current profiles required to suppress the growth of tearing modes could be achieved by ECH. Accordingly, this possibility was investigated theoretically [12] and experimentally [13]. These investigations are ongoing with the aim of developing efficient and effective suppression techniques for the ITER [14].

The first experimental results, reported by Hoshima *et al.* [13], were obtained on the JFT-2M tokamak, whose major radius $R_o = 131$ cm and whose minor radii are $a = 35$ cm and $b = 53$ cm. Although JFT-2M is designed to produce discharges with noncircular cross sections, the experimental results reported in Ref. [13] utilized a conventional circular cross-sectional discharge with a limiter radius of 34 cm. The toroidal magnetic field on axis was limited to $B_o \leq 1.4$ T. Microwave power was supplied by two gyrotrons, one operating at 59.75 GHz (cw) and the other at 59.90 GHz (cw). The maximum microwave power that could be coupled into the plasma was 250 kW. This power in the linearly polarized TE_{11} mode was launched at an angle of 82° with respect to the major radius from two horn antennas located on the equatorial plane of the torus. The electric field of the input power was oriented perpendicular to the static magnetic field, **B**, so as to excite X-modes propagating from the low-field side to the second-harmonic resonance surface. The radial location of the resonance surface was varied in small increments by changing the toroidal magnetic intensity, B_o. The plasma density was maintained below a maximum value of $2.2 \times 10^{13}\,cm^{-3}$ to prevent the right-hand cutoff from blocking the access to the resonance surface. The electron temperature ranged from 100 to 300 eV in these experiments, for which the plasma current was typically around 210 kA, giving a limiter safety factor, $q_a \approx 3$. For line-averaged densities around $1 \times 10^{13}\,cm^{-3}$ and with $2.8 < q_a < 3.3$, the plasma exhibited a robust level of MHD activity but did not disrupt. The poloidal and toroidal mode numbers of the MHD activity ("Mirnov oscillations") were determined by an array of magnetic probes to be $m = 2$ and $n = 1$. These probes measured the perturbed magnetic fields associated with the magnetic islands resulting from the unstable tearing mode. The modes oscillated at frequencies around 2.5 kHz under the usual experimental conditions. When 120 kW of microwave power was injected into the plasma with the second-harmonic resonance surface placed precisely at the optimal location, the amplitude of the MHD oscillations was greatly reduced within a time of 50 ms. The frequency of the remaining low-level oscillations was observed to increase slightly. Associated changes in the internal inductance of the plasma column occurred over a longer time, indicating that the suppression of the MHD activity resulted from local changes in the radial plasma current profile. Suppression of the MHD activity resulted only if the position of the resonance surface was at $r/a = 0.70 \pm 0.03$, corresponding to a radial interval of 2 cm. The MHD activity was found to be enhanced if the resonance surface were placed at $r/a < 0.6$.

The width of the radial interval within which the microwave power is deposited is governed by the relativistic, Doppler-shifted resonance condition: $\omega - k_{||}v_{||} - 2\Omega = 0$. Although the electrons in the JFT-2M plasma were only weakly relativistic, it is useful to use the relativistic form of the electron gyrofrequency: $\Omega(r) = eB(r)/\gamma m$. We can then express the second-harmonic resonance condition at the radial position r in the following form, $2eB(r)/m = \gamma(\omega - k_{||}v_{||})$; and setting $x = v_{||}/c$ and $y = v_{\perp}/c$, we have

$$y^2/2 = [2\Omega(r)/\omega - (1 + x^2/2)(1 - nx\cos\theta)]/(1 - nx\cos\theta), \tag{9.11}$$

where $n = kc/\omega$ and $\cos\theta = k_{||}/k$ and thus the product $n\cos\theta = n_{||}$. For the JFT-2M experimental conditions that yielded the optimum suppression of the MHD activity, we conclude that the cold-electron second-harmonic resonance is at $r/a = 0.70$, and therefore $2\Omega(r = 23.8\,\text{cm})/\omega = 1$. At any other radial position, r,

$$2\Omega(r)/\omega = (R_o + 23.8\,\text{cm})/(R_o + r). \tag{9.12}$$

For $n\cos\theta > 0$, the Doppler shift lowers the frequency of the RF electric field in the rest frame of the electron, that is, the frequency is "downshifted." Conversely, for $n\cos\theta < 0$, the frequency is "upshifted." For $r/a > 0.70$, the resonance condition requires a progressively greater degree of downshifting as r increases which, in turn, requires increasing values of $v_{||}$. At some radius, there will not be a significant number of electrons with the required parallel velocity and no resonant transfer of energy will take place. This situation is illustrated in Figure 9.1.

Solid lines in the figure represent the loci in velocity space where the second-harmonic resonance condition is satisfied for radii of (reading left to right) 22.8, 23.0, 23.2, 23.4, and 23.6, for which the frequency is upshifted; and 24.0, 24.2, 24.4, and 24.6, for which the frequency is downshifted. The surfaces of constant v/c corresponding to energies of 100 and 300 eV are shown as dotted lines and dashed lines, respectively. In this figure, the value of $n\cos\theta$ has been set at $n_{||} = 0.16$. Since the value of the index of refraction for these waves under these experimental conditions is estimated at $n = 0.866$, this value of $n\cos\theta$ corresponds to an angle of 79°, or roughly 3–4° away from the central ray. The beam divergence was given as 9°, so this ray is in the outer part of the microwave beam where $k_{||}/k$ and thus the Doppler shift is stronger than for the central ray. Clearly, since the electron temperature in JFT-2M was less than 300 eV in these experiments, there are few plasma electrons able to experience second-harmonic resonance at a radius greater than $r = 24.6$ cm. Note that although the Doppler shift is symmetrical in $v_{||}$, the relativistic effect is not; the relativistic increase in the electron mass always reduces the local electron gyrofrequency, increasing the frequency mismatch on the low-field side and decreasing it on the high-field side. Indeed, for the radius at 23.6 cm, Figure 9.1 shows that the relativistic effect alone could permit resonance for sufficiently energetic electrons. We therefore anticipate a wider radial interval for resonance on the high-field side as compared to the low-field side.

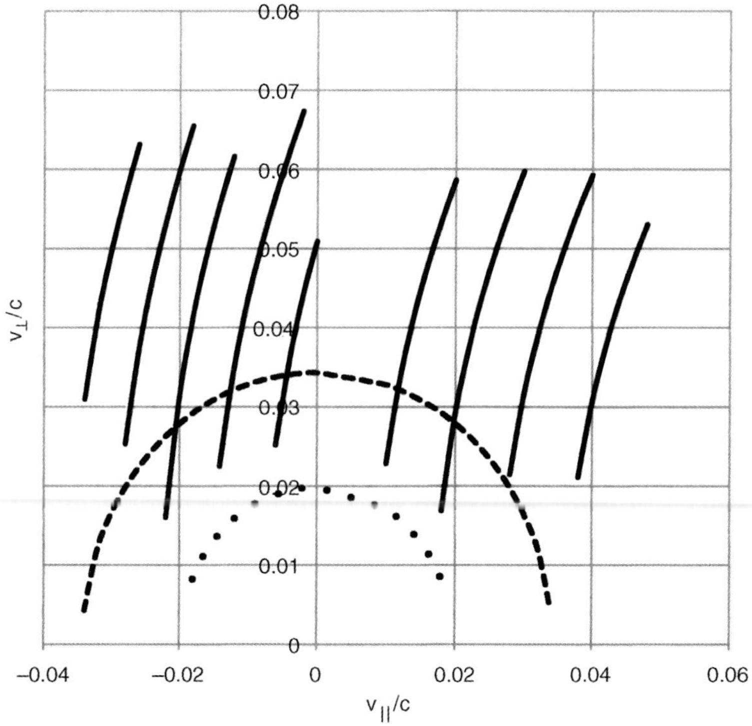

Figure 9.1 Resonance conditions in the JFT-2M disruption avoidance experiments: Solid vertical lines (reading left to right) are for minor radii of: r = 22.8, 23.0, 23.2, 23.4, 23.6, 24.0, 24.2, 24.4, and 24.6 cm. The dashed line is the locus where the electron energy is 300 eV and the dotted line is the locus where the electron energy is 100 eV. The horizontal axis is $v_{||}/c$, the vertical axis is $v_{\perp}/c$, and the parallel index of refraction is $n_{||} = 0.16$.

This plot demonstrates that as the microwaves approach the (cold-electron) resonance surface from the low-field side, electrons with $v_{||} \gg v_{\perp}$ will be heated first while electrons with $v_{||} \approx v_{\perp}$ will only experience resonance nearer the cold-electron resonance surface. For example, 300 eV electrons with $v_{||} \gg v_{\perp}$ can resonate with the RF electric fields at $r \approx 24.5$ cm, whereas 300 eV electrons with $v_{||} \approx v_{\perp}$ will experience resonance at $r \approx 24.3$ cm. On the high-field side of the cold-electron resonance surface, 300 eV electrons can be resonant at $r \approx 22.9$ cm if $v_{||} \gg v_{\perp}$, but at $r \approx 23.2$ cm if $v_{||} \approx v_{\perp}$. For an isotropic distribution of 300 eV electrons in velocity space, the effective thickness of the resonance zone for $n \cos\theta = 0.16$ would be around 12 mm. For 100 eV electrons, the thickness of the resonance zone would be only 7 mm. Hoshino *et al.* [13] estimate the experimental thickness of the resonance zone to be between 6 and 12 mm. The volume of the resonance zone would then range from 7.5×10^4 to $1.5 \times 10^5\ \text{cm}^3$. The dependence of the thickness on the value of $n \cos\theta$ is displayed in Figure 9.2; the Doppler shift becomes negligible for $n \cos\theta \ll v/c$.

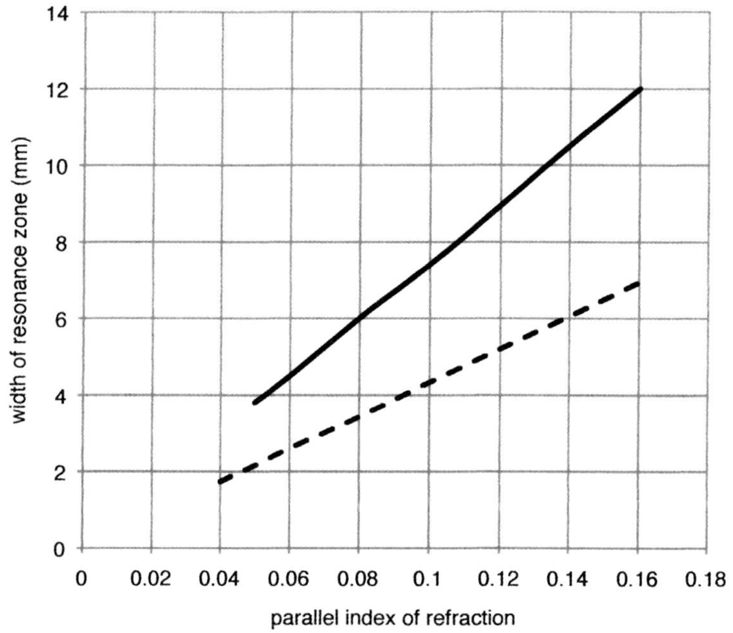

Figure 9.2 Radial thickness of the resonance zone in the JFT-2M experiments versus the parallel index of refraction, $n_{||}$.

The earliest theoretical studies [12] of ECH stabilization of MHD tearing modes predicted that RF power densities of 1 W cm^{-3} would be adequate to stabilize low-order tearing modes through local flattening of the plasma current profile. We can use the experimental data from Hoshino *et al.* [13] to estimate the power density deposited in the resonance zones for electron temperatures of 100–300 eV. The optical depth for second-harmonic X-modes propagating perpendicular to the magnetic field was given by Antonsen and Manheimer [15]:

$$\tau = (2\pi/3)(L\omega/c)\left[\left(3-2\omega_{pe}^2/\omega^2\right)/\left(1-2\omega_{pe}^2/\omega^2\right)\right]$$
$$\times\left\{\left[4\left(1-\omega_{pe}^2/\omega^2\right)^2-1\right]/\left(3-4\omega_{pe}^2/\omega^2\right)\right\}^{3/2} T_e/mc^2.$$

Hoshino *et al.* [13] report only the line-averaged electron density; but if we assume a parabolic radial dependence, we can infer that the density at the cold-electron resonance surface is around 8.3×10^{12} cm^{-3} giving $\omega_{pe}^2/\omega^2 \approx 0.2$. If $T_e \approx 100$ eV, the optical depth is then $\tau \approx 2.49$ and roughly 92% of the incident right-hand circularly polarized power is absorbed in the resonance layer. If $T_e \approx 300$ eV, the optical depth is 7.49 and virtually all of the incident right-hand power is absorbed. Recall that the polarization of the X-mode at the cold-electron resonance surface is given by

$$E_-/E_+ = -\left[K_{11}/n^2 - \cos^2\theta - (\sin^2\theta)/2\right]/\left[(\sin^2\theta)/2\right] \approx -1.414.$$

The right-hand circularly polarized component is then about two-third of the incident microwave power. For the low-temperature cases, we can thus estimate the absorbed resonant power (per unit volume) to be around 1 W cm^{-3}, while for the higher temperature cases, the corresponding value would be around 0.6 W cm^{-3}. Thus, the inferred experimental values are in good agreement with theoretical expectations.

9.4
Electron Cyclotron Current Drive

The specific mechanism by which localized heating suppresses the tearing mode in the JFT-2M experiments was not determined. Raising the electron temperature in the resonance zone is expected to increase the current density there, since the electrical conductivity varies as $T_e^{3/2}$. This seems the most likely stabilizing mechanism in these JFT-2M studies. Subsequent experimental investigations have attempted to explore the effect on tearing modes of currents driven directly by electron cyclotron heating, now generally referred to as Electron Cyclotron Current Drive (ECCD). It has been suggested [16] that this could be a much more efficient way of avoiding tearing modes by maintaining the current profile in a stable regime.

From Figure 9.1, it is clear that microwave power launched from the low-field side with $n_{||} = 0.16$ first encounters resonant 300 eV electrons at a radius of 24.6 cm, where the electrons with $v_{||} \gg v_{\perp}$ experience the strongest RF electric fields. If the optical depth is large enough, 300 eV electrons with negative values of $v_{||}$ and thus with resonance surfaces only for $r < 23.8$ cm will encounter much weaker RF electric fields. This asymmetry in $v_{||}$ can drive net current through a mechanism first described theoretically by Fisch and Boozer [17]. In this low-field launch scenario, resonant electrons with $v_{||} > 0$ will gain greater perpendicular energy than those with $v_{||} < 0$, since the heating rate varies as $|E_-|^2$ which, in turn, varies as $\exp(-\tau)$. These more energetic electrons will require longer period to thermalize and isotropize through Coulomb collisions since the Coulomb scattering rate varies as v^{-3}, and a net current can result.

For a somewhat more quantitative conceptual description of this mechanism, we consider a small volume in velocity space, Region 1, at a point **r** within a particular narrow resonance zone. The steady-state electron density (in velocity space) is determined by the balance between resonant heating and velocity–space relaxation due to Coulomb scattering and can be described at least conceptually through an equation of the following form:

$$\partial f_1/\partial t = -P(\mathbf{r}, \mathbf{v})/\Delta W_{\perp} + \nu_{rlx1}(f_{1\,amb} - f_1) = 0, \tag{9.13}$$

where $P(\mathbf{r},\mathbf{v})$ is the RF power (per unit volume) absorbed in Region 1 at the point **r** in the specified resonance zone by electrons of velocity **v**, $\Delta W_{\perp}$ is the average energy increment due to ECH, ν_{rlx1} is the Coulomb relaxation rate in Region 1, and $f_{1\,amb}$ is the velocity–space density in the neighborhood surrounding Region 1. Electrons from Region 1 enter Region 2, where $v_{||,2} \cong v_{||,1}$ and $v_{\perp,2} = v_{\perp,1} + (2\Delta W_{\perp}/m)^{1/2}$. In Region 2, the condition for steady-state balance takes the form

$$\partial f_2/\partial t = P(\mathbf{r}, \mathbf{v})/\Delta W_{\perp} - \nu_{rlx\,2}(f_2 - f_{2\,amb}) = 0. \tag{9.14}$$

The resulting contribution to the steady-state current density is

$$\begin{aligned}\delta j_{||} &= -ev_{||}(f_2 - f_{2\,\mathrm{amb}} + f_1 - f_{1\,\mathrm{amb}}) = ev_{||}[P(\mathbf{r},\mathbf{v})/\Delta W_{\perp}](1/\nu_{\mathrm{rlx}1} - 1/\nu_{\mathrm{rlx}2}) \\ &= (ev_{||}/\nu_{\mathrm{rlx}1})[P(\mathbf{r},\mathbf{v})/\Delta W_{\perp}][1-(E_2/E_1)^{3/2}] \\ &= (ev_{||}/\nu_{\mathrm{rlx}1})[P(\mathbf{r},\mathbf{v})/\Delta W_{\perp}]\{1-[(E_1+\Delta W_{\perp})/E_1]^{3/2}\} \\ &\approx -(3/2)(ev_{||}/\nu_{\mathrm{rlx}})[P(\mathbf{r},\mathbf{v})/E],\end{aligned} \tag{9.15}$$

where we have dropped the subscripts in the final equation. The heated electrons relax in energy and pitch angle at a rate which is the sum of the electron–electron and electron–ion scattering rates. From the discussion of Coulomb scattering in Chapter 6 and keeping in mind that charge neutrality requires $Z_i n_i = n_e$, one finds that [17]:

$$\delta j_{||}/P(\mathbf{r},\mathbf{v}) = -(3/2)ev_{||}v/\{[(1/4\pi)(e^2/\varepsilon_o)^2 n_e \ln\Lambda/m](5+Z_i)/2\}, \tag{9.16}$$

where we set $\nu_{\mathrm{rlx}}E = (5+Z_i)\nu_{\mathrm{Coul}}(mv^2/2) = (1/4\pi)(e^2/\varepsilon_o)^2 n_e \ln\Lambda/mv(5+Z_i)/2$. These contributions to the parallel current can be evaluated at each point in the resonance zone that is illuminated by microwave power using Fokker–Planck models of the electron dynamics implemented in geometrical optics codes. The individual contributions are then integrated to obtain the total driven current, I. The efficiency of this current drive mechanism is proportional to the electron temperature and can be increased if the microwave power is absorbed mainly by energetic electrons so as to minimize the rate of Coulomb relaxation. In addition, the optical depth must be large enough to ensure that the power is absorbed primarily by electrons with the desired sign of $v_{||}$. Note that if the heated electrons enter a part of velocity space where they are trapped in the mirror-like region of the toroidal magnetic field, the resulting rapid bounce motion will prevent them from carrying unidirectional current. However, their absence from the passing-particle regions of velocity space will actually constitute a net current in the opposite direction [18]. The trapped-passing boundary in the JFT-2M experiments (the "loss cone") is a cone whose half angle is 57°, from which it is clear that 300 eV electrons resonating at r = 24.2 cm could become trapped with relatively small increments in their perpendicular energy. If the resonance zone were moved to smaller minor radii, where the mirror ratio is smaller and the loss-cone angle is larger, the retrograde current associated with trapping of heated electrons could be reduced. Sophisticated computer codes are now in routine use at all of the installations where ECCD is being investigated experimentally to aid both in optimizing the current drive and in interpreting the results of the experiments.

The experimental support for this theory of ECCD is now quite extensive. For example, in the JT-60U tokamak [19], noninductive currents as large as 185 kA were driven by 1.3 MW of 110 GHz microwave power. The salient features of these experiments are given below.

The major radius of JT-60U is around 3.3 m, but in the present experiments, the magnetic axis was at R = 340 cm. The nominal minor radius is around 1 m, but the plasma cross section is noncircular, as indicated in Figure 9.3, taken from Ref. [18].

The 110 GHz power was coupled into O-modes propagating at an angle of 20° with respect to a major radius, so that $k_{||}/k = \cos 70° = 0.342$, and at an angle of roughly

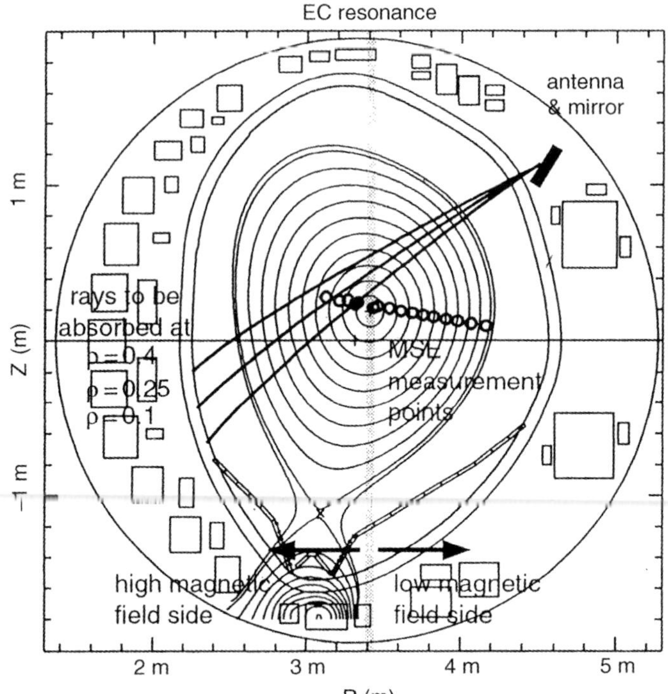

Figure 9.3 Geometry of the microwave launch in the JT-60U ECCD experiments.

$35°$ relative to the equatorial plane of the discharge. This poloidal angle could be varied to have the point of intersection of the microwave beam with the (cold-electron) resonance surface take on specified values, namely, $\rho \equiv r/a = 0.1$, 0.25, and 0.4. The toroidal magnetic field at the magnetic axis of the discharge was set at the cold-electron resonance value. The line-averaged electron density was maintained at the relatively low value of $8 \times 10^{12}\,\mathrm{cm}^{-3}$ to minimize the so-called bootstrap current in order to facilitate evaluation of the ECCD current. Central electron temperatures ranged from 5 to 10 keV. Isolation of the ECCD current is made difficult in practice by several unavoidable inductive and noninductive phenomena and is made possible in these experiments by an array of sophisticated diagnostic and analytical techniques. Magnetic measurements including motional Stark effect polarimetry were used to construct a temporal sequence of computed MHD equilibria, joined in time by a spline fit.

For the case in which $\rho = 0.25$, the authors report both the total deduced ECCD current, 185 ± 111 kA, as well as the radial profile of the current. These experimental results are then compared with theoretical predictions made with a geometrical optics code that utilizes a Fokker–Planck model of the electron dynamics. The results of the computations agree well with the experimental results and, in particular, predict total ECCD currents of 157 kA. Most of this, 136 kA, is localized in a radial interval that is 14 cm wide.

We can understand these results in an approximate way and see how they relate to the theoretical model of EECD by first examining the loci in velocity space of resonance surfaces in the radial interval beginning with the magnetic axis, $R_o = 340\,\text{cm}$, and extending outward to a place where there are negligible numbers of resonant electrons. As in our discussions of the JFT-2M experiments, the mildly relativistic, Doppler-shifted resonance condition is $eB(R)/\gamma m = \omega - k_{\|}v_{\|}$, which we again write as

$$\Omega(R)/\omega = 340\,\text{cm}/R = \left(1 + 0.5\,u_{\|}^2 + 0.5\,u_{\perp}^2\right)(1 - n_{\|}u_{\|}),$$

where $u_{\|} = v_{\|}/c$ and $u_{\perp} = v_{\perp}/c$. The authors cite a value of $n_{\|} \sim 0.5$, but because the density is low ($\omega_{pe}^2/\omega^2 \sim 0.05$), the O-mode index of refraction is close to unity and we therefore have set $n_{\|} = 0.33$ in constructing Figure 9.4.

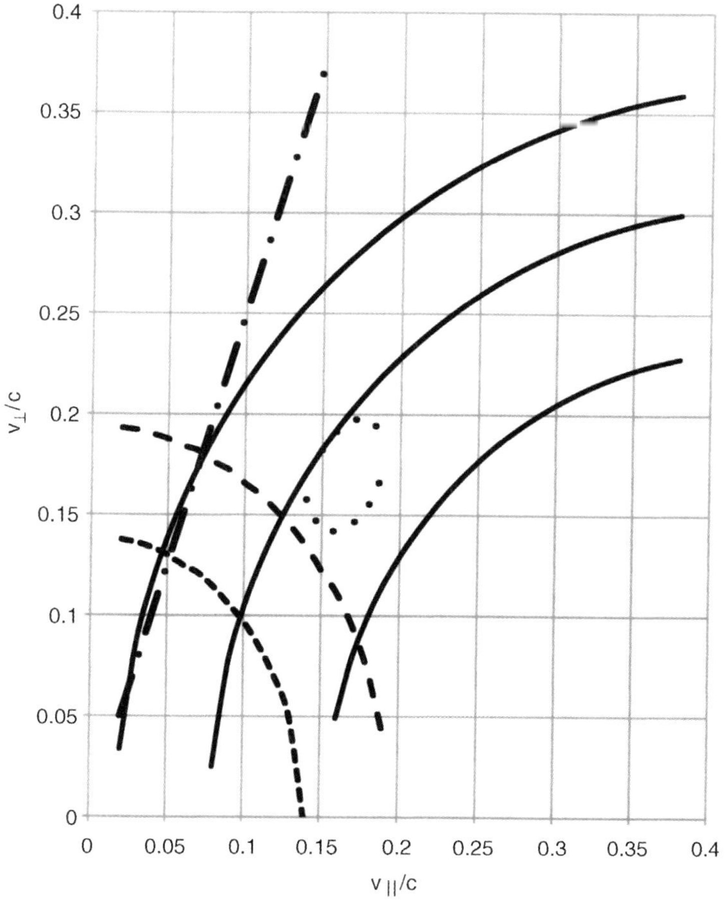

Figure 9.4 Resonance conditions for the ECCD experiments in JT-60U: solid curved lines are for radii of (reading from left to right) $R = 342$, 348, and 354 cm. The magnetic axis is at $R_o = 340\,\text{cm}$ and the parallel index of refraction is $n_{\|} = 0.33$. The long dash line is the locus where the electron energy is 10 keV and the short dash line is the locus where the electron energy is 5 keV. The boundary between trapped and passing electrons is shown as the dash-dot line. The elliptical area enclosed by dots is the assumed region of optimum conditions for current drive.

Note that the loss cone for $\rho = 0.25$ is also shown in Figure 9.4, and we see that any power absorbed by electrons near the resonance surface at $R = 342$ cm is likely to contribute mainly to the retrograde current as discussed earlier. The optical depth in these experiments is large enough that most of the incident microwave power will be absorbed before the beam reaches this resonance surface. We therefore expect most of the current to be driven by absorption of microwave power that takes place in the radial interval for which $344\,\text{cm} < R < 354\,\text{cm}$. If we approximate the poloidal flux surfaces by circles centered on the magnetic axis, the resulting power deposition is inside the surface on which $\rho = 0.38$. Since the width of the microwave beam is not reported, it is not possible to determine a more detailed deposition profile. Nonetheless, we can make a rough estimate of the total ECCD current driven by the 1.3 MW incident O-mode power using the earlier theoretical result:

$$j_{||} = (P/V)(3eu_{||}umc^2)/[(1/4\pi)(e^2/\varepsilon_o)^2 n_e \ln\Lambda(5+Z_i)]. \tag{9.17}$$

We approximate the current density by the ratio of the total current to the cross-sectional area of the flux tube within which the power is absorbed to write

$$I/A = j_{||} = (P/2\pi RA)(3eu_{||}umc^2)/[(1/4\pi)(e^2/\varepsilon_o)^2 n_e \ln\Lambda(5+Z_i)], \text{ or}$$
$$I = (P/2\pi R)(3eu_{||}umc^2)/[(1/4\pi)(e^2/\varepsilon_o)^2 n_e \ln\Lambda(5+Z_i)]. \tag{9.18}$$

At this point, we must resort to some hopefully plausible speculation to decide on appropriate values of $u_{||}$ and u. We assume that the diffusion coefficient describing the velocity-space flux resulting from the fundamental O-mode resonance is proportional to v^2. The gradient in velocity space is proportional to $v\exp(-v^2/\alpha^2)$, where α is the thermal speed. The maximum flux in velocity space would then be expected to occur for speeds that are larger than the thermal speed by $(3/2)^{1/2}$, and plausibly in the neighborhood of equal parallel and perpendicular velocities. The indicated region in velocity space is shown dotted in Figure 9.4, assuming the electron temperature is 10 keV. Using the published value of $Z_i = 3$, we would then estimate the ECCD current to be around 129 kA, somewhat smaller but similar to the value obtained by the Fokker–Planck code. The maximum absorption would appear to be at the resonance surface near $R = 350$ cm, or near the poloidal flux surface at $\rho = 0.33$. This, too, is in reasonable agreement with the experiment.

As we have seen earlier, the efficiency of ECCD, as measured by the ratio I/P, is predicted to be proportional to the energy of the electrons participating in the process. The question then naturally arises as to whether ECH could also be used in conjunction with ECCD to create a minority population of relativistic electrons in specified regions of tokamaks similar to the relativistic-electron plasmas created in many magnetic-mirror experiments. These more energetic electrons would then be expected to increase the ECCD efficiency in proportion to their energy relative to the bulk population. It appears that it should indeed be possible to form such a minority population of relativistic electrons in tokamaks, provided appropriate heating strategies were adopted [20]. As was discussed in Chapter 8, it is necessary that resonance surfaces for the relativistic electrons exist in the desired confinement volume and that this volume be illuminated with microwave power that is preferentially absorbed by the relativistic electrons. To illustrate a conceptual approach to satisfying these two requirements, we consider again the 110 GHz ECCD experiments in JT-60U, where the cold-electron resonance surface was at R = 340 cm, and stipulate that the relativistic electrons should occupy the flux tube with minor radii between 25 and 35 cm. To achieve preferential absorption of microwave power by the energetic electrons, one could illuminate the specified flux tube with O-modes propagating perpendicular to the magnetic field at a frequency near the second harmonic of the cold-electron gyrofrequency at a major radius of roughly 365 cm. In the JT-60U experiments, this would require power at close to 200 GHz. To heat electrons in the specified flux tube, the power would need to be confined within a horizontal beam some 10 cm high by quasioptical confocal mirrors, where it could heat electrons up to energies around 100 keV.

References

1 P.C. Efthimion, V. Arunasalam, and J.C. Hosea, *Phys. Rev. Lett.* **44**, 396 (1980).
2 K. Hoshino *et al.*, *J. Phys. Soc. Jpn.* **54**, 2503 (1985).
3 J. Wesson, ed., *Tokamaks*, third edition, p. 711 ff, Clarendon Press, Oxford (2004).
4 B. Lloyd *et al.*, *Nucl. Fusion* **31**, 2031 (1991).
5 D.R. Whaley *et al.*, *Nucl. Fusion* **32**, 757 (1992).
6 H.W. Drawin, Euratom Report EUR-CEA-383, Fontenay-aux-Roses (1967).
7 E.G. Harris, *J. Nucl. Energy* **C2**, 138 (1961).
8 I.B. Bernstein, *Phys. Rev.* **109**, 10 (1958).
9 See, for example, G.E. Guest and R.A. Dory, *Phys. Fluids* **8**, 1853 (1965), as well as Donald A. Gurnett and Amativa Bhattacharjee, *Introduction to Plasma Physics*, Cambridge University Press, New York (2005), Section 8.2.4.
10 For a convenient summary of the Coulomb collision rates, see David L. Book *Revised and Enlarged Collection of Plasma Physics Formulas and Data*, Naval Research Laboratory, Washington, D.C. (1977) p. 53.
11 A.H. Glsser, H.P. Furth, and P.H. Rutherford, *Phys. Rev. Lett.* **38**, 234 (1977).
12 V. Chan and G. Guest, *Nucl. Fusion* **22**, 272 (1982); Y. Yoshioka, S. Kinoshita, and T. Kobayashi, *Nucl. Fusion* **24**, 565 (1984); E. Westerhof and W.J. Goedheer, *Plasma Phys. Control Fusion* **30**, 1691 (1988).
13 K. Hoshino *et al.*, *Phys. Rev. Lett.* **69**, 2208 (1992); D.A. Kislov *et al.*, *Nucl. Fusion* **37**, 339 (1997).
14 F. Leuterer *et al.*, *Nucl. Fusion* **43**, 1329 (2003) and references cited therein.
15 T.M. Antonsen, Jr., and W.M. Manheimer, *Phys. Fluids* **21**, 2295 (1975).
16 E. Westerhof, Requirements on heating or current drive for tearing mode stabilization by current profile tailoring, *Nucl. Fusion* **30**, 1143 (1990).
17 N.J. Fisch and A.H. Boozer, *Phys. Rev. Lett.* **45**, 720 (1980), see also, N.J. Fisch, *Rev. Mod. Phys.* **59**, 175 (1987).
18 T. Ohkawa, "Steady state operation of tokamaks by rf heating," General Atomics Report GA-A13847 (1976).
19 T. Suzuki *et al.*, *Plasma Phys. Control Fusion* **44**, 1 (2002).
20 G.E. Guest, R.L. Miller, and C.S. Chang, *Nucl. Fusion* **27**, 1245 (1987).
21 C.C. Petty *et al.*, *Nucl. Fusion* **41**, 551 (2001).

Exercises

9.1. *A comprehensive study of ECH-assisted start-up in large tokamaks was carried out in DIII-D [4]. The basic parameters were $R_o = 1.67$ m, $a = 0.67$ m, $B_o \leq 2$ T, and 1.1 MW RF power at 60 GHz.*

(a) *Assuming the working gas to be hydrogen, use the point model to estimate the electron temperature if the gas pressure is $p_o = 4 \times 10^{-5}$ Torr and the connection length, L, is 1000 m.*

(b) *Under these conditions, what RF power density would be required to sustain an average density of $n_e \approx 2 \times 10^{12}\ \mathrm{cm}^{-3}$?*

(c) *Where would be upper hybrid resonance be for this density if $B_o = 2$ T?*

9.2. *ECCD experiments in DIII-D [21] employed 110 GHz power launched at an angle of 26° from radial with the polarization chosen to couple to X-modes. On the magnetic axis, the field strength was 2 T, the density was $2 \times 10^{13}\ \mathrm{cm}^{-3}$, and the electron temperature was 3 keV.*

(a) *Determine the radial interval in which the RF power is deposited.*

(b) *Estimate the noninductive current that could be driven by 1 MW of ECCD power.*

(c) *If the RF power were launched at 13°, where would be power be deposited?*

9.3. *In order to heat energetic electrons preferentially in DIII-D, RF power at 110 GHz is launched perpendicular to the magnetic field in the equatorial plane with polarization chosen to couple to O-modes. On the magnetic axis, the field strength was 2 T.*

(a) *Where will this power be absorbed by 5 keV electrons?*

(b) *What is the maximum energy for which electrons can resonate with this power at the second harmonic of their gyrofrequency?*

(c) *What is the minimum energy for which electrons can resonate with this power at the third harmonic of their gyrofrequency?*

10
The ELMO Bumpy Torus

It was recognized early in the controlled fusion program that charged particles could be confined in a simple toroidal magnetic trap if the individual coils used to generate the toroidal field were spaced sufficiently far apart, relative to the minor radius of the torus, to make the magnetic field vary significantly along each magnetic line of force. In such a "bumpy" toroidal magnetic field, local gradients in magnetic intensity give energetic charged particles a poloidal drift motion which balances the $\nabla B \times \mathbf{B}$ drift inherent in ideal toroidal magnetic fields. Kadomtsev [1] analyzed the single particle orbits in such a bumpy torus, and subsequently Gibson *et al.* [2], Morozov and Solov'ev [3], and others developed a rigorous picture of single-particle confinement in bumpy torii using adiabatic constants of the motion as well as numerical studies of particle orbits. Although these studies showed good confinement properties for the bumpy torus, it was also clear almost from the outset that plasmas confined in a simple bumpy torus were likely to be unstable to large-scale interchange instabilities. Kadomtsev [4] gave a general stability criterion for these modes in an ideal scalar-pressure plasma confined in a magnetic trap in which all of the magnetic lines of force close on themselves, as is the case for the bumpy torus. Kadomtsev's criterion predicted that any negative pressure gradients at the outer edge of the plasma, which are unavoidable in a confined plasma, would lead to the unstable growth of interchange perturbations. As a result of these dire predictions, there was only limited interest in experimental tests of the bumpy torus concept.

Then, almost a decade after that first interest in the bumpy torus, the experiments with ECH in simple magnetic mirrors discussed in Chapter 8 demonstrated that high-beta, hot-electron plasmas could be created and maintained in stable steady state in these configurations, which were also predicted to be unstable to interchange instabilities. As we have seen earlier, the relativistic electrons formed an annulus centered on the midplane of the magnetic mirror and in some instances, most notably in the ELMO device, their diamagnetic currents were strong enough to create a significant depression or "magnetic well" in the center of the annulus. These hot-electron rings appeared to offer the possibility of modifying the magnetic field at the outer edge of a bumpy torus sufficiently to stabilize the plasma in the interior with respect to interchange modes. If that proved to be the case, plasmas confined in such stabilized bumpy torus traps would offer an attractive prospect for steady-state

Electron Cyclotron Heating of Plasmas. Gareth Guest

ISBN: 978-3-527-40916-7

operation of a device with a conveniently large aspect ratio, well suited for auxiliary heating by energetic neutral or molecular-ion beams, for example. A relatively small experimental device, the ELMO Bumpy Torus (EBT-I) [5], was constructed to test this concept. In 1973, this initial EBT device successfully demonstrated the stable, steady-state confinement of a toroidally circulating plasma, provided that the hot-electron rings had high enough energy densities. The EBT employed electron cyclotron heating for altogether novel purposes and its inclusion here provides a different perspective on the physics underlying ECH as well as an opportunity to discuss a number of aspects of classical plasma confinement in toroidal traps which were germane to the EBT by the virtue of the observed quiescent stability of its plasma.

As was described earlier, a simple bumpy torus is predicted to be unstable to interchange instabilities. The fundamental supposition underlying the EBT concept is that the high-beta relativistic-electron annuli similar to those observed in the ELMO experiments can reverse the radial gradient of the specific magnetic volume, $\mathbf{U} = \int d\ell/B$, and provide stable confinement. The basis for this supposition is shown schematically in Figure 10.1.

Figure 10.1(a) shows an artist's conception of how $\mathbf{U}$ varies with the radius, r, for the vacuum magnetic fields (dashed) and with the ELMO rings present. In Figure 10.1(b), the corresponding radial profiles of magnetic intensity are sketched together with a plasma pressure profile that satisfies Kadomtsev's interchange stability criterion. It should be stressed that the conditions under which the ELMO rings themselves are stable are not fully understood, but most theoretical and experimental evidence to date indicate that some form of "line-tying" is responsible for the observed stability. The empirical facts are that the rings are completely stable if the ambient gas pressure is high enough and that the stable rings form in radial regions where there is good electrical contact along magnetic lines of force with conducting walls of the vacuum chamber.

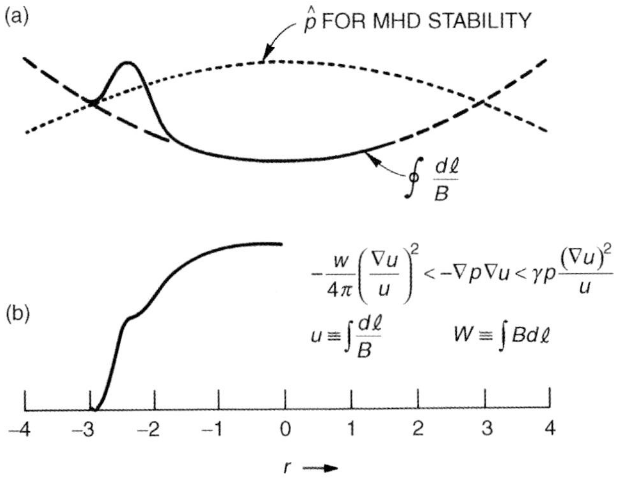

Figure 10.1 A highly schematic representation of the conceptual basis for interchange stabilization in an ELMO Bumpy Torus.

10.1 The Canted Mirror Experiments

Prior to construction of the first EBT device experiments were carried out in the Canted Mirror facility [6] to determine whether the ELMO rings of adequate beta could be formed in the nonaxisymmetric bumpy torus configuration. These experiments provided a clear affirmative answer to that question and, in addition, permitted an experimental study of the pitch-angle distribution of some of the relativistic electrons, resulting from ECH and forming the high-beta annulus. It is that aspect of these experiments that will be described in this section.

The Canted Mirror, shown schematically in Figure 10.2, was a simple magnetic-mirror configuration with a mirror ratio on axis of 2 : 1, but with the unique capability that the magnetic field coils could be pivoted ("canted") about parallel axes to resemble one sector of a bumpy torus.

Plasma confined in this magnetic mirror was heated with up to 3 kW (cw) of fundamental resonance 10.6 GHz power and up to 800 W (cw) of 35.7 GHz upper off-resonant heating power. In the axisymmetric "uncanted" configuration, the resulting plasma contained stored energies up to around 16 J with hot-electron temperatures between 700 and 800 keV and with maximum electron energies around 4 MeV. The main features of the heating geometry in the uncanted configuration can be seen in the magnetic field plots displayed in Figures 10.3(a) and (b).

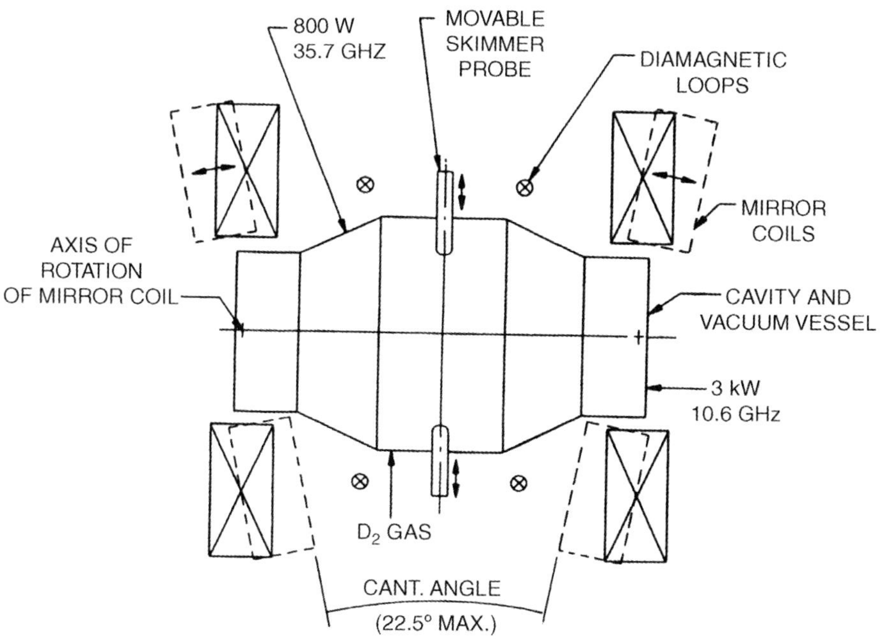

Figure 10.2 Schematic representation of the Canted Mirror Facility.

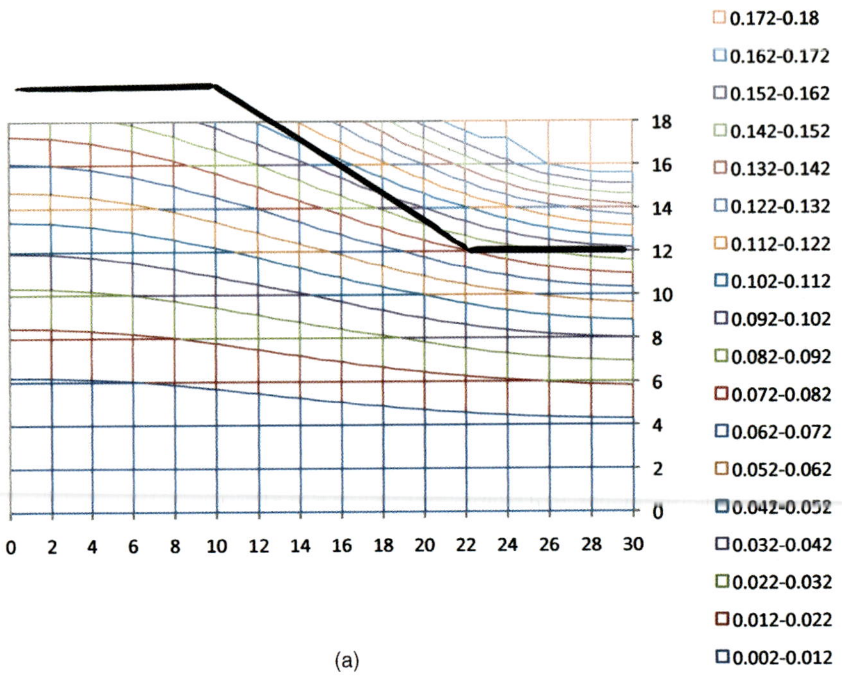

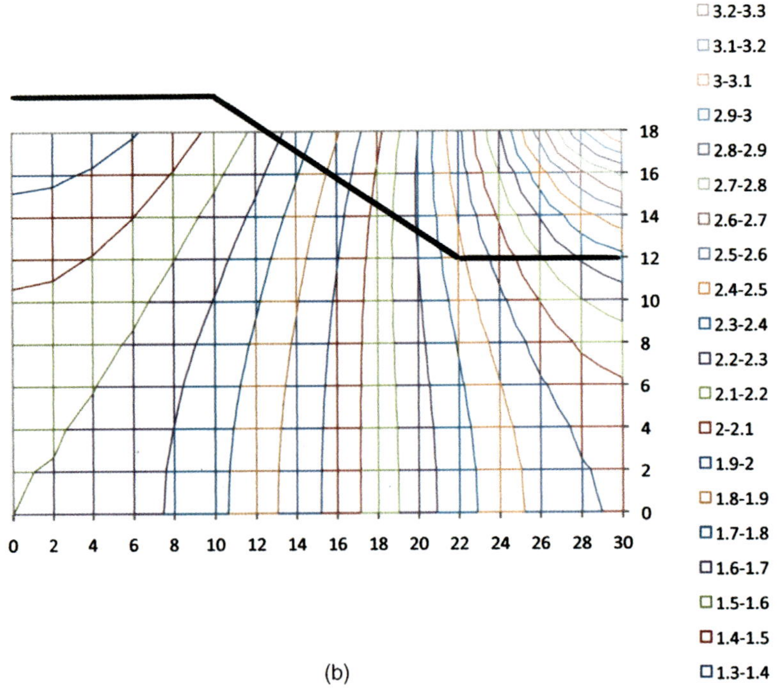

Figure 10.3 (a) Flux plot of the magnetic fields in the (uncanted) Canted Mirror Facility. (b) Mod-B contours in the (uncanted) Canted Mirror Facility.

The flux plot in Figure 10.3(a) shows that the outermost magnetic line of force that just passes the corner of the vacuum chamber intercepts the midplane at a radius of 17 cm. The mod-B contours shown in Figure 10.3(b) show that the mirror ratio along this field line is approximately 3 : 1. As was the case in the TPM experiments discussed in Chapter 8, the corner of the vacuum chamber functions as a limiter and effectively fixes this flux surface as the position of the outer surface of the plasma.

When the magnetic field coils are canted, a toroidal curvature is introduced into the magnetic lines of force and the electron guiding centers will drift in azimuth on circular paths whose centers are shifted off the magnetic axis by an amount, Δ, that depends on the electron's pitch angle, $\xi = (v_{||}/v_{\perp})^2$. Experiments were carried out using two probes, separated in azimuth by 180°, as indicated in Figure 10.4.

The inner probe was fixed at a radial position just inside the surface of the plasma, and the flux of energetic electrons striking the tip of the inner probe was measured as the outer probe was inserted progressively deeper into the plasma. Typical experimental results are shown by the solid curve in Figure 10.3(d).

Electrons whose guiding centers are on a drift surface that intersects both probes satisfy $\eta \equiv r_{outer}/r_{inner} = 1 - 2\Delta(\xi)/r_{inner}$, where $\xi = (v_{||}/v_{\perp})^2$ is to be evaluated at the position of the inner probe. We assume that the probe signal is a measure of the flux, F, of energetic electrons striking its tip:

$$F(\eta) = \int d\xi \int d\varepsilon\, J(\varepsilon, \xi) f_o(\varepsilon, \xi) v_\theta(\varepsilon, \xi), \tag{10.1}$$

where $J(\varepsilon,\xi)$ is a suitable Jacobian. The change in the flux, F, as the outer probe is inserted further into the plasma is then related to the distribution function through

$$dF/d\eta = -d\xi/d\eta \int d\varepsilon\, J(\varepsilon, \xi) f_o(\varepsilon, \xi) v_\theta(\varepsilon, \xi), \tag{10.2}$$

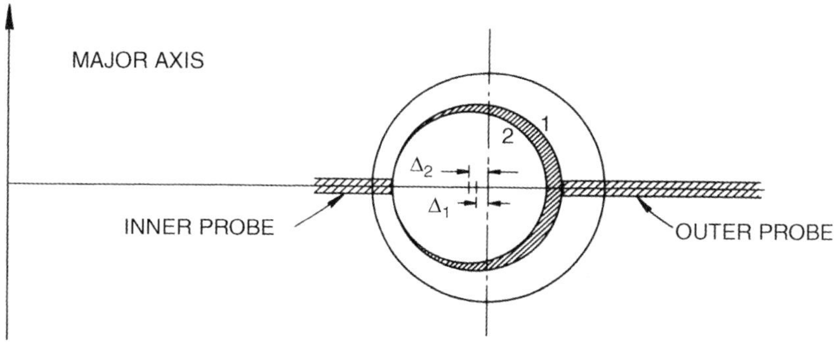

Figure 10.4 The probe geometry used to measure the pitch-angle distribution of electrons at the edge of the hot-electron plasma in the Canted Mirror Facility.

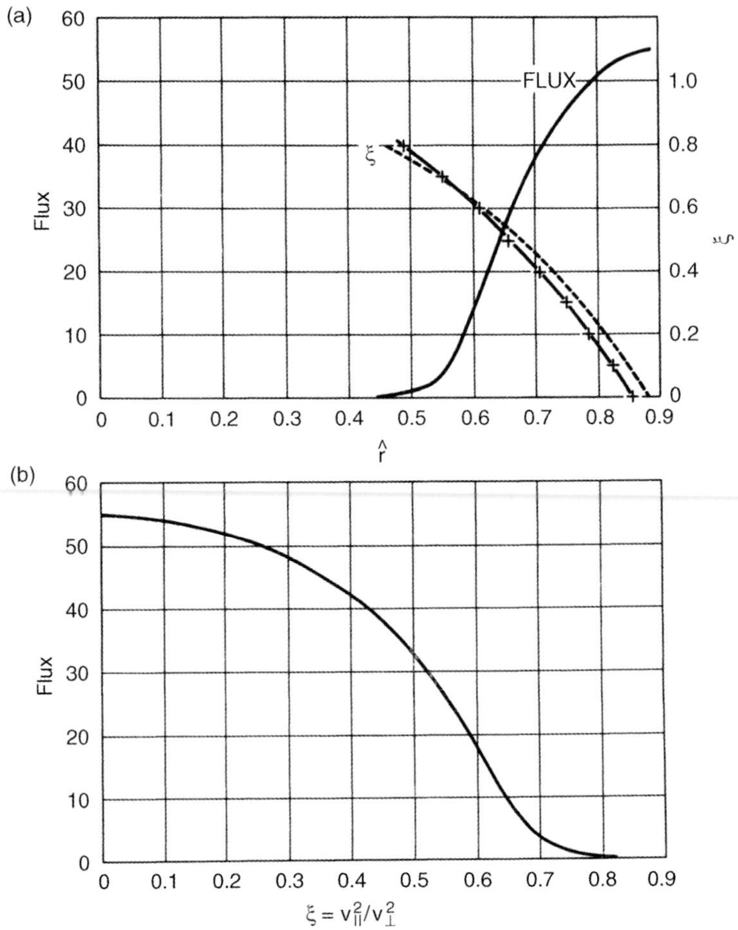

Figure 10.5 (a) Experimental values of the flux to the inner probe as the outer probe is inserted progressively deeper into the hot-electron plasma. (b) The corresponding values of flux to the inner probe versus the hot-electron pitch angle.

which we write as

$$N(\xi) = -(dF/d\eta)(d\eta/d\xi)/\langle v_\theta(\xi)\rangle = -(dF/d\xi)/\langle v_\theta(\xi)\rangle. \tag{10.3}$$

From the guiding-center drift analysis of Chapter 3, we have

$$\Delta(\xi)/r_{inner} = (2/k_o R_t)(M-1)^{-1}[I_o(k_o r_{inner})/I_1(k_o r_{inner})]$$
$$(1+2\xi)\{K(k^2)/[2E(k^2)-K(k^2)]\},$$

where $k^2 = \xi/(M-1)$. Note that for $\xi = 0$

$$\Delta/r_{inner} = (2/k_o R_t)(M-1)^{-1}[I_o(k_o r_{inner})/I_1(k_o r_{inner})] \equiv (\Delta/r_{inner})_o,$$

and $\Delta(\xi)/r_{inner}$ increases monotonically with ξ up to the value where $2E - K = 0$. The first electrons to be intercepted as the outer probe is inserted are those with $\xi = 0$, and thus the value of $(\Delta/r_{inner})_o$ can be determined experimentally. We can therefore use the following two convenient forms for unfolding the experimental data:

$$\begin{aligned}\Delta(\xi)/r_{inner} &= (\Delta/r_{inner})_o(1+2\xi)\{K(k^2)/[2E(k^2)-K(k^2)]\} \\ &\equiv (\Delta/r_{inner})_o f_2(\xi, M)\end{aligned} \tag{10.4}$$

and

$$\begin{aligned}\langle v_\theta(\xi)\rangle^{-1} &= \langle v_\theta(0)\rangle^{-1}(1+\xi)\{K(k^2)/[2E(k^2)-K(k^2)]\} \\ &\equiv \langle v_\theta(0)\rangle^{-1} f_1(\xi, M).\end{aligned} \tag{10.5}$$

The experiments cited in Ref. [6] were carried out with a cant angle of 15°. The resonance frequency is at a mirror ratio of 1.06 relative to the magnetic intensity on the midplane and at the axis, B(0,0). The tip of the inner probe was 16.8 cm from the machine axis. Using the experimental results shown by the solid curve in Figure 10.5(a), we obtain $(\Delta/r_{inner})_o$ from the end point, $\eta = 0.88$, and then by using $f_2(\xi,M)$ with $M = 3$, we construct the dashed curve showing ξ versus η. The flux can then be plotted versus ξ to obtain the curve shown in Figure 10.5(b). The derivative of this curve together with the function $f_1(\xi,M)$ then yields $N(\xi)$ shown in Figure 10.6.

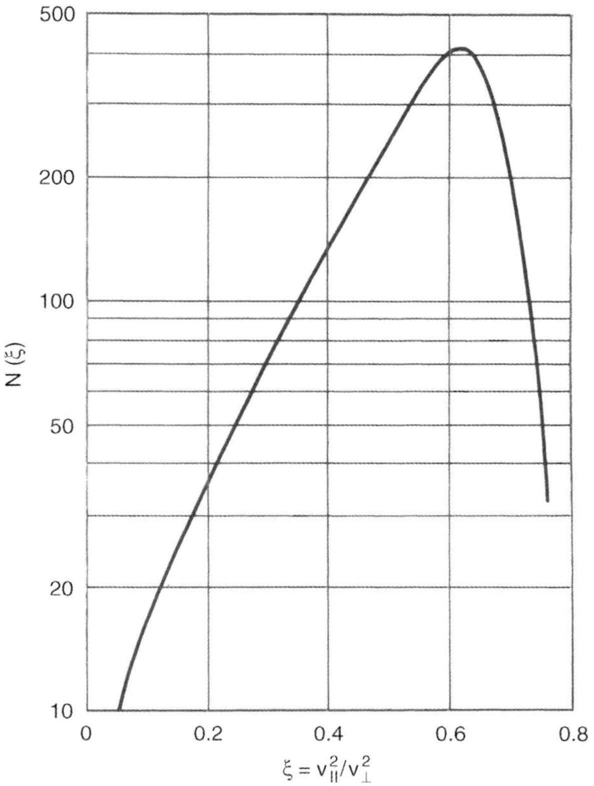

Figure 10.6 The distribution of hot electrons in pitch angle deduced from these measurements.

The strong single peak in the pitch-angle distribution indicates that most of the energetic electrons are turning at a point where the magnetic intensity is 1.6 times the value at the inner probe. Since for the present experiments, the magnetic intensity at the inner probe is roughly 0.74 B(0,0), we conclude that the hot electrons at the outer edge of the annulus are turning inside the modulus-B surface where $B_t \approx 1.18B(0,0)$. If we assume an approximate correction for the finite beta modification to the magnetic intensity, with $\beta \approx 20\%$, we would conclude that the electrons in this outer edge of the annulus were turning inside the fundamental resonance surface, $B_{res} = 1.06B(0,0)$, as anticipated.

10.2
Experiments in EBT-I

The EBT-I device, shown in Figure 10.7, was designed to permit plasma to be generated, heated, and stabilized by ECH using microwave power at frequencies of 10.6 GHz and 18 GHz.

The basic structure consisted of 24 identical canted mirrors with 2 : 1 mirror ratios connected to form a complete bumpy torus. The major radius was 150 cm and the minor radius at the midplane of each sector was 25 cm. The maximum steady-state magnetic field strength on the midplane of each sector was 5 kG requiring 6 MW of DC magnetic field power. The total plasma volume was roughly 0.5 m^3. The microwave power available for experiments initially consisted of 30 kW (cw)

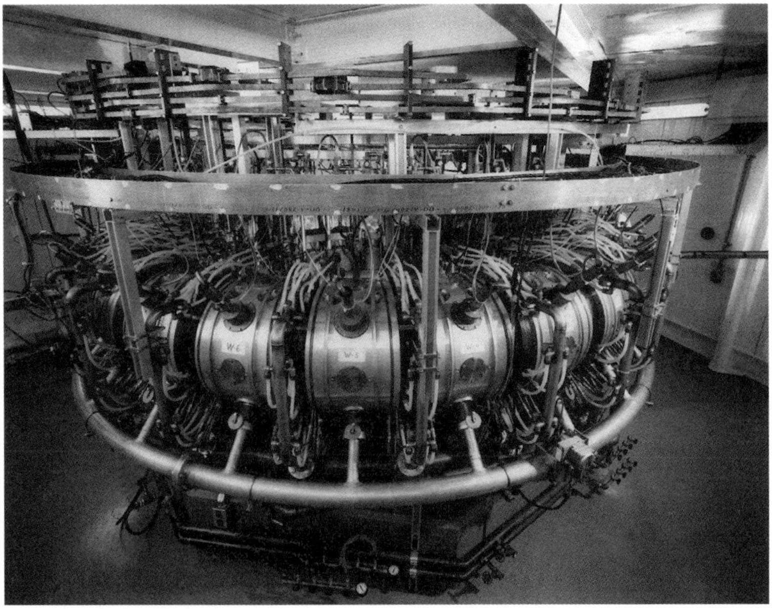

Figure 10.7 The EBT-I Facility.

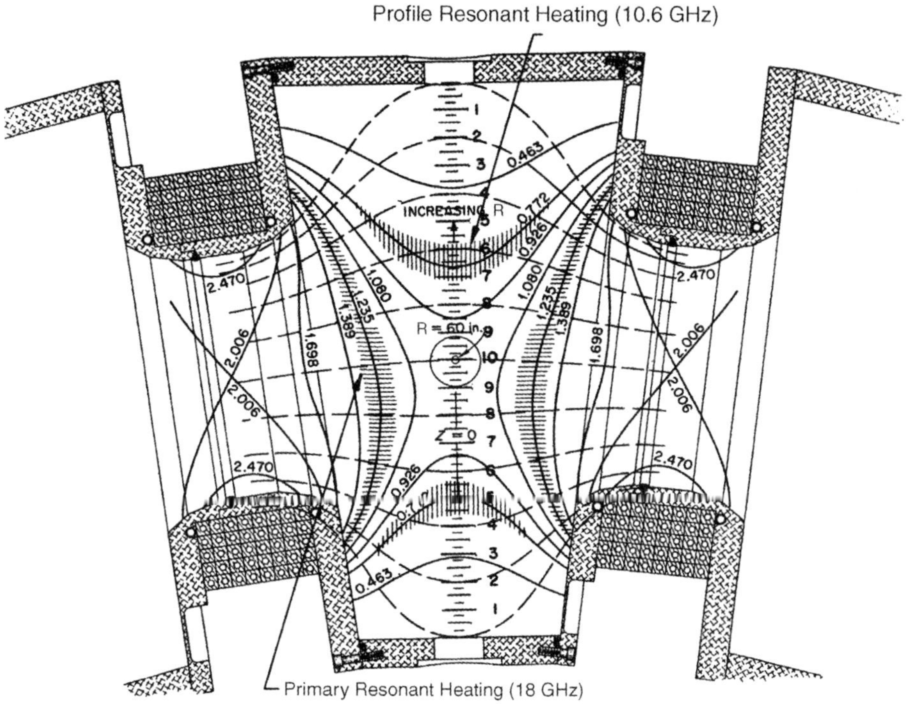

EBT Mod-B Contours (–), Flux Lines (---), and Heating Geometry (/////)

Figure 10.8 The ECH heating geometry in EBT-I.

of 10.6 GHz and 60 kW (cw) of 18 GHz. The heating geometry is shown in Figure 10.8.

The 18 GHz power was provided for heating at the fundamental resonance surface as well as at the second-harmonic resonance where it also heated the high-beta annuli. The 10.6 GHz power was absorbed at fundamental resonance in a region adjacent to the inner surface of the second-harmonic 18 GHz resonance surface, thereby augmenting the heating of the annuli. The microwave power was launched into the plasma at the midplane of the inside wall of the vacuum chamber in each sector using hybrid couplers oriented to couple to O-modes in the plasma. The microwave power was observed to be absorbed with an efficiency that approached 100%. It was experimentally verified that the high-beta annuli were formed at the 18 GHz second-harmonic resonance surfaces.

Three distinct and highly reproducible modes of operation were found, depending on the applied microwave power level and the ambient neutral gas pressure. The different modes were clearly distinguished by fluctuation levels in the line-integrated electron density, $\delta(\int n_e d\ell)$ as measured with a 75 GHz microwave interferometer, by the stored energy in the high-beta annuli, $W_\perp$ and by the ambipolar potential, ϕ. At the highest gas pressures, the plasma temperature and stored energy are both low; ϕ is small and generally positive relative to the cavity wall. Although no gross instabilities are seen, large amplitude density fluctuations are observed in the frequency range suggestive of drift wave phenomena. These fluctuations observed on the microwave interferometer provide a convenient way of identifying this mode of operation, designated the C-mode. In this high-pressure C-mode of operation, most of the microwave power is dissipated in the generation of cold plasma and very few energetic electrons are produced.

As the pressure is reduced, the diamagnetic stored energy, $W_\perp$, increases significantly; and when $W_\perp$ exceeds a critical value, the electron density fluctuation level drops abruptly to very low values, as shown in Figure 10.9.

In this regime of operation, designated the T-mode, both the toroidally confined plasma and the high-beta annuli are free of gross instabilities and the electron and ion

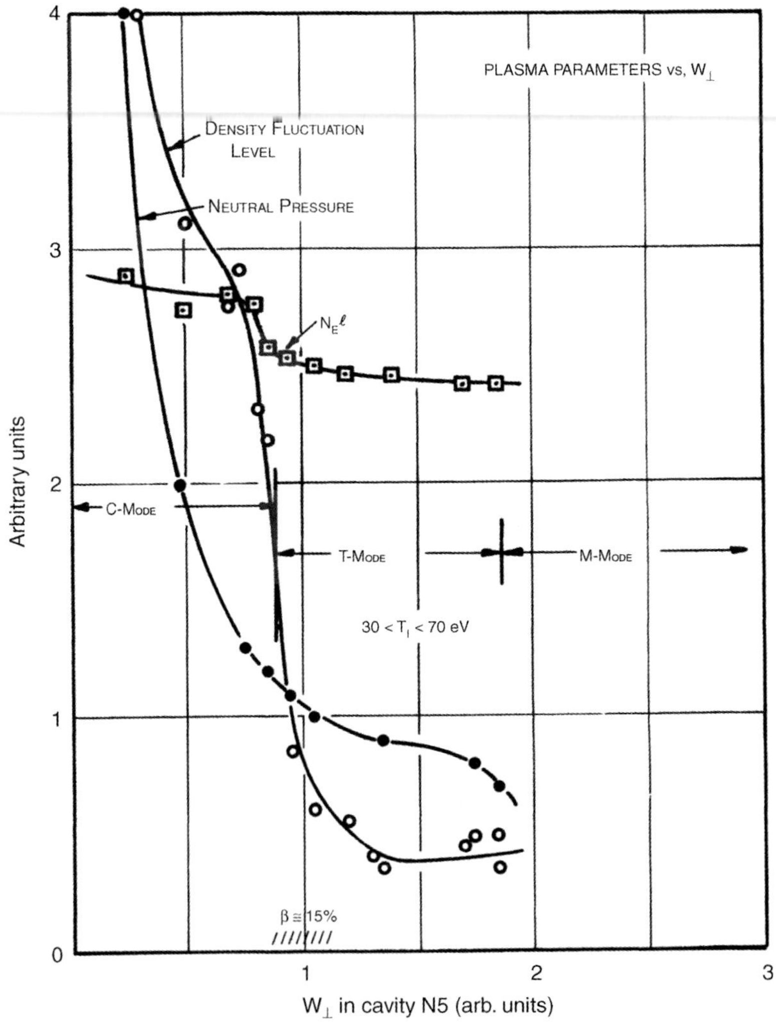

Figure 10.9 Experimental results showing the transition from the C-mode to the T-mode in EBT-I.

temperatures of the toroidal plasma increase by more than an order of magnitude. A positive space potential $\phi \sim 100$ V is measured at the position of the annuli, whereas a negative potential is observed in the interior of the plasma. Typically, the annuli have beta values around 15% at the transition from the C-mode to the T-mode, in reasonable agreement with the estimated value needed to reverse the radial gradient in the specific flux volume, $\mathbf{U} = \int d\ell/B$.

If the pressure is lowered further, a second abrupt transition is encountered in which the stored energy increases to values corresponding to $\beta \geq 50\%$, while the potential at the surface of the plasma becomes large and negative and may exceed 10^4 V. Although the annuli appear to remain stable, the toroidal plasma supports large amplitude density fluctuations. This regime of operation, designated the M-mode, is chiefly of interest because of the high-energy density and large electric fields that it exhibits. In what follows, we summarize observations made in the quiescent T-mode of operation.

The line-averaged electron density measured by the 75 GHz microwave interferometer ranges from $2\text{–}5 \times 10^{13}\ \text{cm}^{-2}$; and since the plasma diameter is around 18 cm in the midplane of each sector, the average electron density is estimated to be between 1×10^{12} and $2 \times 10^{12}\ \text{cm}^{-3}$. The local electron temperature, $T_e(r)$, has been measured directly using Thompson scattered laser light; but at these low densities, the intensity of the scattered light is only marginally detectable. Steady-state operation of the plasma allows data from multiple laser pulses to be accumulated, and in this way, electron temperatures between 130 and 200 eV were measured during relatively low-power heating with roughly 20 kW of microwave power. The variation of electron temperature with ambient hydrogen gas pressure was measured using soft x-ray spectroscopy, and typical results are shown in Figure 10.10. The corresponding ion temperatures were measured by charge-exchange analysis with results as shown in Figure 10.11.

The ambipolar potential was measured using a rubidium beam probe developed in collaboration with Colestock [7]. In this technique, singly charged rubidium ions are injected into the plasma where they become doubly ionized. Analysis of their orbits then permits a direct spatially resolved measurement of the ambipolar potential, $\phi(x,y)$. An example is shown in Figure 10.12.

The potential is typically around $+100$ V positive near the surface of the plasma and -100 V near the center line of the torus. The resulting inward-directed radial electric field enhances electron confinement and provides electrostatic confinement of most of the plasma ions. A spline fit to a large number of potential measurements reveals the closed, bowl-shaped nature of the potential surfaces in the T-mode of operation, when magnetic field errors have been properly compensated. The annuli typically have densities in the $1\text{–}4 \times 10^{11}\ \text{cm}^{-3}$ with temperatures around 250 keV. The corresponding values of beta are in the range of 10–40%.

The parameters of the plasma in the quiescent T-mode of operation can be interpreted at least qualitatively in terms of neoclassical transport theory; initially formulated for stellarators and bumpy torii by Kovrizhnikh [8] and subsequently developed extensively by many others. The detailed effect on ion confinement of the ambipolar potential observed in EBT has not yet been fully investigated, but the

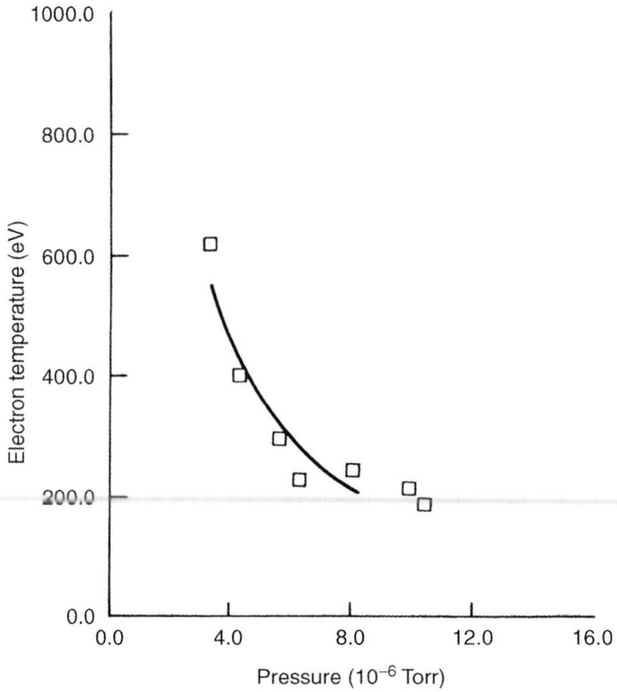

Figure 10.10 T_e versus pressure in the T-mode.

EBT experiments provide some useful phenomenological insights into the electron transport processes. In its most basic form, neoclassical transport theory predicts that the flux of particles and heat in quiescent plasma will be governed by the pitch-angle dependence of the spatial properties of the drift orbits of the guiding centers. The resulting transport will have a diffusive character with a diffusion coefficient of the following form:

$$D \sim [\langle(\delta x)^2\rangle \nu_\theta]\{1/[1+(\nu_\theta/\Omega_p)^2]\},$$

where once again δx is the shift of the guiding-center drift orbit relative to the ring axis of the torus, $r\Omega_p$ is the poloidal drift speed, and $\nu_\theta = \tau_\theta^{-1}$ is the rate at which the particle velocity is scattered in angle by Coulomb collisions. Generally speaking,

$$\langle(\delta x)^2\rangle \approx (rR_{co}/R_t)^2(1+qER_{co}/T)^{-2}$$

and

$$\Omega_p \approx (T/qBrR_{co})(1+qER_{co}/T).$$

The Coulomb scattering rate is the inverse of the Spitzer value cited earlier:

$$\tau_\theta = 25.8\sqrt{\pi}(\varepsilon_o/q^2)^2(n\ln\Lambda)^{-1}\sqrt{m}\,T^{3/2}.$$

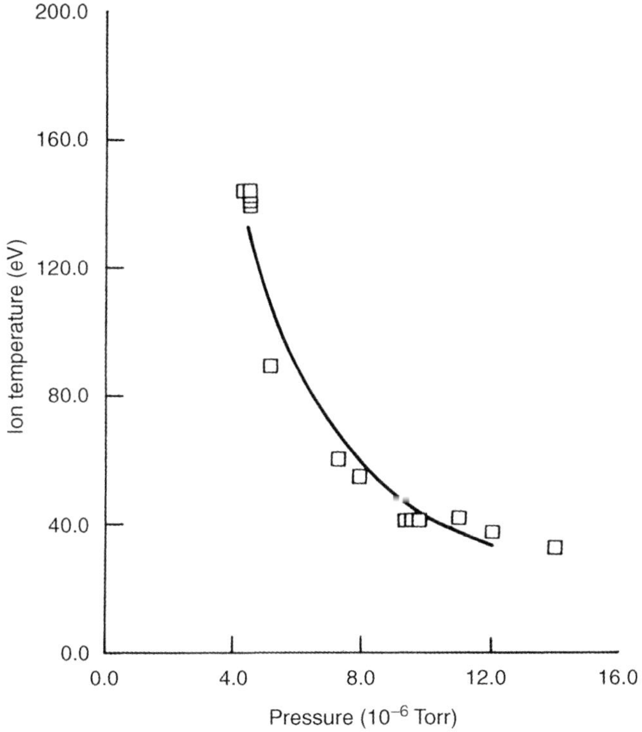

Figure 10.11 T_i versus pressure in the T-mode.

For 400 eV electrons with a density of $2 \times 10^{12}\,cm^{-3}$ near the plasma surface in EBT-I, we estimate that $\langle(\delta x)^2\rangle \approx 0.3\,cm^2$, $\tau_\theta = 10^{-4}\,s$, and $\Omega_p \approx 10^5\,s^{-1}$. The corresponding diffusion coefficient is $D \approx 3 \times 10^3\,cm^2\,s^{-1}$; and the electrons are in the collisionless regime since $(\nu_\theta/\Omega_p)^2 \ll 1$. The transport times estimated from this crude picture of neoclassical transport are then in rough agreement with the experimental values and given by

$$\tau \sim a^2/2D \approx \tau_\theta/2\{[R_t/R_{co}(a)](1 + qER_{co}/T)\}^2 \sim 10\,ms.$$

Perhaps of greater significance is the empirical observation that the confinement time increases with electron temperature roughly as $T_e^{3/2}$, in agreement with the temperature dependence of neoclassical transport in the "collisionless" regime. The data indicating this temperature dependence are shown in Figure 10.13, where the product of electron temperature and energy confinement time is plotted against the ratio of the Coulomb collision frequency to the poloidal precession frequency, (ν_θ/Ω_p), proportional to $T_e^{-5/2}$.

In order to explore the transport implications of these experimental results from EBT-I, we express x, the shift in the guiding-center drift surfaces, in the following symbolic form:

$$x(r, \theta_v, W) = (r/R_t)W/[W\langle\nabla B/B\rangle + q\langle\nabla\phi\rangle],$$

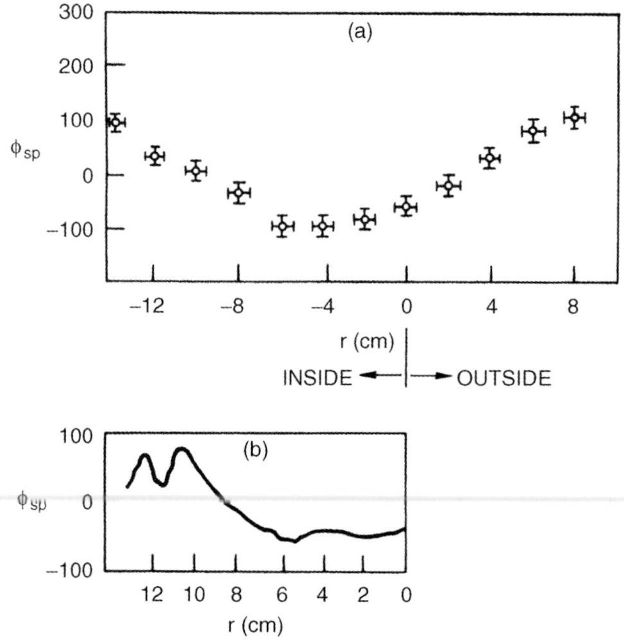

Figure 10.12 The ambipolar potential in the T-mode.

where $\theta_v = \sin^{-1}(v_\perp/v)$ is a measure of the velocity-space pitch angle, W is the particle energy, and $\langle\,\rangle$ indicates the average over a bounce or transit period. As a notational convenience, we denote the average radial (ambipolar) electric field as $E_r(r,\theta_v) = \langle\nabla\phi\rangle$. Similarly, the average radius of curvature of the magnetic lines of force is

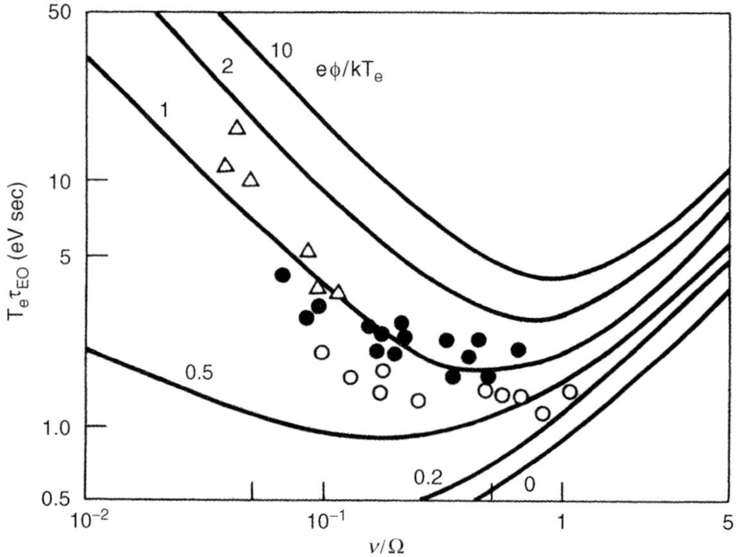

Figure 10.13 $T_e\tau_E$ versus ν/Ω_p in the T-mode.

denoted by $R_c(r,\theta_v) = -\langle \nabla B/B \rangle^{-1}$. Since the net poloidal drift is the vector sum of the $\nabla B \times \mathbf{B}$ and $\mathbf{E} \times \mathbf{B}$ drifts, and since the radial electric field in the EBT-I experiments is directed inward toward the toroidal ring axis, there is an ion energy at which the two drifts will cancel and the net poloidal drift will vanish. This energy, W_t, marks the transition from electrostatic confinement to magnetic confinement and provides a loss mechanism for ions that can ensure quasineutrality. Symbolically, this transition energy is given by $W_t = -qE_rR_c$. Our expression for the shift in the center of the guiding-center drift surfaces now takes the form

$$x(r, \theta_v, W) = -(rR_c/R_t)W/(W-W_t).$$

Clearly, the transition energy depends on the position, r, and the pitch angle, θ_v, but these dependences are relatively weak for well-trapped particles and we will ignore them for the present qualitative discussion.

We now imagine that the ions and electrons experience a large number of small, random changes in energy and pitch angle, resulting from Coulomb collisions and any operative heating processes. The corresponding changes in the centers of the guiding-center drift surfaces will be given by

$$\delta x = (\partial x/\partial \theta_v)\delta\theta_v + (\partial x/\partial W)\delta W,$$

which can readily be written in the following form:

$$\delta x = (rR_c/R_t)(1-W/W_t)^{-2}\{[(W/W_t)^2(\partial \ln R_c/\partial \theta_v) + (W/W_t)(\partial \ln E/\partial \theta_v)]\delta\theta_v - \delta W/W_t\}.$$

If the ambipolar potential is constant along the magnetic lines of force and the mirror ratio is not too large, $\partial \ln E/\partial \theta_v$ is small and will be neglected for the present qualitative discussion. Moreover, since the electric field in EBT-I is directed radially inward so that the transition energy for electrons is negative, we may assume that the electrons will diffuse mainly as a result of pitch-angle scattering:

$$\delta x_e \approx (rR_c/R_t)(1-eER_c/T_e)^{-2}(\partial \ln R_c/\partial \theta_v)\delta\theta_v.$$

For the plasma ions whose energies are less than the transitional energy, the dominant contribution to transport may come from the heating term:

$$\delta x_i \approx -(rR_c/R_t)(1-W/W_t)^{-2}\delta W/W_t \approx -(r/R_t)(1-W/W_t)^{-2}\delta W/eE.$$

Evidentially, as the ions are heated and W approaches W_t, δx_i will become large and radial transport will become increasingly rapid. Note that for these low-energy-electrostatically confined ions, the radial diffusion associated with heating will transport ions in proportion to the energy gained:

$$\Delta x(W) = -(rR_c/R_t)(1-W/W_t)^{-2}\Delta W/W_t.$$

Since the ions are lost if $\Delta x(W)$ exceeds the plasma radius, we can obtain a rough estimate of the maximum energy which these electrostatically confined ions can achieve; namely, $W_{max} \approx W_t(R_t/R_c)(1 - W_{max}/W_t)^2$. For typical operating conditions in EBT-I, this gives $W_{max} \approx 0.7\, W_t$.

One will immediately recognize the parallels between this mechanism and the ambipolar transport mechanism discussed in Chapter 6. In EBT-I, the electrons are magnetically confined with an apparently neoclassical lifetime τ_e, and the ion lifetime must equal that of the electrons to maintain electrical neutrality. The ambipolar (radial) electric field takes on the value required for the two lifetimes to be equal. A theoretical formula for the neoclassical diffusion coefficient has been derived by Kovriznykh [8] and is given by:

$$D_n = (1/6)(v_{th}/\Omega_{pol})^2 v_\theta/[1 + (v_\theta/\Omega_p)^2].$$

The scaling of electron lifetime with the parameters of the EBT is then given by

$$\tau_e \sim a^2/(2D_{ne}) \sim (R_t/R_c)\tau_{\theta e},$$

and it appears that the electron lifetime in EBT is enhanced over the Coulomb scattering time by a (large) factor, (R_t/R_c)

References

1 B.B. Kadomtsev, in *Plasma Physics and the Problems of Controlled Thermonuclear Reactions* (M.A. Leontovich and J. Turkevich,eds.), Vol. 3, p. 340 ff, Pergamon, Oxford (1959).

2 G. Gibson, W.C. Jordan, and E.J. Lauer, *Phys. Rev. Lett.* **4**, 217 (1960).

3 A.I. Morozov and L.S. Solov'ev, in *Reviews of Plasma Physics* (M.A. Leontovich, ed.), Vol. III, pp. 267–272, Consultants Bureau, New York (1966).

4 B.B. Kadomtsev, in *Plasma Physics and the Problems of Controlled Thermonuclear Reactions* (M.A. Leontovich and J. Turkevich, eds.), Vol. 4, p. 450, Pergamon, Oxford (1960).

5 R.A. Dandl *et al.*, "Plasma confinement and heating in the ELMO Bumpy Torus (EBT)," in *Plasma Physics and Controlled Nuclear Fusion Research* (Proceedings of the Fifth Conference, Tokyo, 1974) II, IAEA, Vienna, p. 141 ff (1975).

6 R.A. Dandl, H.O. Eason, P.H. Edmonds, A.C. England, G.E. Guest, C.L. Hedrick, J.T. Hogan, and J.C. Sprott,in *Proceedings of the Fourth International Conference on Plasma Physics and Controlled Thermonuclear Reactions, Madison, Wisconsin*, Vol. 2, p. 607. IAEA, Vienna (1971). See also, G.E. Guest, C.L. Hedrick, and J.T. Hogan, *Phys. Fluids* **15**, 1159 (1972).

7 P.L. Colestock, K.A. Connor, R.L. Hickok, and R.A. Dandl, *Phys. Rev. Lett.* **40**, 1717 (1978).

8 L.M. Kovrizhnykh, *Zh. Eksp. Teor. Fiz.* **46**, 877 (1969). [English Translation: *Sov. Phys. JETP* **19**, 475 (1969)].

Exercises

10.1. *It has been suggested that the observed stability of the hot-electron annulus in ELMO and similar hot-electron plasmas may be due to the rapid poloidal drift speed of the relativistic electrons, provided it is much faster than the growth rate of potentially unstable fluctuations.*

(a) Derive an approximate expression for the growth rate of curvature-driven flute modes in a plasma consisting of low-temperature ions and high-temperature electrons.

(b) Estimate the hot-electron temperature, for which this growth rate is much less than the hot-electron poloidal frequency.

10.2. *As a crude model of the diamagnetic modifications of the vacuum magnetic field brought about by a high-beta relativistic-electron annulus, we adopt the following ad hoc* model of the kinetic pressure, $p = (3/2)nkT$, in the annulus:

$$p = p_a \exp\{-[(r - r_a)/\Delta r]^2\}, \text{ for } |z| \leq L_a$$
$$= 0, \text{ for } |z| > L_a.$$

In the MHD picture, the radial pressure gradients in the annulus are balanced by diamagnetic currents whose densities are given by $j \times B = \Delta p$.

(a) Derive expressions for the azimuthal diamagnetic current densities in the inner and outer surfaces of the annulus.

(b) What are the maximum values of these currents and where do they occur?

(c) Derive an approximate expression for the total diamagnetic field at the center of the annulus, $r = r_a$ *and* $z = 0$.

(d) Estimate the resulting diamagnetic fields as a function of beta, $\beta = 2\mu_o p/B^2$, *and the dimensions of the annulus.*

10.3. *In the "collisional" regime of plasma confinement in a simple bumpy torus, the poloidal drift frequency and the (pitch-angle) scattering time satisfy the condition that* $\Omega_p \tau_\theta < 1$. *Thus, the poloidal drift is too slow to compensate for the toroidal drift,* $v_t u_y$, *where the major axis of the torus is in the y-direction. The random walk step size is then given by* $\delta y = v_t \tau_\theta$, *where* v_t *is the guiding-center drift speed due to the toroidal curvature. Show that the resulting diffusion coefficient varies as* B^{-2}, *i.e., the same B dependence as classical diffusion across a uniform magnetic field.*

11
ECH Applications to Space Plasmas

In this chapter, we consider the possible use of electron cyclotron heating (ECH) for active experiments in space as well as for basic terrestrial studies of plasmas that may be helpful in understanding some astrophysical phenomena. The illustrative active experiments to be considered here are aimed at achieving control of energetic particle fluxes from the Earth's radiation belts, an undertaking sometimes referred to as "Radiation Belt Remediation." The basic terrestrial plasma studies envisioned in this chapter emphasize plasma processes of interest in astrophysics that can be expected to take place in the high-energy-density plasma media that can be produced and sustained in stable steady state using ECH. Plasmas such as those produced in ELMO have substantial populations of relativistic electrons with energies in excess of the thresholds for electron–positron pair production as well as the electrodissociation of deuterons discussed earlier in this volume. The content of this chapter is necessarily more speculative, as compared with the previous three chapters in which well-diagnosed experiments provided the basis for theoretical interpretations. Nonetheless, it will be possible to introduce several unique aspects of ECH that should prove to be important in future ECH applications in these two areas.

11.1
Active Experiments in Space

Achieving active control of the flux of energetic particles from the Earth's radiation belts has been of considerable interest for many years, particularly in the United States and in the former Soviet Union. There are natural sources such as strong solar storms that create a population of energetic charged particles in the magnetosphere, which have sometimes caused damage to satellites and electrical power networks. An even greater concern has been the possible creation of dense populations of damaging energetic electrons and ions by the intentional detonation of nuclear weapons in the radiation belts as a way of neutralizing satellites intended for defense purposes.

One conceptual approach to pumping energetic charged particles out of the Earth's radiation belts was made almost 25 years ago by Trakhtengerts [1]. In this approach,

Electron Cyclotron Heating of Plasmas. Gareth Guest

ISBN: 978-3-527-40916-7

which he called the "Alfvén Maser," collective oscillations in the magnetospheric plasma are driven to large amplitudes through the injection of suitable populations of hot electrons together with synchronous modulation of the ionospheric plasma density at the foot of the chosen magnetic flux tube. In some respects, this process resembles the triggering of whistler instabilities, described in Section 8.3. The Alfvén Maser concept raises several issues that can be addressed in terms of basic ECH theory; it also suggests some experimental strategies that have not been fully investigated, as well as some unresolved theoretical issues, and therefore, the present discussion must be regarded as somewhat speculative.

The main technical difficulty that must be overcome if adequate densities of ionospheric electrons are to be heated to even moderately relativistic energies arises from the nearly uniform magnetic field in the resonant interaction region where the RF fields resonate with the electron gyrofrequency and its harmonics. As discussed earlier, the relativistic increase in the mass of the heated electrons limits the increase in electron energy to values that may be inadequate to pump the Alfvén Maser. This heating limit is usually referred to as a relativistic limit cycle or, more recently, a relativistic heating gap [2]. The term "gap" refers to the difference between the maximum energy that permits resonance at a particular gyroharmonic and the minimum energy that permits resonance at the next higher harmonic. There have been studies of the use of multiple heating frequencies to bridge the relativistic heating gaps and achieve stochastic heating over a wide energy range. A more speculative approach that has yet to receive similar attention is the possible use of frequency-modulated ECH power (FMECH) to extend the duration of a single-pass resonance and thereby achieving the desired final energy in a single-electron transit of the resonance zone. FMECH also offers the possibility of bunching the electrons in gyrophase as well as in energy, although this has not been demonstrated for the oblique illumination that would characterize ionospheric heating. If phase bunching could be achieved, the resulting energetic electrons may be especially efficient in triggering the desired collective response in the radiation belts. Note that the precipitation of electrons into the ionosphere will increase the conductivity at the foot of the flux tube and provide positive feedback, which could conceivably yield large amplitude bursts of wave energy and ejected particles.

To establish the basic heating geometry we recall the geomagnetic field model described briefly in Chapter 2, following Alfvén and Fälthammar [3]. In terms of a magnetic scalar potential, χ, the magnetic field is given in spherical polar coordinates by $\mathbf{B} = -\nabla\chi$ with $\chi = \mathbf{a}\cdot\mathbf{r}/r^3 = a \sin\lambda/r^2$. The strength of the dipole is $a = 8.1 \times 10^{15}$ Tm^3 and the longitudinal angle λ is measured from the equatorial plane, $\lambda = \pi/2 - \theta$. The resulting components of the geomagnetic field are then: $B_\phi = 0$, $B_r = B_p \sin\lambda$, and $B_\lambda = -(B_p/2)\cos\lambda$, where $B_p = 2a/r^3$. The magnetic lines of force are specified by $\phi = \text{constant}$ and $r = r_{eq}\cos^2\lambda$. Here, r_{eq} is the distance from the origin to the point of intersection of the magnetic field line with the equatorial plane. Along a line of force, the magnetic intensity varies as

$$B = \left(a/r_{eq}^3\right)\left[(1 + 3\sin^2\lambda)^{1/2}/\cos^6\lambda\right]. \tag{11.1}$$

If s measures the distance along a magnetic line of force, we have

$$(1/B)(dB/ds) = (3 \tan \lambda/r)[(8-5 \cos^2 \lambda)/(4-3 \cos^2 \lambda)](1+4 \tan^2 \lambda)^{-1/2}. \tag{11.2}$$

As a specific illustrative example, we will utilize the HIPAS facility in Alaska [4], where $\lambda = 64.87^\circ$ and $(1/B)(dB/ds) \approx 3/r$. For a resonant interaction region 100 km above the Earth's surface at the location of HIPAS, $r = 6{,}478$ km and thus, $(1/B)(dB/ds) \approx 4.6 \times 10^{-4}\,\text{km}^{-1}$. If the total length of the interaction region illuminated by a given antenna array is 10 km, for example, then over this region, the magnetic intensity varies by $\Delta B/B \approx 0.0046$ and the magnetic field can be considered as essentially uniform over the interaction region. Note that the distance along a magnetic line of force from a resonant interaction region at longitude λ and its conjugate point is given in the present dipole model by

$$L = 2\int (ds/d\lambda)d\lambda = \sqrt{3}r_{eq}\{\sin \lambda(1/3 + \sin^2 \lambda)^{1/2}$$

$$+ (1/3)\ln[\sin \lambda + (1/3 + \sin^2 \lambda)^{1/2}] - (1/6)\ln(1/3)\}, \tag{11.3}$$

where the limits of integral are from 0 to λ. In our HIPAS example, with the resonant interaction region at an altitude of 100 km, $r_{eq} \approx 3.6 \times 10^4$ km and $L \approx 9.8 \times 10^4$ km. The remaining basic parameters characterizing the resonant interaction region 100 km above HIPAS are: $B = 5.54 \times 10^{-5}$ T, $f_{ce} = 1.55$ MHz, and $c/f_{ce} = 193$ m. The cutoff densities for fundamental and second-harmonic resonance are 3×10^4 and 12×10^4 electrons/cm^3, respectively. For comparison, the maximum day-time electron density in the E-layer (100–120 km) is around 12×10^4 electrons/cm^3 and the night-time density is around 1.8×10^3 electrons cm^{-3} [5].

In Eq. (5.4), a term was included for the rate of change of the phase factor, $\nu'(t_{res})$, arising from the relativistic change in the mass of the heated electrons: $\nu'(t_{res}) = \Omega(dW/dt)/\gamma mc^2$. This effect will limit the duration of resonance if the magnetic field is essentially uniform, as is the case under consideration here. If $\nu'(t_{res})$ is not compensated by a corresponding change in the wave frequency, ω, electrons will initially gain energy until their gyrophase has slipped by a large enough value to cause the sign of dW/dt to reverse. The electrons will then loose energy until the gyrophase reaches a value for which dW/dt is once again positive. This relativistic limit cycle will repeat until some external condition is altered. We can gain some semiquantitative insights into the implications of this relativistic effect using results from Chapter 5.

Equation (5.5) gave the following expression for the effective duration of resonance, t_{eff}, defined as the time interval within which the gyrophase would change by $\pm\pi/4$: $t_{eff} \cong 2[\pi/|2\nu'(t_{res})|]^{1/2}$. If we now substitute the rate of change due to relativistic detuning: $\nu'(t_{res}) = -\Omega(dW/dt)/\gamma mc^2$, approximate the heating rate in the heating phase of the relativistic limit cycle by $dW/dt \approx \Delta W_\perp/t_{eff}$ and replace Ω by $2\pi f$, we have

$$f\, t_{eff} \approx \gamma\, mc^2/\Delta W_\perp. \tag{11.4}$$

Thus, the duration of resonance is inversely proportional to the energy increment as one might have expected. The energy increment in one transit of the resonance zone was given by Eq. (5.13):

$$\Delta W_\perp = -e\sqrt{2}|E_-|v_\perp(0)J_o^2(k_\perp\rho)t_{eff}\cos\phi_{res} + (e^2/\gamma m)|E_-|^2 J_o^2(k_\perp\rho)t_{eff}^2.$$

Defining $K \equiv e|E_-|\lambda J_o(k_\perp\rho)ft_{eff}/\gamma mc^2$, where $\lambda = c/f$ is the free-space wavelength, and substituting from Eq. (11.4), we can rewrite Eq. (5.13) in the following compact form:

$$K^2 - \sqrt{2}[v_\perp(0)/c]J_o(k_\perp\rho)\cos\phi_{res}K = \Delta W_\perp/\gamma mc^2 = 1/(f\,t_{eff}). \qquad (11.5)$$

The particular choice of $\cos\phi_{res} = 0$ displays the basic scaling with RF electric field strength: $K^2 = 1/ft_{eff}$ becomes $1/ft_{eff} = [e|E_-|\lambda J_o(k_\perp\rho)/\gamma mc^2]^{2/3}$, so that

$$\Delta W_\perp = \gamma mc^2\left[e|E_-|\lambda J_o(k_\perp\rho)/\gamma mc^2\right]^{2/3}. \qquad (11.6)$$

The variation of $\Delta W_\perp$ with the two-third power of the effective RF electric field strength has been demonstrated in numerical studies by Ginet and Heinemann [6]. To realize this energy increment in practice would require that the length of the resonance zone illuminated by RF power be matched to the distance traveled by a heated electron in the time for a single transit of resonance, t_{eff}.

In principle, it is possible to compensate for the relativistic detuning by employing frequency modulation of the RF power (FMECH) [7]. For fundamental resonance in a uniform magnetic field, the necessary rate of change of the RF frequency is given by $d\omega/dt = -\Omega(dW/dt)\gamma mc^2$. For realizable power sources, the maximum frequency deviation rate is limited by the effective bandwidth of the RF system, Δf:

$$df/dt \le (3/2)(\Delta f)^2. \qquad (11.7)$$

The maximum heating rate for which the resonance condition, $\nu = 0$, can be maintained is therefore limited by the available bandwidth:

$$(dW/dt)_{max} \approx (3/2)\gamma mc^2 f(\Delta f/f)^2. \qquad (11.8)$$

The maximum change in the electron energy for which resonance can be maintained is similarly limited by the bandwidth of the RF system:

$$(\Delta W)_{max} \approx \gamma mc^2|\Delta f/f|. \qquad (11.9)$$

These limitations determine the maximum energy increment, the optimum RF electric field strength, and the duration of resonance for a given bandwidth. For ideal resonance, $v_-(t)$, and thus the heating rate, dW/dt, increases linearly with time whereas the electron energy increases quadratically with time. At a time $t = t_{max}$, the heating rate will reach the limit set by Eq. (11.8), and the condition for ideal resonance can no longer be satisfied. For optimum choice of the RF electric field strength, we require the energy at this time to satisfy Eq. (11.9). We employ the linearized description of the heating from Chapter 5 to make this picture more quantitative. The right-hand circularly polarized component of the velocity of an electron acted on by a right-hand circularly polarized RF electric field was obtained in

Chapter 5 by integrating the equation of motion. The resulting expression was

$$v_-(t) = v_-(0)\exp\left(-i\int\Omega dt\right) + A_-\exp\left(-i\int\Omega dt\right)\sum J_n(k_\perp\rho)\exp[i(n\phi_o - \rho\sin\phi_o)]$$
$$\times\int dt\exp\{i\int dt[k_{||}v_{||} - \omega + (n+1)\Omega]\},$$

where $A_- \equiv -(e/\gamma m)|E_-|$. For fundamental resonance, $n = 0$ and $k_{||}v_{||} - \omega + \Omega = 0$, so that the linear time dependence of $v_-(t)$ is clearly evident:

$$\begin{aligned} v_-(t) &= [v_-(0) + A_-J_o(k_\perp\rho)\exp(-ik_\perp\rho\sin\phi_o)t]\exp\left(-i\int\Omega dt\right) \\ &= [v_-(0) + A_{eff}t]\exp\left(-i\int\Omega dt\right). \end{aligned} \quad (11.10)$$

The heating rate averaged over a wave period is given by $(1/2)\mathrm{Re}(-e\mathbf{E}^*\cdot\mathbf{v}) \approx (1/2)\mathrm{Re}(-eE_-^*v_-)$, resulting in the following expression for the time-dependent fundamental-resonance heating rate:

$$\begin{aligned} (dW/dt) &= -e|E_-|J_o(k_\perp\rho)\cos(\phi_o - k_\perp\rho\sin\phi_o)v_\perp(0)/2\sqrt{2} \\ &\quad + (e^2/\gamma m)[|E_-|J_o(k_\perp\rho)]^2 t/2. \end{aligned}$$

The electron energy at time t is thus given by

$$\begin{aligned} W(t) &= W(0) - e|E_-|J_o(k_\perp\rho)\cos(\phi_o - k_\perp\rho\sin\phi_o)v_\perp(0)t/2\sqrt{2} \\ &\quad + (e^2/\gamma m)[|E_-|J_o(k_\perp\rho)]^2 t^2/4. \end{aligned}$$

In order to display the implications of the heating limitations based on the RF bandwidth, we reconsider only the deterministic, gyrophase-independent term and write for the maximum heating rate and energy increment as

$$(dW/dt)_{max} = (3/2)\gamma mc^2 f(\Delta f/f)^2 \approx (e^2/\gamma m)[|E_-|J_o(k_\perp\rho)]^2 t_{max}/2 \quad (11.11)$$

and

$$\Delta W_{max} = \gamma mc^2|\Delta f/f| \approx (e^2/\gamma m)[|E_-|J_o(k_\perp\rho)]^2(t_{max}/2)^2. \quad (11.12)$$

The optimum values of the duration of resonance and the RF electric field strength are thus given by

$$f\,t_{max} = 3(\Delta f/f)^2\left[e|E_-|\lambda J_o(k_\perp\rho)/\gamma mc^2\right]^{-2} \quad (11.13)$$

and

$$e|E_-|\lambda J_o(k_\perp\rho)/\gamma mc^2 \approx (3/2)(|\Delta f/f|)^{3/2}. \quad (11.14)$$

Since $(\Delta W)_{max} = \gamma mc^2|\Delta f/f|$, Eq. (11.14) indicates once again that the maximum energy increment is proportional to the two-third power of the RF electric field:

$$(\Delta W)_{max}/\gamma mc^2 \approx \left[(2/3)e|E_-|\lambda J_o(k_\perp\rho)/\gamma mc^2\right]^{2/3}.$$

Apart from the factor of $(2/3)^{2/3}$, this is the same result as the more approximate value from Eq. (11.6).

11.2
Laboratory Experiments of Astrophysical Significance

The understanding of astrophysical phenomena is likely to depend to a significant degree on the understanding of basic processes taking place in plasmas. This understanding, in turn, must ultimately be derived from closely linked theoretical and experimental studies of a wide range of plasma media. ECH has proven to be an effective means for creating many different plasma media, but especially the steady-state relativistic-electron plasmas that can provide experimental and theoretical access to plasmas capable of exhibiting a unique regime of plasma phenomena. Among these are the steady-state creation and confinement of populations of electron–positron pairs and the electron dissociation of deuterium nuclei, as well as the generation of various types of double layers. With the potential use of ECH in mind, we include here an overview of the production of relativistic-electron plasmas by ECH with an emphasis of the fundamental processes that could be manipulated for optimizing this production. Many of the technical details of this overview have been discussed in earlier chapters, but they will be placed in a broader context here to provide something of a prescription for efficient ECH generation of relativistic-electron plasmas.

The conceptual approach to be discussed here is based in part on the anticipation that under typical circumstances, a significant number of collisionless electrons can accumulate in the superadiabatic or null-heating limit cycles, as discussed in Chapter 5. By suitable restoration of the stochastic heating of these electrons before they can generate turbulence, many more of them can be heated to relativistic energies. The use of multiple-frequency ECH (MFECH), which has shown great promise in increasing the efficiency of relativistic-electron generation, has been interpreted in just this way; but other approaches might also be effective.

Broadly speaking, the fundamental processes governing the generation of relativistic-electron plasmas are those that are involved in providing a steady source of ion–electron pairs, confining the resulting plasma without excessive turbulence and heating the electrons into the MeV energy range. For definiteness, we will use magnetic-mirror confinement in discussing these processes, but much of the discussion will be applicable to other configurations as well. To that end, we consider a mirror-confined electron cyclotron heated plasma created by ionization of ambient gas and identify fundamental aspects of the heating and confinement in the schematic representation shown in Figure 11.1.

The arrows in the figure represent fluxes of electrons in energy and in space. The energy fluxes indicate the effect of competition between heating and cooling, whereas the spatial fluxes indicate the loss of confinement. The equilibrium distribution of electrons in energy is governed by the net result of these competing processes. Note

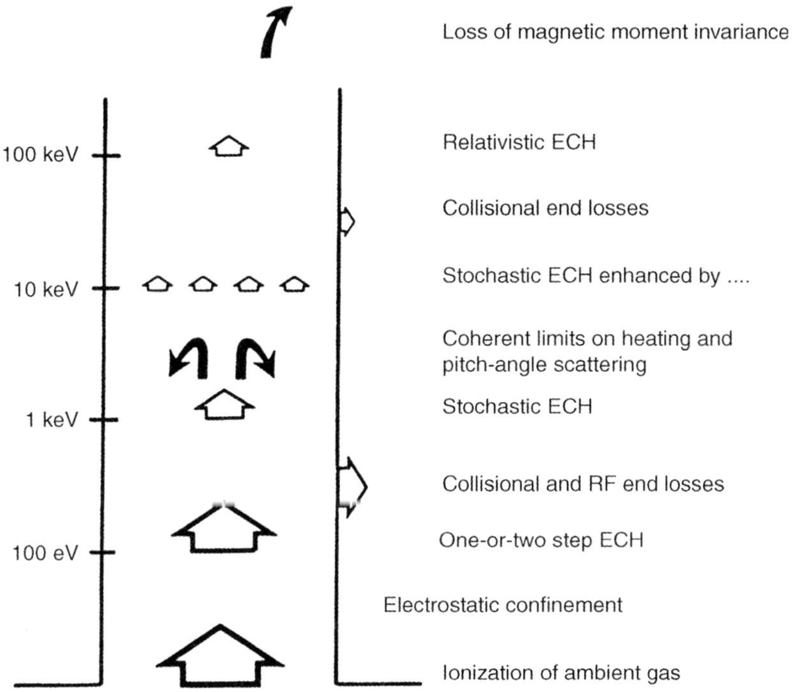

Figure 11.1 A schematic model of the fundamental ECH processes in a magnetic mirror.

that the energy dimension increases upward in the figure, starting at 10 eV for the newly created ion–electron pairs and ending somewhere above 10^5 eV where nonadiabatic losses may become dominant in magnetic mirrors of modest size and magnetic field strength. The scale is somewhat arbitrary, but is based on our earlier classification of groups of electrons according to the energy dependence of their fundamental kinetic and dynamical processes. Despite its arbitrariness, this scheme provides a useful framework for identifying the elements of heating and confinement that can be manipulated to optimize the steady-state ECH production of relativistic-electron plasmas.

In the first energy decade shown in the figure, between 10 and 100 eV, electrons are confined by an ambipolar electrostatic field whose magnitude is governed by the conditions for quasineutrality, discussed in detail in Chapter 6. These electrons make a number of bounce orbits before escaping and therefore encounter ECH resonance a number of times in this energy interval. Under typical ECH conditions, only a few transits of the resonance surfaces are required to heat electrons to energies above the ambipolar potential (see "One-or two-step ECH" in Figure 11.1). Electrons heated above the ambipolar potential barrier can escape promptly if they are in the loss cone or by scattering into the loss cone if they are initially outside of it. Since the rate of collisional scattering decreases with the cube of the electron speed, the loss of electrons at this phase of the process will be reduced if the ambipolar potential is increased, for example, by inhibiting the outward flux of ions through anisotropic ion heating or by replacing some of the cold ions by well-confined hot ions. Note that any change that reduces the ambipolar potential or, more particularly, causes it to reverse sign is likely to cause a reduction in efficiency or a loss of the hot-electron population altogether unless additional steps are taken to prevent rapid loss of low-energy electrons in this first energy interval.

The second energy decade identified in Figure 11.1 is typically characterized by a Coulomb collision rate that is high enough to randomize the electron gyrophase between successive transits of the resonance surfaces. These electrons will therefore be heated stochastically and diffuse in energy and pitch angle. The mirror reflection points of those diffusing upward in energy will accumulate around the resonance surfaces, whereas those diffusing downward in energy may reach the loss cone and escape. The accumulation at the resonance surfaces can elevate the temperature anisotropy to levels at which instabilities can occur if the resonance surfaces are too near the mirror midplane, although measures for mitigating this effect such as UORH have been well established. Choosing resonance surfaces too near the loss cone will inevitably cause higher levels of RF driven losses.

In the third energy interval identified in Figure 11.1, from 10^3 to 10^4 eV, the Coulomb scattering rate of electrons is so slow under typical conditions that gyrophase randomization between successive transits of the resonance surfaces may no longer obtain. Stochastic heating may then cease and the electrons may accumulate in the superadiabatic or null-heating limit cycles, as discussed in Chapter 5. From the point of view of preserving the source of electrons that are to be heated to relativistic energies, it appears that the elimination of these limit cycles can provide the basis for a useful optimization mechanism. These electrons are nearly collisionless and are therefore well confined by magnetic mirrors. Moreover, the pitch angles of these electrons are restricted to relatively narrow ranges by the combined action of the ECH power and the static magnetic field of the mirror. These electrons can be released from the limit cycles and stochastically heated with high efficiency if suitable perturbing forces are applied. One such perturbing force that can be especially efficient is multiple-frequency ECH (MFECH), to be elaborated upon shortly. Other interactions may also be used to destroy the coherence underlying these limit cycles, such as the automatic second-harmonic heating that occurs if the second-harmonic resonance surfaces interact the magnetic flux surfaces on which the limit cycles obtain. This mechanism was active in TPM, the Canted Mirror, and EBT, but its efficacy could not be isolated as clearly as that of MFECH, as demonstrated, for example, in experiments carried out in the SM-1 facility [8]. In these experiments, the energy stored in relativistic electrons was increased by a factor of $5\times$ when the same amount of RF power was supplied at four different frequencies relative to that which resulted when the same power was coupled at a single frequency.

SM-1 was a simple magnetic mirror with a 2.2 : 1 mirror ratio on axis. The distance between the mirror "throats" was 71 cm. RF power was supplied by four klystron amplifiers at frequencies 8.3, 9.2, 9.4, and 10.1 GHz, respectively, and launched through hybrid junctions oriented to couple to O-modes. The klystron amplifiers could be driven at several frequencies within a bandwidth of 50 MHz. Very extensive diamagnetic measurements were made and matched to computed equilibriums to establish the equilibrium properties of the high-beta annuli generated in SM-1; but the main focus here is on the remarkable results achieved by MFECH that are summarized in Figure 11.2.

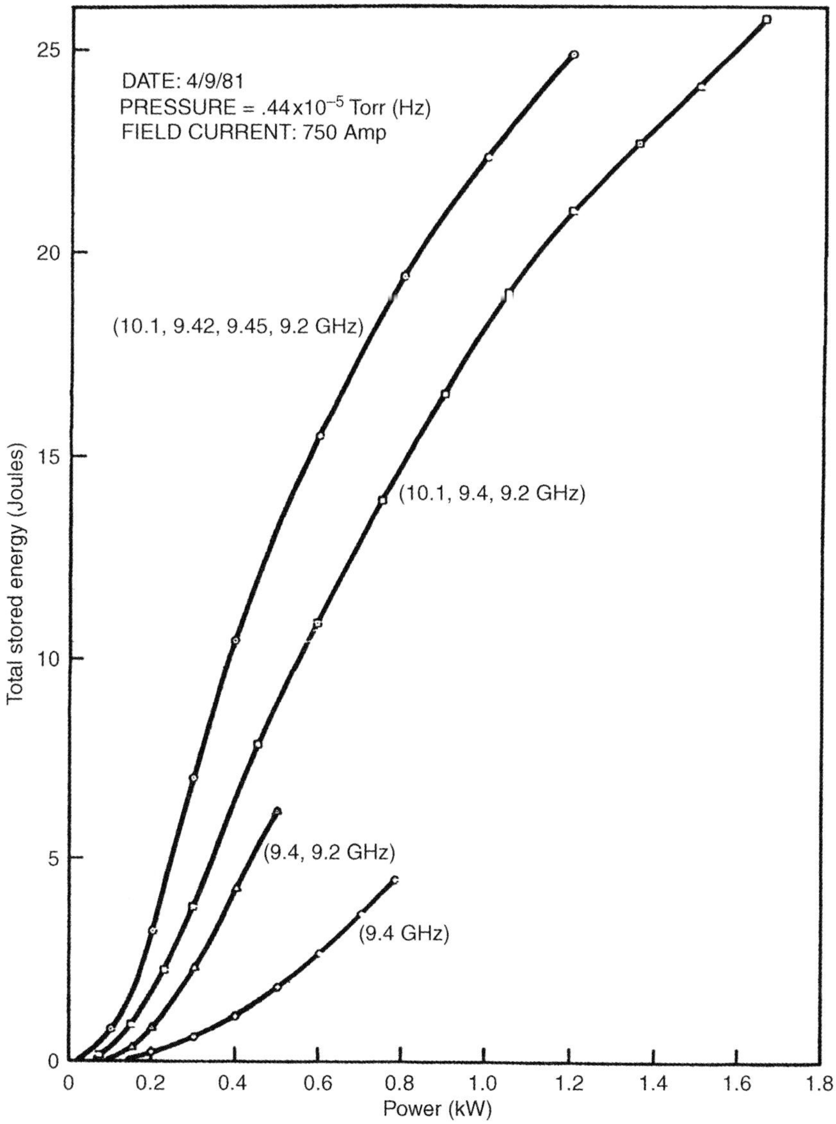

Figure 11.2 Dependence of total stored energy on total heating power for four different compositions of the heating power spectrum. The total power is always distributed equally among all of the active heating frequencies.

In Figure 11.2, the measured values of the perpendicular energy, $W_\perp$, stored in the hot-electron population are plotted as functions of the total heating power. In all the four cases shown here, the total power is divided equally among the frequencies listed for each case. When the power was supplied at a single frequency (9.4 GHz), the stored energy increased quadratically with power up to a limit around 5 J at 800 W, at which point the equilibrium was lost. When the power was divided equally between two frequencies (9.4 and 9.2 GHz), $W_\perp$ increased by a factor of three to four times higher than that was achieved with an equal power at a single frequency. With three or four frequencies, the stored energy could reach 25 J, and total powers as high as 1200 W could be used, compared with a maximum of 800 W for the single-frequency case. Experiments with two-frequency heating in which the frequency difference between the two sources was varied continuously revealed a clear optimum frequency separation of 40 MHz, which is approximately equal to the bounce frequency of 10 keV electrons in SM-1. With single-frequency heating, the escaping current carried by the electrons in this energy interval exhibited low-frequency ($\sim$30 kHz) fluctuations. These fluctuations were strongly suppressed with three or four frequency heating. It was not possible to identify the stabilization mechanism responsible for the suppression of these fluctuations, but the increased density of hot electrons resulting from MFECH is a likely candidate.

The fourth energy decade, 10^4–10^5 eV, in Figure 11.1 has been identified with renewed stochastic heating, since the gyroradius of electrons in this decade is large enough to yield significant heating rates at the second and higher harmonics of the electron gyrofrequency. It is in this energy interval that upper off resonant heating or other forms of preferential heating can be employed with excellent effect to heat the more energetic electrons. Fokker–Planck models of the resulting equilibrium distribution function have been quite successful in describing the kinetic processes in this interval [9]. These same studies have also modeled the losses caused by of the breakdown of adiabatic invariance, in good agreement with experiments.

A useful prescription for the efficient ECH generation of relativistic-electron plasmas is thus available for application to the study of basic plasma processes in these unusual plasma media.

11.3 Nonlinear Dynamical Ambipolar Equilibria

As we have seen earlier, the absorption of whistler waves launched in the high-field region can be localized in a narrow zone near the resonance surface, especially in over dense plasmas. Such strong local heating can generate a substantial electron-pressure gradient that must be balanced by an electrostatic field. The resulting structure can take the form of a phase-space vortex or a compound charge layer anchored to the resonance surface [10]. Such structures may exhibit a threshold for instability if the electron distribution contains two distinct populations of electrons: those that are trapped in the electrostatic ambipolar potential and those that are streaming through this potential.

In the course of experimental studies of plasmas created and sustained by whistler wave heating, several observations have been noted that raise questions regarding the basic properties of the plasma equilibria. For example, significant differences have been observed in the electron energy distribution when cw heating was replaced by trains of pulses with duty cycles around 25%. The electron temperature achieved with this form of pulsed heating was substantially higher than with cw heating at the same value of peak power. Since it was demonstrated that the cw heated plasma was subject to repetitive bursts of instabilities at frequencies near the heating frequency it is plausible that the use of pulse trains of heating power permits classical mechanisms to allow ordered groups of electrons to relax before the threshold for instability is reached. In this section we wish to consider the dynamical properties of the ambipolar potential that must arise if the plasma is to achieve the necessary quasineutrality that was discussed earlier. There we typically considered cases where two symmetrically located resonance surfaces were illuminated by microwave power. The ambipolar potential could reasonably be assumed to extend more or less uniformly along the magnetic lines of force between the resonance surfaces and decrease beyond these surfaces where the field lines intercepted metallic walls. If the mean-free-paths for atomic processes such as charge exchange and recombination are greater than the distance to the nearest material wall, the potential drop will be largely confined to the conventional sheath separating the plasma from the wall. In contrast, if a single resonance surface is illuminated, as would be the case in the kinds of active experiments in space discussed earlier in this chapter, the ambipolar potential is expected to reach a maximum positive value where the microwave power is absorbed and decrease along the magnetic lines of force at a rate that depends on the rate at which ions and electrons are lost by transport as well as atomic processes.

In this section we will consider a dynamical model of such an ambipolar potential containing a mixture of trapped and untrapped electrons and explore the conditions under which it might be possible to employ strong local absorption of whistler-wave power to generate a stable electrostatic trap for electrons. Such a trap could permit trapped electrons to experience multiple transits through the resonance surface and thereby be heated more efficiently at low power than would be possible if the electrons made only a single transit of the resonance surface. We assume that the dominant phenomena affecting the collective dynamics of the equilibria are electrostatic and therefore governed by Poisson's equation: $\nabla \cdot \varepsilon_o \mathbf{E} = -\nabla \cdot \varepsilon_o \nabla\phi = -\varepsilon_o \nabla^2\phi = \rho$. Here ρ is the electric charge density and ϕ is the electrostatic potential. In a strongly magnetized plasma with cylindrical symmetry, the equilibrium conditions along the magnetic lines of force can be described approximately by the one-dimensional limit of Poisson's equation:

$$d^2\phi/dx^2 = -\rho/\varepsilon_o = -\sum(q_s/\varepsilon_o)\int dv\, f_{os}(v).$$

The summation is over all species of charged particles. Since $\varepsilon = mv^2/2 + q\phi$ is a constant of the motion, the equilibrium distribution functions can be taken to be functions of ε. Note that the speed is given by $v = [2(\varepsilon - q\phi)/m]^{1/2}$ so that the derivative $d\varepsilon/dv = [2m(\varepsilon - q\phi)]^{1/2}$. If we now consider the function $V(\phi)$ defined as the following integral from $\varepsilon = q\phi$ to $\varepsilon = \infty$:

$$\varepsilon_o V(\phi) = -\int d\varepsilon\, f_o(\varepsilon)[2(\varepsilon - q\phi)/m]^{1/2}.$$

Clearly,

$$\varepsilon_o dV/d\phi = q\int d\varepsilon\, f_o(\varepsilon)[2m(\varepsilon - q\phi)]^{-1/2} = q\int d\varepsilon\, f_o(\varepsilon) dv/d\varepsilon = \rho,$$

so that $\rho/\varepsilon_o = dV/d\phi$. Poisson's equation can therefore be written as

$$d^2\phi/dx^2 + dV/d\phi = 0.$$

If we now multiply this equation by $(d\phi/dx)$ we have

$$(d\phi/dx)d^2\phi/dx^2 + (d\phi/dx)dV/d\phi = 0$$

$$(d/dx)[(d\phi/dx)^2/2] + dV/dx = 0, \text{ or}$$

$$(d\phi/dx)^2/2 + V = \text{constant}.$$

Note that Poisson's equation can now be written as

$$d[(d\phi/dx)^2/2]/d\phi = d(E^2/2)/d\phi = -dV/d\phi = -\rho/\varepsilon_o = 2(e/\varepsilon_o)(n_e - n_i), \tag{11.15}$$

which is the form we will employ in the following. The function, $V(\phi)$, is generally designated as the "pseudo potential."

One of the simplest techniques for obtaining self-consistent, nonlinear dynamical descriptions of equilibria of the sort under consideration here employs the so-called water-bag model [11] to obtain explicit forms for $n(\phi)$ that can be integrated to obtain $(d\phi/dx)^2$ and $\phi(x)$. The water-bag model represents the equilibria as a number of sharply bounded regions in phase space within which the distribution function has a constant value and Liouville's theorem is explicitly satisfied. For two groups of electrons, the phase-space orbits may have the form shown in Figure 11.3. The distribution of electrons in speed is assumed to be constant within boundaries

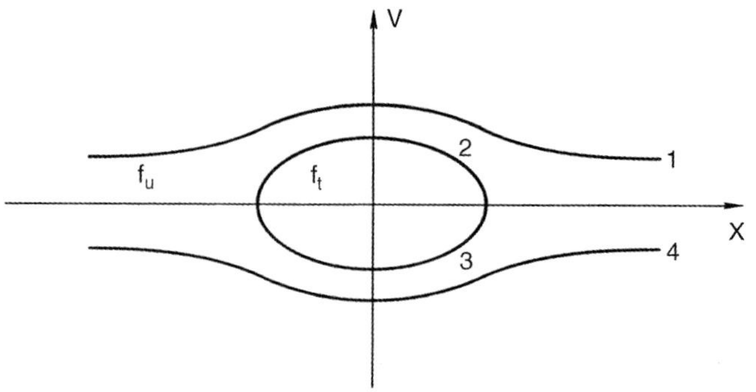

Figure 11.3 Phase-space boundaries for a rudimentary water-bag model of the dynamic, self-consistent equilibrium containing trapped and untrapped electrons.

determined by the constants of the motion. In the present illustrative case two such regions are identified, corresponding to electrons that are trapped in the self-consistent electrostatic potential and those that are untrapped and stream through the potential. The associated distribution functions have values f_t and f_u, respectively. The four boundaries shown in the figure are described by energy conservation:

$$mv_j^2/2 - e\phi = \varepsilon_j = \text{constant} = \pm mu_j^2/2, \tag{11.16}$$

where the index j takes on the values 1, 2, 3, and 4. Since v_2 vanishes at the turning points, $\varepsilon_2 \langle 0$ if $\phi \rangle 0$, as is the case here. Additionally, since v_2 and v_3 vanish at the same position, we have $\varepsilon_2 = \varepsilon_3$. For the present case we will adopt the simplifying assumption that the equilibrium is symmetric so that $\varepsilon_1 = \varepsilon_4$, and the electron density is then given by

$$\begin{aligned} n_e &= 2f_u v_1, \quad \text{for} \quad e\phi + \varepsilon_2 \le 0, \text{ and} \\ n_e &= 2f_u v_1 + 2(f_t - f_u)v_2, \quad \text{for} \quad e\phi + \varepsilon_2 \ge 0. \end{aligned}$$

If we introduce the Heaviside unit step function, $U(e\phi + \varepsilon_2)$, we can write this more compactly as

$$n_e = 2f_u v_1 + 2(f_t - f_u)v_2 U(e\phi + \varepsilon_2). \tag{11.17}$$

As a notational convenience we use the maximum value of the electrostatic potential, ϕ_o, to define the following three dimensionless variables:

$$\begin{aligned} \xi &\equiv \phi/\phi_o \\ \zeta_1^2 &\equiv mu_1^2/2e\phi_o = \varepsilon_1/e\phi_o \\ \zeta_2^2 &\equiv mu_2^2/2e\phi_o = -\varepsilon_2/e\phi_o. \end{aligned} \tag{11.18}$$

In term of these variables, the speeds on the boundaries are given by

$$\begin{aligned} v_1 &= (2e\phi_o/m)^{1/2}(\zeta_1^2 + \xi)^{1/2} \text{ and} \\ v_2 &= (2e\phi_o/m)^{1/2}(\xi - \zeta_2^2)^{1/2}. \end{aligned} \tag{11.19}$$

Our expression for the electron density now becomes

$$n_e = 2(2e\phi_o/m)^{1/2}\left[f_u(\zeta_1^2 + \xi)^{1/2} + (f_t - f_u)(\xi - \zeta_2^2)^{1/2} U(\xi - \zeta_2^2)\right]. \tag{11.20}$$

If the plasma ions are regarded as a spatially uniform distribution of infinitely massive positive charges, we can set $n_i = $ constant. In the center of the plasma, where $\xi \equiv \phi/\phi_o = 1$, we require charge neutrality, $n_i = n_e(\xi = 1)$, and thus

$$n_i = 2(2e\phi_o/m)^{1/2}\left[f_u(\zeta_1^2 + 1)^{1/2} + (f_t - f_u)(1 - \zeta_2^2)^{1/2}\right]. \tag{11.21}$$

The square of the electric field, $E^2(\phi)$, can now be determined by integrating Poisson's equation, Eq. (11.15), using the above expressions for the electron and ion densities. In terms of the dimensionless variables, Poisson's equation then becomes the following:

$$\begin{aligned} d\,E^2(\xi) = (2e\phi_o/\varepsilon_o)\{&2(2e\phi_o/m)^{1/2}\left[f_u(\zeta_1^2 + \xi)^{1/2}\right. \\ &\left. + (f_t - f_u)(\xi - \zeta_2^2)^{1/2} U(\xi - \zeta_2^2)\right] - n_i\}d\xi. \end{aligned} \tag{11.22}$$

At the surface of the plasma, where $\xi = 0$, we impose the additional boundary condition that $E(\xi = 0) = 0$. Under these conditions, $E^2(\xi)$ is given by

$$E^2(\xi) = (2e\phi_o/\varepsilon_o)\{(4/3)(2e\phi_o/m)^{1/2}\left[f_u(\zeta_1^2+\xi)^{3/2} - f_u\zeta_1^3 + (f_t - f_u)(\xi - \zeta_2^2)^{3/2}U(\xi - \zeta_2^2)\right] - n_i\xi\}. \quad (11.23)$$

We also require the electric field to vanish at the center of the plasma, so that $E(\xi = 1) = 0$, which fixes the value of the ion density:

$$n_i = (4/3)(2e\phi_o/m)^{1/2}\left[f_u(\zeta_1^2+1)^{3/2} - f_u\zeta_1^3 + (f_t - f_u)(1-\zeta_2^2)^{3/2}\right].$$

Since this ion density must also satisfy the electrical neutrality condition, we have

$$A \equiv (f_t - f_u)/f_u = \left[(\zeta_1^2+1)^{1/2}(\zeta_1^2+1/2) - \zeta_1^3\right]/\left[(1-\zeta_2^2)^{1/2}(\zeta_2^2 - 1/2)\right]. \quad (11.24)$$

Our expression for $E^2(\xi)$ can now be rewritten in the following form:

$$E^2(\xi) = (2n_i e\phi_o/\varepsilon_o) \times \{\left[(\zeta_1^2+\xi)^{3/2} - \zeta_1^3 + A(\xi-\zeta_2^2)^{3/2}U(\xi-\zeta_2^2)\right]/\left[(\zeta_1^2+1)^{3/2} - \zeta_1^3 + A(1-\zeta_2^2)^{3/2}\right] - \xi\} \equiv E_o^2F(\xi). \quad (11.25)$$

In order to ensure that $E^2(\xi) \geq 0$ for $0 \leq \xi \leq 1$, it is necessary to require that $dE^2/d\xi \geq 0$ for $\xi \to 0$. This condition is satisfied if $\zeta_1 \geq (\zeta_1^2+1)^{1/2} + A(1-\zeta_2^2)^{1/2}$.

Since we have required $E^2(\xi)$ to vanish at $\xi = 0$ and $\xi = 1$, we can expand $E^2(\xi)$ in the neighborhood of both of these points:

$$E^2(\xi) \cong E_o^2F'(0)\xi, \quad \text{as } \xi \to 0, \quad \text{and}$$

$$E^2(\xi) \cong E_o^2F''(1)(1-\xi)^2, \quad \text{as } \xi \to 1.$$

Note also that $E(\xi) = -d\phi/dx = -\phi_o d\xi/dx = \pm E_o\sqrt{F}(\xi)$, so that

$$dx = \pm x_o d\xi/\sqrt{F}(\xi),$$

where $x_o \equiv -\phi_o/E_o$. If we let $x = 0$ be the point at which $\xi = 0$, we can obtain the spatial profile of the potential by integrating this last expression using the local approximation, valid as $\xi \to 0$:

$$x \cong 2x_o\xi^{1/2}/\sqrt{F'}(0),$$

or

$$\phi \cong \phi_o F'(0)(x/2x_o)^2.$$

Similarly, near $\xi = 1$,

$$dx = -x_o d\ln(1-\xi)/\sqrt{F''}(1)/2, \text{ so that if } \xi_1 < \xi_2 = 1,$$

$$x_2 - x_1 = x_o\sqrt{2/F''}(1)\ln[(1-\xi_1)/(1-\xi_2)].$$

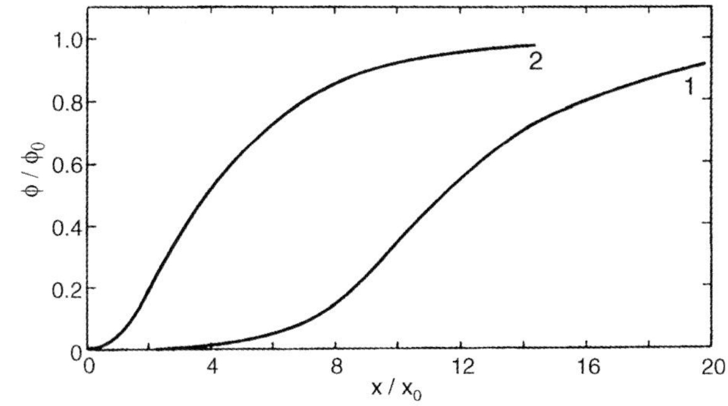

Figure 11.4 Profiles of the self-consistent ambipolar potentials for the two cases described in the text.

This model has been used to construct the equilibrium potential profiles shown in Figure 11.4. The parameters defining the two cases are as follows.

Case	ζ_1^2	ζ_2^2	n_t/n_u	A	$F'(0)$	$F''(1)$
1	1	0.2	0.92	−0.4678	0.0042	0.0924
2	0.2	0.1	1.72	−0.7345	0.1218	0.1739

Note that $E_o^2 = 2n_i e\phi_o/\varepsilon_o$ and thus the scale length, $x_o = \phi_o/E_o = (\varepsilon_o\phi_o/2en_i)^{1/2}$.

The sheath-like structure of the plasma surface is evident from Figure 11.4: the potential rises within a distance of $O(10x_o)$ to values approaching its maximum value, ϕ_o. The details of the sheath-like edge profile depend on the relative density of trapped and untrapped electrons; but in any event, the model leads to equilibria whose lengths are infinite. That is, the length is not determined by charge balance but rather by particle and heat balance. It appears that the conditions for charge balance in this dynamical model of the equilibrium determine whether or not equilibria exist at all. This rudimentary three-component model containing trapped and untrapped electrons and fixed ions indicates that the untrapped electrons must have energies greater than a critical minimum value that depends on the relative density of trapped and untrapped components. Otherwise the local density of electrons near the surface will be too small to satisfy Poisson's equation near the plasma surface while simultaneously satisfying charge neutrality in the interior of the plasma. The condition for the existence of equilibria corresponds to the physical requirement that there be no net positive space charge at the plasma surface.

References

1 V. Yu. Trakhtengerts, Active Experiments in Space, Symposium at Alpbach, 24–28 May, 1983, ESA SP-195, July 1983, pp. 67–74.

2 G.E. Guest, R.L. Miller, and C.S. Chang, *Nucl. Fusion* **27**, 1245 (1987).
3 H. Alfvén and C.-G. Fälthammar, *Cosmical Electrodynamics*, Clarendon Press, Oxford, pp. 3–6 (1963).
4 A.Y. Wong, "HIPAS Status Report – March 1–May 15, 1986," June 1986, Center for Plasma Physics and Fusion Engineering, University of California, Los Angeles.
5 Y.L. Alpert and D.S. Fligel, *Propagation of ELF and VLF Waves Near the Earth*, Consultants Bureau, New York (1970).
6 G.P. Ginet and M.A. Heinemann, *Phys. Fluids* **B2**, 700 (1990).
7 R.A. Dandl and G.E. Guest, *Phys. Rev. Lett.* **50**, 970 (1983).
8 B.H. Quon, R.A. Dandl, W. DiVergilio, G.E. Guest, L.L. Lao, N.H. Lazar, T.K. Samec, and R.F. Wuerker, *Phys. Fluids* **28**, 1503 (1985).
9 G.E. Guest and R.L. Miller, *Nucl. Fusion* **28**, 419 (1988).
10 see, for example, R.H. Berman, D.J. Tetreault, T.H. Dupree, and T. Boutros-Ghali, *Phys. Rev. Lett.* **48**, 1249 (1982); Thomas H. Dupree, *Phys. Fluids* **25**, 277 (1982); Thomas H. Dupree, *Phys. Fluids* **26**, 2460 (1983); Robert H. Berman, David J. Tetrault and Thomas H. Dupree, *Phys. Fluids* **26**, 2437 (1983); and references cited in these works.
11 see, for example, H.L. Berk, C.E. Nielsen, and K.V. Roberts, *Phys. Fluids* **13**, 980 (1970);Hans L. Pécseli, in *Proceedings of the Second Symposium on Plasma Double Layers and Related Topics*, R. Schrittwieser and G. Eder, editors, July 5/6, 1984, Innsbruck, Austria, (University of Innsbruck, Innsbruck, Austria, a984), pp. 81–117 and references cited therein.

Exercises

11.1. ***For the simple dipole model of the geomagnetic field, estimate the following properties of single electrons:***

- ***(a) The bounce time of electrons heated to (perpendicular) energies of 5 keV in a resonance zone 100 km above HIPAS.***
- ***(b) The $\nabla B \times B$ drift speed of these same electrons at the equatorial plane.***
- ***(c) The maximum electron energy for adiabatic invariance on the geomagnetic field line through HIPAS.***

11.2. ***Estimate the RF power flux required to heat these electrons to 5 keV:***

- ***(a) with fixed frequency power and***
- ***(b) with frequency-modulated power with a 2% bandwidth.***

11.3. ***The radius of curvature of field lines in SM-1 is given by*** **$rR_c = 600\,\text{cm}^2$. If the magnetic field at a radius of 10 cm on the midplane is about 0.2 T, estimate the poloidal drift frequency and the frequencies of curvature-driven flute modes for electrons of 10-keV energy. Assume that the ions have a negligible temperature.**

12
Some Aspects of Microwave Technology

It is not appropriate to close the treatise without adding some brief remarks regarding the aspects of microwave technology that are important in electron cyclotron heating (ECH). These remarks, however, are in no way intended to substitute for an in-depth review of microwave technology that would be far beyond the scope of the present work. Instead, the aim is to highlight some issues that may prompt the interested reader to delve more deeply into the vast literature on this subject. We rather arbitrarily separate the field into frequencies less than 35 GHz, where the technology is well established for the most part, and frequencies in the 100–170 GHz range, where the development of high-power continuous wave (cw) sources is ongoing and the technology is advancing rapidly as the power becomes available at multimegawatt levels.

12.1
Low-Frequency Technology: Sources

At frequencies below 35 GHz, there is a wide range of cw microwave power sources. For instance, klystron amplifiers, such as those employed in the SM-1 experiments, have been widely used at frequencies up to 18 GHz and cw magnetrons have been used at frequencies up to 9 GHz. TWT amplifiers employing coupled cavity, slow-wave structures have been used for ECH at multikilowatt cw power output levels at frequencies up to 55 GHz. Amplifier power output devices are emphasized in the text since they offer some advantages over oscillators by virtue of the inherent simplicity and flexibility of power output control and the possibility of a wide range of time-dependent waveforms that they afford. Although few systematic studies of heating with different waveforms have been reported, these have demonstrated beneficial effects on some aspects of the heating process. Additionally, the power output of a system containing multiple amplifiers can readily be controlled by simultaneous adjustment of a common low-level drive power source.

Electron Cyclotron Heating of Plasmas. Gareth Guest

ISBN: 978-3-527-40916-7

12.2 Low-Frequency Transmission Systems

In the frequency range below 35 GHz transmission systems for most laboratory plasma facilities have generally employed dominant mode waveguide. Bends may be incorporated in these waveguide transmission systems to permit considerable flexibility in the location of the RF sources relative to the plasma device. Waveguide vacuum window availability in this frequency range has presented few problems, since the window designs used on the microwave output tubes are generally applicable for use at equivalent power levels on the fusion device itself. Window location is flexible since it may be located at an arbitrary distance from the plasma device and can be shielded from plasma bombardment by interposing waveguide bends. A complication arises when the waveguide passes through a magnetic field where the microwave frequency matches the local electron gyrofrequency. In order to prevent breakdown and arcing in such regions, it has been common practice to locate the waveguide vacuum window in a region where the magnetic field exceeds the resonant electron gyrofrequency while maintaining pressurized waveguide operation through the resonance region.

12.3 Low-Frequency Coupling Techniques

Impedance matching is an especially important consideration at the power levels of interest to most ECH experiments, but because the plasma parameters can vary significantly in time, it is difficult to apply traditional methods of impedance matching using networks of passive elements. It is helpful to minimize local plasma coupling in the near field of the waveguide coupling aperture and exploit the polarization-changing reflections at the walls of the vacuum chamber to convert power initially coupled to weakly absorbed O-modes into X-modes that are then rapidly damped. Since O-modes propagate readily through the underdense plasma, near-field interaction is minimized and significant plasma coupling occurs only after subsequent reflections from the chamber walls. Under these circumstances, the impedance presented to the incident wave is nearly equal to the impedance of free space. Impedance matching considerations entering into the design of the coupling aperture are therefore similar to those for a radiating antenna. The simplest aperture suitable for high-power use is an open-ended waveguide terminating flush with the interior wall of the vacuum chamber. A widely used and very effective method of coupling from dominant mode waveguide employs quadrature-type waveguide hybrid junctions (four-port hybrid couplers) terminating in two identical radiating apertures at the cavity wall. In this approach, the combination of the network properties of the hybrid junction and physically symmetrical connection of the output ports produces high-order cancellation of mismatch effects caused by discontinuities as well as by near-field interaction with the plasma. If the feed is oriented with the RF electric field parallel to the magnetic field so as to couple to O-modes, the near-field effects are further minimized.

12.4 High-Frequency Power Sources

The development of the gyrotron oscillator has made possible the implementation of ECH on large tokamaks and stellarators by providing megawatt power levels at millimeter wavelengths for pulse lengths that are approaching and in some instances achieving what is, in effect, continuous operation. Because of the importance of the gyrotron to contemporary fusion research, we present here a brief introduction to the theory of its operation and suggest the interested reader to take advantage of detailed reports [1] that are issued regularly to update the technical community on the rapid progress in this area.

A highly schematic representation of a gyrotron is shown in Figure 12.1.

A thin annular beam of electrons is generated at the cathode of a magnetron-type electron gun and is accelerated by the potential difference applied to the anode, which is typically around 80 kV. The electrons flow along a static magnetic field whose intensity increases gradually, adiabatically converting parallel kinetic energy into perpendicular kinetic energy. The electrons then pass into an interaction region where a fraction of their perpendicular kinetic energy is transformed into the energy of electromagnetic radiation. This transformation comes about because of relativistic changes in the electron gyrofrequency, $\Omega = eB/\gamma m$.

As discussed in a uniform magnetic field in Chapter 11, the rate of change of the phase of an electron relative to the phase of an imposed electromagnetic field is governed by the relativistic change in the gyrofrequency:

$$d\phi/dt \equiv \nu = \Omega + k_{||}v_{||} - \omega \approx \nu(t_{res}) + \nu'(t_{res})(t - t_{res}) + \text{higher order terms},$$

where $\nu'(t_{res}) = -[\Omega(dW/dt)/\gamma mc^2]_{res}$ is the dominant term in the case of a uniform magnetic field and $dW/dt \approx -eE_\perp v_\perp \cos\phi$. If the initial phase of an electron is in the range for deceleration, $\pi/2 > \phi > -\pi/2$ and the heating rate, $dW/dt < 0$, the electron

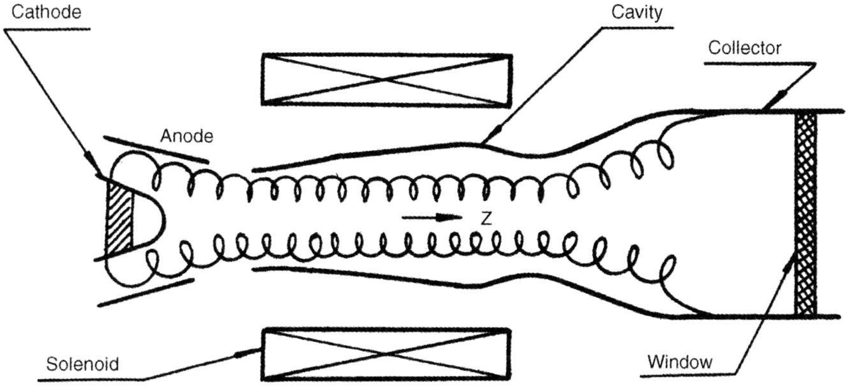

Figure 12.1 A highly schematic representation of a generic gyrotron.

looses energy, the gyrofrequency increases, and $d\phi/dt > 0$. The initial phase thus increases with time until it reaches $\pi/2$, after which the heating ceases. Conversely, if $\pi/2 < \phi < 3\pi/2$, the heating rate is positive and $d\phi/dt < 0$. The initial phase will therefore decrease in time until it, too, reaches $\pi/2$. When the resulting gyrophase bunch is formed in the decelerating phase of the wave, electrons do net work on the wave and amplify it. In practice, $\omega - \Omega$ is small but positive to keep electron bunches in the retarding phase.

To produce coherent radiation, the contribution from the electrons must reinforce the original emitted radiation in the oscillator, and the bunching mechanism must create electron density variations comparable in size to the wavelength of the imposed radiation. The number of electron gyro orbits required for efficient bunching and deceleration of electrons can be large. Since the annular electron beam is quite thin in the radial dimension, and since the beam excites azimuthally rotating electromagnetic waves in an axially symmetric cavity, all of the beamlets comprising the beam interact identically with the wave.

The interaction region can be considered as an open resonator formed by an axially symmetric open waveguide with smoothly varying wall radius. In the central, constant radius section, a mode is excited at a frequency close to the cutoff value, so that $k_{||} \sim 1/L$ is very small. The radius is tapered downward toward the cathode to create a space in which the waves are evanescent. The radius is tapered upward toward the collector to permit the amplified wave to propagate into an internal mode converter that yields a Gaussian beam of microwave radiation in a mode that can be transported efficiently to the plasma device.

The remaining megawatts of electron beam power must be absorbed without damaging components, and this is accomplished in different ways. In some gyrotrons, the beam is deposited over a large surface area by magnetically sweeping the beam across the collecting surface. Other gyrotrons employ depressed collectors to recover the beam energy.

One of the more recent developments that have made continuous operation possible is the use of synthetic diamond discs for output windows on the gyrotron. These discs are made by chemical vapor deposition and have outstanding properties for this application: high thermal conductivity (roughly four times that of copper) and low absorptivity of microwave power ($\sim$0.2%). Edge cooling is adequate to extract the energy absorbed by the diamond from the transmitted microwave beam. The photograph in Figure 12.2 shows a 110-GHz 1-MW gyrotron manufactured by the Communications and Power Industries.

12.5
High-Frequency Transmission Systems

Power from the gyrotrons is generally transmitted to the plasma device using low-loss, windowless, evacuated transmission lines that are formed from circular corrugated waveguide for propagation in the HE_{11} mode. This mode, existing only in corrugated waveguides, has exact linear polarization that can be selected

Figure 12.2 Photograph of a 110-GHz, 1-MW, 10-s Pulse Gyrotron, Type VGT-8110 manufactured by the Communications and Power Industries.

for coupling to O-modes or X-modes in the plasma. The lines will typically have a number of mitre bends, each of which may result in the loss of around 1% of the power. The overall efficiency of a 100-m-long transmission line can exceed 80%.

12.6
High-Frequency Couplers

It is now a common practice to launch the high-frequency power-using mirrors that can steer the beam in both the poloidal and toroidal directions. Polarization-changing mirrors, mounted on the inside wall of a torus, can also be used to convert O-modes launched from the low-field side of a torus into X-modes for injection from the high-field side.

References

1 An excellent resource for current information and references to the literature is available from Manfred Thumm, Forschungszentrum Karlsruhe GmbH, Postfach 3640, 76021 Karlsruhe, Germany.

13
Frequency Modulated Electron Cyclotron Heating (FMECH)

For most of the situations considered in previous chapters the duration of resonance, t_{eff}, was determined either by the spatial gradients of the magnetostatic field or, in the case of heating in a locally uniform magnetic field, by the relativistic change in the electron mass and gyrofrequency. The frequency of the microwave power was usually assumed to be constant, although in Chapter 11 we employed frequency modulation to extend the duration of resonance by compensating for the relativistic change in the electron gyrofrequency. The maximum extent of this compensation was limited in a very straightforward way by the bandwidth of the microwave power source. In the present chapter we will explore further some of the potential uses of microwave power with time-varying frequency and, in particular, the possible use of FMECH to achieve preferential interactions with electrons having selected parallel velocities.

As we have seen, in general, the duration of resonance is determined by the time-history integral of $\nu = \Omega + k_{||}v_{||} - \omega$, the rate at which the electron gyrophase changes relative to the (Doppler shifted) wave frequency: $t_{eff} = \mathrm{Re}\int dt \exp(i\int dt\ \nu)$. At a resonance $\nu = 0$, $t_{eff} = t$, and the duration of resonance increases with time as long as the resonance condition is satisfied. In Chapter 5 we proposed an *ad hoc* definition of t_{eff} as the interval in time throughout which the phase difference remained within $\pm\pi/4$ relative to its value at resonance: $|\phi(t_{eff}) - \phi_{res}| \leq \pi/4$. We used this rather arbitrary definition to illustrate for mirror-confined electrons the dependence of t_{eff} on the electron pitch angle in velocity space. To do this we expanded $\nu(t)$ in a Taylor series about the instant of resonance:

$$\nu(t) = \nu_{res} + \nu'_{res}(t - t_{res}) + \nu''_{res}(t - t_{res})^2/2! + \cdots . \quad (13.1)$$

Here the primes indicate derivatives with respect to time. Since $\nu_{res} = 0$, the phase evolves in time according to

$$\phi(t) = \phi_{res} + \nu'_{res}(t - t_{res})^2/2 + \nu''_{res}(t - t_{res})^3/3! + \cdots . \quad (13.2)$$

If the intervals prior to resonance and after resonance are designated, respectively, by δt_- and δt_+ and, for $\nu'_{res} \neq 0$, given by $|\nu'_{res}|\delta t_\pm^2/2 = \pi/4$, then the duration of resonance in our *ad hoc* model is $t_{eff} = \delta t_- + \delta t \cong 2\pi/|\nu'_{res}|)^{1/2}$. If we now include the

Electron Cyclotron Heating of Plasmas. Gareth Guest

ISBN: 978-3-527-40916-7

possibility that the microwave frequency can vary in time, then

$$v'_{res} = [\Omega v_{||} d\ln B/dz - \Omega (dW/dt)/\gamma mc^2 + k_{||}(dv_{||}/dt) - d\omega/dt]_{res}. \quad (13.3)$$

For the moment we will neglect the relativistic and Doppler terms and approximate the duration of resonance as

$$\begin{aligned} t_{eff} &= \left(2\pi/|v'_{res}|\right)^{1/2} \cong \left\{2\pi/\left|\left[\Omega v_{||} d\ln B/dz - d\omega/dt\right]_{res}\right|\right\}^{1/2} \\ &= \left(2\pi/\left|\left[\Omega d\ln B/dz\right|_{res}\right)^{1/2}\left(|v_{||} - v_{res}|\right)^{-1/2}, \end{aligned} \quad (13.4)$$

where $v_{res} = (d\omega/dt)/\Omega d\ln B/dz$. If the frequency deviation rate is such that $v_{||} = v_{res}$, the duration of resonance is given as before by $t_{eff} = t_{max}$, where t_{max} is the maximum time for which the resonance condition can be satisfied. As we shall see in the following section the duration of resonance for electrons with $v_{||} \approx v_{res}$ can be more than 5 times greater than the duration for electrons having the same speed but moving in the opposite direction, as suggested by Eq. (13.4).

The possibility of interacting preferentially with electrons having a particular velocity was utilized in an approach to ECH current drive in tokamaks that exploits this potential [1]. In this approach, pulsed, fixed-frequency ECH, resonant near the maximum magnetic field, is used to create a collisionless group of barely trapped electrons. FMECH resonant near the minimum magnetic field is then used to displace some of these collisionless electrons into regions of velocity space where they can pass freely around the torus and thereby contribute to the toroidal current. In order to achieve a net current the FMECH must interact preferentially with the electrons moving in the same direction as the bulk electrons that carry the main toroidal current ("co-streaming" electrons) to yield a significant toroidal asymmetry. In the next section we consider in a more rigorous way the degree to which the interaction with co-streaming electrons can be enhanced relative to the interaction with the counter-streaming electrons (those moving in the opposite direction).

13.1 Achievable Values of Toroidal Asymmetry

Before undertaking a detailed analysis of the toroidal asymmetry that can be achieved with FMECH, we first sketch a heuristic picture that gives a rough estimate of the achievable toroidal asymmetry and displays the elements of the optimization process that will then be examined with fewer arbitrary assumptions in what follows. If we choose the frequency deviation rate to match the parallel velocity of the co-streaming electrons the maximum time for which resonance can be maintained will be estimated by assuming that the bandwidth $\Delta f = \Delta\omega/2\pi$ is swept by the FMECH at the maximum frequency deviation rate:

$$t_{max} \approx 2\pi\Delta f/(df/dt) \geq 2\pi\Delta f/[(3/2)(\Delta f)^2 \approx 4\pi/(3\Delta f). \quad (13.5)$$

Our *ad hoc* model for the duration of resonance for the counter-streaming electrons gives

$$t_{\text{eff, counter}} = \left(2\pi/\left|v'_{\text{res}}\right|\right)^{1/2} \cong \left[2\pi/\left|-2d\omega/dt\right|_{\text{res}}\right]^{1/2} \cong 1/\left(\sqrt{3}\Delta f\right). \tag{13.6}$$

The ratio of the FMECH interaction with co-streaming electrons to that with counter-streaming electrons is then given under these assumptions by

$$t_{\max}/t_{\text{eff, counter}} \approx 4\pi/\sqrt{3} = 7.3,$$

as was suggested by the simple $(|v_{||} - v_{\text{res}}|)^{-1/2}$ dependence of t_{eff} cited earlier.

We now consider a more detailed and rigorous analysis that seeks to optimize the toroidal asymmetry without making such arbitrary assumptions as those made above. To achieve the maximum possible toroidal asymmetry it is necessary to maximize the rate of change of the phase of the counter-streaming electrons while prolonging the resonance of co-streaming electrons to the fullest extent permitted by the bandwidth of the FMECH power source. We first consider the maximum duration of resonance that can be maintained for the co-streaming electrons by employing FMECH. In the rest frame of a trapped electron the gyrofrequency varies in time according to

$$\Omega/\Omega_o = [(M_t + 1)/2] - [(M_t - 1)/2]\cos 2\,\omega_b t. \tag{13.7}$$

Here ω_b is the bounce frequency and we have taken $t = 0$ to be the time at which the electron is at $z = 0$. In order for the resonance condition to be satisfied we require

$$d\omega/dt = (M_t - 1)\omega_b\Omega_o \sin 2\,\omega_b t. \tag{13.8}$$

But since $d\omega/dt = 2\pi df/dt \leq 3\pi(\Delta f)^2$, it is not possible to provide the frequency deviation rate required for resonance for times greater than a maximum time, $t_{\max}$, given by

$$\sin 2\,\omega_b t_{\max} = 3\pi(\Delta f)^2/[(M_t - 1)\omega_b\Omega_o]. \tag{13.9}$$

At time $t = t_{\max}$ we shall require that the FMECH frequency is as high as the bandwidth permits:

$$f(t_{\max}) = f_c + (\Delta f/2), \tag{13.10}$$

where f_c is the center frequency of the FMECH power source. If the instantaneous frequency deviation rate is matched to the electron gyrofrequency for the duration of the resonance, Δt, we have

$$\begin{aligned} f(t_{\max}) &= f(t_{\max} - \Delta t) + \Delta f(\Omega_o/2\pi)\{[(M_t + 1)/2] - [(M_t - 1)/2]\}\cos 2\omega_b t_{\max} \\ &= (\Omega_o/2\pi)\{[(M_t + 1)/2] - [(M_t - 1)/2]\}\cos 2\omega_b(t_{\max} - \Delta t) + \Delta f, \end{aligned}$$

or

$$\cos 2\omega_b(t_{\max} - \Delta t) = \cos 2\omega_b t_{\max} + (\Delta f/f)[2M_{\text{res}}/(M_t - 1)]. \tag{13.11}$$

Note that we have set $f = M_{res}(\Omega_o/2\pi)$ as a notational convenience. Clearly the optimum situation is obtained if $\Delta t = t_{max}$, in which case

$$\cos 2\omega_b t_{max} = 1 - (\Delta f/f)[2M_{res}/(M_t - 1)], \tag{13.12}$$

provided that the necessary frequency deviation rate is maintained:

$$\sin 2\omega_b t_{max} = 3\pi(\Delta f)^2/[(M_t - 1)\omega_b \Omega_o] = (3/2)(f/\omega_b)(\Delta f/f)^2[M_{res}/(M_t - 1)]. \tag{13.13}$$

For relatively narrow bandwidths, $\Delta f/f \sim O(10^{-2})$, the optimum conditions will be characterized by small values of $\omega_b t_{max}$ and we can approximate the trigonometric functions with the lowest order terms in their Taylor series and obtain two convenient optimizing conditions:

$$\omega_b t_{max} = \{(\Delta f/f)[M_{res}/(M_t - 1)]\}^{1/2} \tag{13.14}$$

and

$$\omega_b t_{max} = (3/4)(f/\omega_b)(\Delta f/f)^2[M_{res}/(M_t - 1)]. \tag{13.15}$$

If the bounce frequency is eliminated between these two conditions we find for the maximum duration of resonance, $t_{max} = 4/(3\Delta f)$.

In order to estimate the corresponding duration of resonance for counter-streaming electrons we model the variation of magnetic intensity along a magnetic line of force in a tokamak by our earlier expression for mirror-like fields:

$$B(z)/B_o = [(M + 1)/2] - [(M - 1)/2]\cos k_o z, \tag{13.16}$$

Where $k_o = 1/Rq$. Here B_o is the minimum magnetic intensity at $z = 0$, and MB_o is the maximum magnetic intensity at $z = \pi Rq$. $R = R_o + r$ is the major radius of the field line at $z = 0$ in the equatorial plane of the tokamak and q is the safety factor on this flux surface. The mirror ratio is $M = (R_o + r)/(R_o - r)$. In this model of the tokamak magnetic field the parallel gradient in B and thus in Ω is given by

$$d\Omega/dz = \Omega_o k_o[(M - 1)/2]\sin k_o z. \tag{13.17}$$

The parallel velocity at any point along the field line is related to the (conserved) speed by adiabatic invariance:

$$v_{||} = \pm v(1 - B/B_t)^{1/2}, \tag{13.18}$$

where $B_t = \varepsilon/\mu = M_t B_o$ is the magnetic intensity at which the trapped electron is reflected by the increasing magnetic field. If we denote the initial point along the field line at which resonance occurs as $z = z_{res}$, where $B = B_{res} = M_{res}B_o$, the counter-streaming electrons will undergo a rate of change of phase given by

$$\begin{aligned}|v'_{\text{res counter}}| &= 2d\omega/dt \\ &= 2\pi v(1 - M_{res}/M_t)^{1/2} f k_o[(M - 1)/M_{res}]\sin k_o z_{res} \le 3\pi(\Delta f)^2.\end{aligned} \tag{13.19}$$

We can now compare the FMECH interaction with co-streaming and counter-streaming electrons under these optimum conditions by evaluating the change in perpendicular energy they experience in a single transit of the resonance:

$$\begin{aligned}\delta W_{\perp} &\cong -eE_{-}v_{\perp res}\mathrm{Re}\left[\exp(i\phi_{res})\int dt\,\exp\left(i\int dt v\right)\right]/2\\ &\cong -eE_{-}v_{\perp res}\mathrm{Re}\left[\exp(i\phi_{res})\int dt\,\exp(iv'_{res}t^2/2)\right]/2\\ &\cong -eE_{-}v_{\perp res}\mathrm{Re}\left\{\exp(i\phi_{res})\left[\int dt\cos v'_{res}t^2/2+i\int dt\sin v'_{res}t^2/2\right]\right\}/2.\end{aligned} \tag{13.20}$$

The two integrals in brackets, whose limits are $t=0$ and $t=t$, are readily expressed in terms of the Fresnel integrals [2], $C(y)$ and $S(y)$. Here, for clarity, we shall consider the case $\phi_{res}=0$ and thus isolate the first term in Eq. (13.20). Then let $y^2=|v'_{res}|t^2/\pi$ so that

$$\int dt\cos v'_{res}t^2/2=(\pi/|v'_{res}|)^{1/2}\int dy\cos\pi y^2/2=(\pi/|v'_{res}|)^{1/2}C(y). \tag{13.21}$$

For the co-streaming electrons $v'_{res}\approx 0$ for $t\le t_{max}$ and the small argument limit of the Fresnel integral applies:

$$(\pi/|v'_{res}|)^{1/2}C(y)\cong(\pi/|v'_{res}|)^{1/2}(\pi/|v'_{res}|)^{-1/2}t_{max}=t_{max}. \tag{13.22}$$

For the counter-streaming electrons $|v'_{res}|=|2d\omega/dt|_{res}$ and $y=(|v'_{res}|/\pi)t\gg 1$. Thus, for the counter-streaming electrons we can take the asymptotic limit of the Fresnel integral:

$$(\pi/|v'_{res}|)^{1/2}C(y)\cong(\pi/|v'_{res}|)^{1/2}(1/2)\cong(\pi/|2d\omega/dt_{res}|)^{1/2}(1/2). \tag{13.23}$$

Finally, we arrive at the following estimate for the toroidal asymmetry that can be achieved with FMECH:

$$\delta W_{\perp co}/\delta W_{\perp counter}\cong t_{max}/(\pi/|2d\omega/dt_{res}|)^{1/2}(1/2)=2t_{max}[2(|d\omega/dt_{res}|)/\pi]^{1/2}. \tag{13.24}$$

With $t_{max}=4/(3\Delta f)$ and $d\omega/dt_{res}\le 3\pi(\Delta f)^2$ we have

$$\delta W_{\perp co}/\delta W_{\perp counter}\cong(8/3)\sqrt{6}=6.5. \tag{13.25}$$

The more detailed optimization thus yields a slightly smaller toroidal asymmetry than our initial crude estimate and provides additional constraints that reflect specific tokamak parameters.

13.2
An Estimate of I/P

The conventional figure-of-merit for noninductively driven current is I/P, where P is the power required to sustain the current I. For the FMECH approach to current drive there are two separate steps to consider: In STEP ONE fixed-frequency ECH is used to generate a small fractional concentration of barely trapped energetic electrons by preferentially heating electrons in the tail of the bulk distribution. The resonance surface for STEP ONE is near the region of maximum magnetic intensity; that is, on the high-field side of the tokamak. We designate the density of these electrons as $n_1(\varepsilon, \theta_v, r)$, where ε is the energy, $\theta_v = \sin^{-1}(v_\perp/v)$ is the pitch angle in velocity space, and r is the minor radius of the flux surface. STEP TWO is the FMECH process described in the preceding section. In STEP TWO, FMECH resonant near the minimum magnetic intensity is used to de-trap some of the co-streaming electrons from the collisionless group, $n_1(\varepsilon,\theta_v,r)$ and so create a source, $S = dn_1/dt$, of current carriers. These will remain in the passing-particle region of velocity space for a time, $\tau(\varepsilon, \theta_v, Zn)$, where Zn is the bulk charge density responsible for (Coulomb) scattering the current-carrying electrons in pitch angle. The resulting current density, j(r), is then given by the following integral over all energies and over the passing-particle pitch angles:

$$j(r) = -e\int d\varepsilon \int d\theta_v\, S(\varepsilon, \theta_v, r)\tau(\varepsilon, \theta_v, Zn)v_{||}. \tag{13.26}$$

The source function, $S(\varepsilon, \theta_v, r)$, is itself given by the following integral over all pitch angles comprising the population of magnetically trapped electrons:

$$S(\varepsilon, \theta_v, r) = \int d\theta_{vi}\, n_1(\varepsilon, \theta_{vi}, r)\nu_{b1}(\varepsilon, \theta_{vi}, r)F(\varepsilon, \theta_v, \theta_{vi}). \tag{13.27}$$

Here $\nu_{b1}(\varepsilon, \theta_{vi}, r)$ is the bounce frequency of the trapped electrons, and the function $F(\varepsilon, \theta_v, \theta_{vi})$ gives the probability that an initially trapped electron will be displaced by the STEP TWO FMECH from its initial pitch angle, θ_{vi}, to a new value, θ_v. We shall somewhat arbitrarily employ a Gaussian form to model this transition probability:

$$F(\varepsilon, \theta_v, \theta_{vi}) = 1/(\Delta\theta\sqrt{\pi})\exp\{-[(\theta_v - \theta_{vi})/\Delta\theta]^2\}. \tag{13.28}$$

The parameter $\Delta\theta$ is thus a measure of the average pitch-angle displacement produced in STEP TWO by the FMECH interaction:

$$\langle(\theta_v - \theta_{vi})^2\rangle \cong (\Delta\theta)^2/2. \tag{13.29}$$

The displacement in pitch angle after a single transit of resonance is related to the change in perpendicular energy as follows: Since $v_\perp^2 = v^2 \sin^2\theta_v$, we have

$$\begin{aligned} &W_\perp = (W_\perp + W_{||})\sin^2\theta_v \text{ which we differentiate as follows,}\\ &\delta W_\perp = \delta W_\perp \sin^2\theta_v + 2(W_\perp + W_{||})\sin\theta_v \cos\theta_v\, \delta\theta_v, \text{ giving}\\ &\delta\theta_v = (M_t - 1)^{1/2}\delta W_\perp/2\varepsilon. \end{aligned} \tag{13.30}$$

In Section 13.1, we found that for co-streaming electrons the maximum duration of resonance that could be achieved with FMECH was given by Eq. (13.15):

$$t_{max} = (3/4)(f/\omega_b^2)(\Delta f/f)^2[M_{res}/(M_t-1)].$$

The corresponding change in the perpendicular energy is then given by

$$\delta W_{\perp max} = -eE_{-}v_{\perp res}\cos\phi_{res}(3/4)(f/\omega_b^2)(\Delta f/f)^2[M_{res}/(M_t-1)]. \qquad (13.31)$$

The FMECH power must be sufficient to displace a significant number of the source electrons into the passing-particle region of velocity space, and the bandwidth of the power source must be adequate to achieve a significant toroidal asymmetry. These requirements can be made somewhat more quantitative if we consider an idealized population of barely trapped ("source") electrons in the form of delta functions:

$$n_1(\varepsilon, \theta_{vi}, r) = n_1(r)\delta(\varepsilon-\varepsilon_1)\delta(\theta_{vi}-\theta_1). \qquad (13.32)$$

The co-streaming source electrons that pass through the de-trapping FMECH interaction with the sign of parallel velocity that sustains the steady-state current will have a transition probability that we denote by F_+:

$$F_+(\varepsilon_1, \theta_v, \theta_1) = 1/(\Delta\theta_+\sqrt{\pi})\exp\{-[(\theta_v-\theta_1)/\Delta\theta_+]^2\}. \qquad (13.33)$$

The source electrons that move in the opposite direction will undergo smaller changes in perpendicular energy and pitch angle:

$$F_-(\varepsilon_1, \theta_v, \theta_1) = 1/(\Delta\theta_-\sqrt{\pi})\exp\{-[(\theta_v-\theta_1)/\Delta\theta_-]^2\}. \qquad (13.34)$$

The net transition amplitude for sustaining the steady-state current is the difference of these: $F_j = F_+ - F_-$; the total transition amplitude resulting in power drain from the microwave power source is the sum of these: $F_p = F_+ + F_-$. Indeed, the power drain (per unit of volume) can be estimated by the following integral over the energy and the (passing-particle) pitch angles:

$$\mathcal{P} \cong \int d\varepsilon \int d\theta_v\, S(\varepsilon, \theta_v, r)\varepsilon = \int d\varepsilon \int d\theta_v \int d\theta_{v1} n_1(\varepsilon, \theta_{v1}, r)v_{b1}(\varepsilon, \theta_{v1}, r)F_p(\varepsilon, \theta_v, \theta_{vi})\varepsilon. \qquad (13.35)$$

Here the integral over θ_v is taken over the passing-particle region of velocity space while the integral over θ_{v1} is taken over the trapped-particle region of velocity space. For the idealized delta-function sources introduced earlier this reduces to the following integral over all passing-particle pitch angles:

$$\mathcal{P} \cong \varepsilon_1 n_1 v_{b1} \int d\theta_v \{[1/(\Delta\theta_+\sqrt{\pi})]\exp\{-[(\theta_v-\theta_1)/\Delta\theta_+]^2\} + [1/(\Delta\theta_-\sqrt{\pi})]\exp\{-[(\theta_v-\theta_1)/\Delta\theta_-]^2\}\},$$

from which we obtain

$$\mathcal{P} \cong \varepsilon_1 n_1 v_{b1}\{2-\mathrm{erf}[(\theta_v-\theta_1)/\Delta\theta_+]-\mathrm{erf}[(\theta_v-\theta_1)/\Delta\theta_-]\}/2, \qquad (13.36)$$

where erf(x) is the error function [3]. As indicated earlier, the current density that is sustained by this power density depends on the lifetime of the current carriers, $\tau(\varepsilon,\theta_v,Zn)$. We shall assume that the dominant loss mechanism is Coulomb scattering of electrons from the passing-particle region to the trapped-particle region of velocity space. Note that STEP TWO displaces some of the source electrons into regions of velocity space that are near the boundary between trapped and passing:

$$\theta_v \cong \theta_1 - \Delta\theta_+ = \theta_{tp} + (\theta_1 - \theta_p - \Delta\theta_+) \le \theta_{tp}. \tag{13.37}$$

If the source electrons are barely trapped, as assumed here, $\theta_1 \ge \theta_{tp}$, the de-trapped electrons are separated from the trapped-passing boundary by an angle less than $\Delta\theta_+$. Thus, Coulomb scattering can lead to the prompt diffusion of half of the de-trapped electrons back into the trapped-particle region of velocity space in a time given by

$$\tau_{j\ \mathrm{prompt}} \approx \tau_{90}(2\Delta\theta_+/\pi)^2 \ll \tau_{90}, \tag{13.38}$$

where τ_{90} is the 90° deflection time. The remaining half of the de-trapped electrons must diffuse in pitch angle across the entire passing-particle region before becoming trapped once again. Their lifetime as current carriers is therefore roughly given by

$$\tau_{j\ \mathrm{delayed}} \approx \tau_{90}(4\theta_{tp}/\pi)^2. \tag{13.39}$$

The net effective lifetime of the current-carrying electrons is therefore roughly given by

$$\tau_j \approx (\tau_{90}/2)(4\theta_{tp}/\pi)^2. \tag{13.40}$$

Our expression for the current density then takes the form

$$j(r) = -e\int d\varepsilon \int d\theta_v\, \tau_j v_{||} \int d\theta_{v1} n_1(\varepsilon, \theta_{v1}, r) v_{b1}(\varepsilon, \theta_{v1}, r) F_j(\varepsilon, \theta_v, \theta_{vi}). \tag{13.41}$$

The integral over θ_v is from $\theta_v = 0$ to $\theta_v = \theta_{tp}$, while the integral over θ_{v1} is from $\theta_{v1} = \theta_{tp}$ to $\theta_{v1} = \pi/2$. For the idealized delta-function distributions of source electrons we have

$$j(r) = (e/2) n_1 v_{b1} v_{||} \tau_{90}(\varepsilon_1)(4\theta_{tp}/\pi)^2 \{\mathrm{erf}[(\theta_1 - \theta_{tp})/\Delta\theta_-] - \mathrm{erf}[(\theta_1 - \theta_{tp})/\Delta\theta_+]\}. \tag{13.42}$$

With sufficient FM bandwidth we can satisfy the following condition:

$$[(\theta_1 - \theta_{tp})/\Delta\theta_+] \ll 1 \ll [(\theta_1 - \theta_{tp})/\Delta\theta_-], \tag{13.43}$$

so that

$$j(r) = e n_1 v_{b1} v_{||} \tau_{90}(\varepsilon_1)(4\theta_{tp}/\pi)^2 \{1 - \mathrm{erf}[(\theta_1 - \theta_{tp})/\Delta\theta_+]\}/2, \tag{13.44}$$

while the associated power drain is

$$\mathcal{P} \cong \varepsilon_1 n_1 v_{b1} \{1 - \mathrm{erf}[(\theta_v - \theta_1)/\Delta\theta_+]\}/2. \tag{13.45}$$

The figure-of-merit follows directly from these two expressions:

$$I/P = (j/2\pi R_o \mathcal{P}) = [ev_{||}\tau_{90}(\varepsilon_1)(4\theta_{tp}/\pi)^2]/(2\pi R_o \varepsilon_1). \quad (13.46)$$

Note that I/P is proportional to ε_1/n, so operation with energetic source electrons is favorable from the standpoint of power requirements. For example, if we consider a case in which $R_o = 150\,\text{cm}$, $Zn = 5 \times 10^{13}\,\text{cm}^{-3}$, and $\theta_{tp} = \pi/4$, and assume that satisfactory de-trapping can be achieved for $\varepsilon_1 = 30\,\text{keV}$, we find that our estimate of the figure-of-merit is roughly $I/P = 0.4\,\text{A/W}$. To estimate the power required we recall that the condition for effective de-trapping can be expressed in the following form:

$$\Delta\theta_- \ll (\theta_1 - \theta_{tp}) \ll \Delta\theta_+,$$

where

$$\Delta\theta_+ = [2\langle(\theta_v - \theta_{vi})^2\rangle^{1/2} \cong [(M_t - 1)/2]^{1/2}\delta W_{\perp max}/(2\varepsilon). \quad (13.47)$$

Since the maximum change in the perpendicular energy for co-streaming electrons was given in Eq. (13.31) as

$$\delta W_{\perp max} = -eE_- v_{\perp res}\cos\phi_{res}(3/4)(f/\omega_b^2)(\Delta f/f)^2[M_{res}/(M_t - 1)],$$

we have

$$\Delta\theta_+ \cong (3/8\sqrt{2})eE_-(v_{\perp res}/\varepsilon)(f/\omega_b^2)(\Delta f/f)^2[M_{res}/(M_t - 1)^{1/2}]. \quad (13.48)$$

By way of illustration, if we set $M = 1.5$ with $M_t = 1.45$, corresponding respectively to pitch angles in velocity space of $\theta_{tp} = 0.96$ and $\theta_1 = 0.98$, and require somewhat arbitrarily that $\Delta\theta_+ = 10(\theta_1 - \theta_{tp})$, we estimate that the necessary FMECH field strength is around 20 V/cm corresponding to a power flux around 100 W/cm^2.

13.3 Generation of the Energetic "Source" Electrons

In this section we recapitulate the main dynamical processes that govern the creation of the super-thermal group of barely trapped electrons in STEP ONE. These electrons comprise the source of current-carrying electrons for STEP TWO, in which some are de-trapped by the FMECH interaction. The creation of a population of energetic source electrons depends on the difference of the velocity dependences of the ECH heating rate ($\sim v^{-1/3}$) and the rate at which the heated electrons are cooled as a result of dynamical friction with the electrons of the bulk plasma ($\sim v^{-1}$). This competition could be effectively treated with a Fokker–Planck analysis of the type discussed earlier; but for now we will find it useful to adopt a more rudimentary approach that makes clear the key physics issues that determine the outcome of this competition. Since the heating rate depends more weakly on the electron speed than the cooling rate, electrons will "runaway" if their speed exceeds the value at which the two rates are equal, $v = v_{crit}$. The density of electrons with speeds greater than v_{crit} can be estimated in terms of error functions if the bulk electrons are described reasonably well by an isotropic Maxwell–Boltzmann distribution.

We first recall our expression from Chapter 5 for the heating rate at the fundamental gyroresonance:

$$dW_\perp/dt = (e^2/m)|E_-|^2 J_o^2(k_\perp \rho) t_{eff}^2 \nu_{coll}. \tag{13.49}$$

For electrons that turn at the resonance surface the duration of resonance was given by

$$t_{eff} = \left\{ \left[3\pi(M-1)/(4\omega_b^2\Omega_o)\right] / \left[\left|\sum(-1)^n n^2 J_{2n}(k_o z_t)\right|\right] \right\}^{1/3}, \tag{13.50}$$

where the summation is from $n=1$ to $n=\infty$ and the frequency at which the electrons encounter the resonance surface is

$$\nu_{coll} = 2\nu_b = \omega_b/\pi. \tag{13.51}$$

The bounce frequency of electrons trapped in a magnetic-mirror field is given by our earlier formula:

$$\omega_b = (k_o v/2)[(M-1)/M_t]^{1/2}\{\pi/2K[(M_t-1)/(M-1)]\}. \tag{13.52}$$

It will sometimes prove convenient to define the expression in braces as $\Theta(M, M_t)$:

$$\Theta(M, M_t) = \{\pi/2K[(M_t-1)/(M-1)]\}, \tag{13.53}$$

where $\Theta(M, M_t) \to 1$ and the electrons turn near the minimum magnetic field, and $\Theta(M, M_t) \to 0$ as $M_t \to M$ and the electrons turn at the maximum magnetic field. Our estimate for the heating rate for electrons turning at the resonance surface is then given by

$$dW_\perp/dt = (e^2/m)|E_-|^2 J_o^2(k_\perp\rho)[(3/4\sqrt{\pi})(M-1)]^{2/3}\left(\omega_b\Omega_o^2\right)^{-1/3} \times \left[\left|\sum(-1)^n n^2 J_{2n}(k_o z_t)\right|\right]^{-2/3} \tag{13.54}$$

The rate at which the heated electrons with speed v_h are cooled by dynamical friction on the bulk electrons is given by [4]

$$dW/dt = \mathbf{F}\cdot\mathbf{v} = -m\Gamma_{hc} n_c[\mathrm{erf}(y) - y\, d\,\mathrm{erf}(y)/dy]\mathbf{v}\cdot\mathbf{v}/v^3. \tag{13.55}$$

Here $y = v_h/\alpha_c$, $\alpha_c = \sqrt{(2T_c/m)}$ is the thermal speed of the bulk electrons, and

$$\Gamma_{hc} = e^4 \ln\Lambda/(2\pi\varepsilon_o^2 m^2). \tag{13.56}$$

The dynamical friction cooling rate is thus given by

$$dW/dt = -\left[e^4 n_c \ln\Lambda/\left(2\pi\varepsilon_o^2 m\alpha_c\right)\right]\{\mathrm{erf}(y)/y - d\,\mathrm{erf}(y)/dy\}. \tag{13.57}$$

The function in braces has a single maximum at $y \cong 1.51$, where its value is 0.525. For values of $y > 2$ the function in braces is given approximately by $1/y$. Since our intention is to heat electrons in the tail of the bulk distribution we can restrict our attention to values of $y > 2$ and take as the cooling rate of the heated electrons

$$dW/dt = -e^4 n_c \ln \Lambda / (2\pi\varepsilon_o^2 m v_h). \tag{13.58}$$

If the heating rate exceeds this cooling rate at some speed v_{crit}, electrons with speeds greater than this value will "runaway" and form the source group. The relative density of these runaway electrons is given by the integral from v_{crit} to ∞ of the bulk electron distribution function:

$$n_1/n_c = \int d^3 v f_o(v).$$

If the distribution function of the thermal bulk electrons is given by the isotropic Maxwell–Boltzmann distribution:

$$f_o(v) = (\alpha^3 \pi^{3/2})^{-1} \exp(-v^2/\alpha^2),$$

the relative density is given by

$$n_1/n_c = (2/\sqrt{\pi}) y_{crit} \exp(-y_{crit}^2) + \mathrm{erfc}(y_{crit}), \tag{13.59}$$

where erfc(y_{crit}) is the complementary error function [3]. As we noted in an earlier chapter, the relative density $n_1/n_c = 0.01$ for $y_{crit} = 2.38$. Since the heating rate is proportional to the microwave power we can estimate the ECH power needed to create a given population of source electrons in a specified bulk plasma.

References

1 R.A. Dandl and G.E. Guest, *Phys. Rev. Lett.* **50**, 970 (1983).

2 Milton Abramowitz and Irene A. Stegun, editors, *Handbook of Mathematical Functions*, Dover Publications, New York (1970), pp. 300–301.

3 Abramowitz and Stegun, op cit, p. 297.

4 Donald A. Gurnett and Amitiva Bhattacharjee, *Introduction to Plasma Physics*, Cambridge University Press, New York (2005), p. 431.

Exercises

13.1 *We wish to create a barely trapped group of 30 keV electrons in a plasma, whose density is 5×10^{13} cm^{-3} and whose electron temperature is 5 keV. At what rate are the 30 keV electrons cooled by dynamical friction from the 5 keV bulk electrons?*

13.2 *The plasma in Exercise 13.1 is confined in a tokamak whose major radius is 150 cm. The barely trapped 30 keV electrons are initially formed near a flux surface whose minor radius is 30 cm. The safety factor on this flux surface is $q = 2$, and the mirror ratio is $M = 1.5$.*

(a) If we choose the resonance surface so that $M_{res} = 1.45$, what is the value of $k_o z_t$ for electrons turning at resonance?

(b) What is the bounce frequency of the 30 keV electrons?

(c) If the ECH power is at a frequency of 140 GHz, what RF field strength, $|E_-|$, will ensure that the heating rate equals the (dynamical friction) cooling rate?

(d) Estimate the relative density of the 30 keV group of electrons.

13.2 *At the start of each sweep in frequency, the FMECH resonance is at the minimum magnetic field on the flux tube of Exercise 2.*

(a) What is the gyrofrequency at this field?

(b) If the bandwidth, $\Delta f/f = 0.01$, estimate the maximum duration of resonance for co-streaming electrons.

(c) During the resonance, $0 \leq t \leq t_{max}$ how far does the 30 keV electron travel in z, $\theta_{poloidal}$ and $\phi_{toroidal}$?

What is the maximum displacement in pitch angle for a co-streaming electron if the FMECH RF electric field strength is 20 V/cm?

Appendix A: Some Useful Physical Constants

Speed of light in vacuum, $c = 2.997\ 925 \times 10^{8}$ m/s
Charge on the electron, $e = 1.602\ 192 \times 10^{-19}$ C
Rest mass of the electron, $m = 9.109\ 558 \times 10^{-31}$ kg
Rest mass of the proton, $M = 1.672\ 614 \times 10^{-27}$ kg
Permittivity of free space, $\varepsilon_o = 8.854\ 188 \times 10^{-12}$ F/m

$$1\ \text{F/m} = 1\ \text{C/(V m)}$$

Permeability of free space, $\mu_o = 4\pi \times 10^{-7}$ H/m

$$1\ \text{H/m} = 1\ \text{N/A}^2$$

Avogadro's number, $N = 6.022\ 141\ 99 \times 10^{23}$/mol
Boltzmann's constant, $k = 1.380\ 650\ 303 \times 10^{-23}$ J/K

Electron Cyclotron Heating of Plasmas. Gareth Guest

ISBN: 978-3-527-40916-7

Appendix B: Formulas from Vector Calculus

The differential operators in orthogonal, curvilinear coordinates, (ξ_1, ξ_2, ξ_3):

The differntial arc length is $ds^2 = (h_1 d\xi_1)^2 + (h_2 d\xi_2)^2 + (h_3 d\xi_3)^2$.

The vector $\mathbf{V}$ is given by $\mathbf{V} = \mathbf{u}_1 V_1 + \mathbf{u}_2 V_2 + \mathbf{u}_3 V_3$.

The gradient of the scalar f is

$$\text{grad } f = \nabla f = (\mathbf{u}_1/h_1)\partial f/\partial\xi_1 + (\mathbf{u}_2/h_2)\partial f/\partial\xi_2 + (\mathbf{u}_3/h_3)\partial f/\partial\xi_3.$$

The divergence of the vector $\mathbf{V}$ is

$$\text{div}\,\mathbf{V} = \nabla\cdot\mathbf{V} = P^{-1}[\partial(PV_1/h_1)/\partial\xi_1 + \partial(PV_2/h_2)/\partial\xi_2 + \partial(PV_3/h_3)/\partial\xi_3], \quad P \equiv h_1 h_2 h_3.$$

The curl of $\mathbf{V}$ is

$$\begin{aligned}\text{curl}\,\mathbf{V} &= \nabla\times\mathbf{V} = (\mathbf{u}_1/h_2h_3)[\partial(h_3V_3)/\partial\xi_2 - \partial(h_2V_2)/\partial\xi_3] + \\ &+ (\mathbf{u}_2/h_3h_1)[\partial(h_1V_1)/\partial\xi_3 - \partial(h_3V_3)/\partial\xi_1] \\ &+ (\mathbf{u}_3/h_1h_2)[\partial(h_2V_2)/\partial\xi_1 - \partial(h_1V_1)/\partial\xi_2]\end{aligned}$$

The differential operators in cylindrical coordinates, (ρ, ϕ, z):

$$\text{grad } f = \nabla f = \mathbf{u}_\rho \partial f/\partial\rho + \mathbf{u}_\phi \rho^{-1}\partial f/\partial\phi + \mathbf{u}_z \partial f/\partial z$$

$$\text{div}\,\mathbf{V} = \nabla\cdot\mathbf{V} = \rho^{-1}\partial(\rho V_\rho)/\partial\rho + \rho^{-1}\partial V_\phi/\partial\phi + \partial V_z/\partial z$$

$$\begin{aligned}\text{curl}\,\mathbf{V} = \nabla\times\mathbf{V} &= \mathbf{u}_\rho(\rho^{-1}\partial V_z/\partial\phi - \partial V_\phi/\partial z) + \mathbf{u}_\phi(\partial V_\rho/\partial z - \partial V_z/\partial\rho) \\ &+ \mathbf{u}_z \rho^{-1}[\partial(\rho V_\phi)/\partial\rho - \partial V_\rho/\partial\phi]\end{aligned}$$

Electron Cyclotron Heating of Plasmas. Gareth Guest

ISBN: 978-3-527-40916-7

$$\nabla \cdot \nabla f = \nabla^2 f = \rho^{-1}\partial(\rho\partial f/\partial\rho)/\partial\rho + \rho^{-2}\partial^2 f/\partial\phi^2 + \partial^2 f/\partial z^2$$

$$\nabla \cdot \nabla \times \mathbf{V} = 0$$

$$\nabla \times \nabla f = 0$$

$$\nabla \times (\nabla \times \mathbf{V}) = \nabla(\nabla \cdot \mathbf{V}) - \nabla^2 \mathbf{V}$$

$$\nabla(fg) = g\nabla f + f\nabla g$$

$$\nabla(\mathbf{U} \cdot \mathbf{V}) = (\mathbf{U} \cdot \nabla)\mathbf{V} + \mathbf{U} \times (\nabla \times \mathbf{V}) + (\mathbf{V} \cdot \nabla)\mathbf{U} + \mathbf{V} \times (\nabla \times \mathbf{U})$$

$$\nabla \cdot (f\mathbf{V}) = \nabla f \cdot \mathbf{V} + f\nabla \cdot \mathbf{V}$$

$$\nabla \times (f\mathbf{V}) = \nabla f \times \mathbf{V} + f\nabla \times \mathbf{V}$$

$$\nabla \cdot (\mathbf{U} \times \mathbf{V}) = (\nabla \times \mathbf{U}) \cdot \mathbf{V} - (\nabla \times \mathbf{V}) \cdot \mathbf{U}$$

$$\nabla \times (\mathbf{U} \times \mathbf{V}) = (\nabla \cdot \mathbf{V})\mathbf{U} - (\nabla \cdot \mathbf{U})\mathbf{V} + (\mathbf{V} \cdot \nabla)\mathbf{U} - (\mathbf{U} \cdot \nabla)\mathbf{V}$$

Appendix C: Properties of Some Mathematical Functions

Complete Elliptic Integrals:

$$K(m) = \int (1 - m \sin^2 \theta)^{-1/2}\, d\theta$$

$$E(m) = \int (1 - m \sin^2 \theta)^{1/2}\, d\theta,$$

where $\theta = 0$ and $\pi/2$ are the limits for both integrals. For $0 \leq m < 1$ and with $m_1 = 1 - m$, K, and E are approximately given by the following power series:

$$K(m) = \left[a_o + a_1 m_1 + a_2 m_1^2\right] + \left[b_o + b_1 m_1 + b_2 m_1^2\right] \ln(1/m_1) + \varepsilon(m)$$

$$\begin{aligned} a_o &= 1.38629\,44, & b_o &= 0.5 \\ a_1 &= 0.11197\,23, & b_1 &= 0.12134\,78 \\ a_2 &= 0.07252\,96, & b_2 &= 0.02887\,29 \quad |\varepsilon(m)| \leq 3 \times 10^{-5}. \end{aligned}$$

$$E(m) = \left[1 + a_1 m_1 + a_2 m_1^2\right] + \left[b_1 m_1 + b_2 m_1^2\right] \ln(1/m_1) + \varepsilon(m)$$

$$\begin{aligned} a_1 &= 0.46301\,51, & b_1 &= 0.24527\,27 \\ a_2 &= 0.10778\,12, & b_2 &= 0.04124\,96 \quad |\varepsilon(m)| \leq 4 \times 10^{-5}. \end{aligned}$$

Bessel Function Relations:

$\exp(ib \sin\theta) = \Sigma J_n(b) \exp(in\theta)$, where the index "n" takes on all integral values from $n = -\infty$ to $n = +\infty$.

$\cos(b \sin\theta) = J_o(b) + \Sigma J_{2n}(b) \cos(2n\theta)$, where the summation index "n" takes on all integral values from $n = 1$ to $n = +\infty$.

$\sin(b \sin\theta) = 2\,\Sigma J_{2n+1}(b) \sin[(2n+1)\theta]$, where the index "n" takes on all integral values from $n = 0$ to $n = +\infty$.

$1 = J_o^2(z) + 2\Sigma J_n^2(z)$, where the index "n" takes on all integral values from $n = 1$ to $n = +\infty$.

$$1 = J_o(z) + 2J_2(z) + 2J_4(z) + 2J_6(z) + \cdots$$

$$J_{n-1}(z) + J_{n+1}(z) = (2n/z) J_n(z)$$

Electron Cyclotron Heating of Plasmas. Gareth Guest

ISBN: 978-3-527-40916-7

$$J_{n-1}(z) - J_{n+1}(z) = 2J'_n(z)$$
$$J'_n(z) = J_{n-1}(z) - (n/z)J_n(z)$$
$$J'_n(z) = -J_{n+1}(z) + (n/z)J_n(z)$$

The "Gaussian Integral": $2\int dx \exp(-x^2) = \sqrt{\pi}$,
and the following integral:
$2\int x dx J_n^2(ax)\exp(-x^2) = \exp(-a^2/2)I_n(a^2/2)$, where the limits on both integrals are $x = 0$ and $x = \infty$.

The Plemelj relation: The limit as $\varepsilon \to 0$ of the integral from $x = -\infty$ to $x = +\infty$, $\int [x - (x_o \pm i\varepsilon)]^{-1} f(x)dx$, is given by $P\int (x - x_o)^{-1} f(x)dx \pm i\pi f(x_o)$. Here $\varepsilon > 0$ and P indicates the principal value integral.

The Plasma Dispersion Function:

$$Z(\zeta) = (1/\sqrt{\pi})\int (z-\zeta)^{-1}\exp(-z^2)dz$$
$$= (1/\sqrt{\pi})P\int (z-\zeta)^{-1}\exp(-z^2)dz + i\sqrt{\pi}\exp(-\zeta^2).$$

where the integrals are from $-\infty$ to ∞ and $\mathrm{Im}\,\zeta > 0$

Asymptotic Expansion:

$$Z(\zeta) = -[1/\zeta^2 + 1/(2\zeta^3) + 3/(4\zeta^4) + \cdots] + i\sqrt{\pi}\exp(-\zeta^2).$$

Small Argument Expansion:

$$Z(\zeta) = -2\zeta + (4/3)\zeta^3 - (8/15)\zeta^5 + \cdots + i\sqrt{\pi}\exp(-\zeta^2).$$

Index

Electron Cyclotron Heating of Plasmas. Gareth Guest

ISBN: 978-3-527-40916-7